D1031634

Emerging Strategies
for
Pesticide Analysis

Library of Congress Cataloging-in-Publication Data

Emerging strategies for pesticide analysis / editors, Thomas Cairns,
  Joseph Sherma.
     p.   cm. -- (Modern methods for pesticide analysis)
     Includes bibliographical references and index.
     ISBN 0-8493-7991-1
     1. Pesticides--Analysis. I. Cairns, Thomas. II. Sherma,
  Joseph. III. Series.
  RA1270.P4E44 1992
  363.17'922--dc20
                                                              92-231
                                                              CIP

Direct all inquiries to CRC Press, Inc., 2000 Corporate Blvd., N.W., Boca Raton, Florida, 33431.

© 1992 by CRC Press, Inc.

International Standard Book Number 0-8493-7991-1

Library of Congress Card Number 92-231

Printed in the United States

Printed on acid-free paper

Emerging Strategies

for

Pesticide Analysis

A Volume in the Series
Modern Methods for Pesticide Analysis

Edited by

Thomas Cairns, Ph.D., D.Sc.
Office of Regulatory Affairs
U.S. Food and Drug Administration
Los Angeles District Office
Los Angeles, California

Joseph Sherma, Ph.D.
John D. and Frances H. Larkin Professor and Head
Department of Chemistry
Lafayette College
Easton, Pennsylvania

CRC Press
Boca Raton   Ann Arbor   London   Tokyo

PREFACE

Between 1963 and 1989, 17 volumes of the highly acclaimed series of books entitled *Analytical Methods for Pesticides and Plant Growth Regulators* were edited by Dr. Gunter (Jack) Zweig and Joseph Sherma for Academic Press. The untimely death of Jack Zweig caused a discontinuity in this educational series of books, which was of practical benefit to pesticide chemists worldwide. Fortunately, CRC Press has agreed to continue this rich heritage of information in the field of pesticides in a new series, *Modern Methods for Pesticide Analysis,* under our editorship. The style and coverage of the series will remain essentially the same, with publication of additional topics related to advances in analytical techniques and improved analytical methods for specific pesticides and pesticide classes as important developments in the field warrant widespread distribution.

In this first volume of the new series, we have collected together chapters on state-of-the-art analytical technologies in the field of pesticide residue analysis written by international experts in their respective areas. In no other field has such intense scrutiny been focused on the development of analyses to protect both the public health and the ecosystem.

In several important respects, this book is a major milestone in an effort to come to grips with what is at once an exquisitely intriguing scientific challenge and a public policy issue of profound importance. No one knows how many miles lie ahead of us in either the scientific or the socio-political aspect of pesticide residue analysis, but this much is certain: the work described in this book represents a quantum leap forward from where the field stood just a few years ago.

It is for historians to decide whether knowledge in the field of pesticide residue analysis compelled a reexamination of public policy regarding public exposure to toxic substances, including pesticides, or if growing public concern goaded science into increasing its efforts related to improved analysis of residues. Whatever the basic reasons, a great number of scientists throughout the world have brought their talents to bear on the issue of reliable multiresidue analysis conducted in a cost-effective manner with minimum hazardous waste generation during the analytical process. The chapters in this book represent that global perspective of modern pesticide residue analysis.

We hope that the reestablishment of this informative series will be welcomed by our colleagues in the field, as well as other scientists who contemplate entry into this analytical area and require a concise, up-to-date source of information and data for evaluation and consideration. We will welcome comments and corrections pertaining to the chapters in this book, as well as suggestions for future topics and volumes in the series.

Thomas Cairns, Ph.D., D.Sc.
Joseph Sherma, Ph.D.

CONTRIBUTORS

J. Paul Aston, Ph.D.
Shell Research Ltd.
Sittingbourne Research Centre
Sittingbourne, England

Damià Barceló, Ph.D.
Centro de Investigacion y
  Desarrolo
C.S.I.C.
Barcelona, Spain

David W. Britton, Ph.D
Shell Research Ltd.
Sittingbourne Research Centre
Sittingbourne, England

Thomas Cairns, Ph.D., D.Sc.
U.S. Food and Drug
  Administration
Mass Spectrometry Service Center
Los Angeles, California

Stephen W. George
E. I. du Pont de Nemours & Co.
Du Pont Agricultural Products
Experimental Station
Wilmington, Delaware

Alan E. Grey, Ph.D
Idaho National Engineering
  Laboratory
Idaho Falls, Idaho

Alan R. C. Hill
Ministry of Agriculture, Fisheries
  and Food
Central Science Laboratory
Hatching Green
Harpenden, England

Patrick T. Holland, Ph.D
Ruakura Agricultural Research
  Center
Ministry of Agriculture and
  Fisheries
Hamilton, New Zealand

Marvin L. Hopper
U.S. Food and Drug
  Administration
Total Diet Research Center
Kansas City, Missouri

Richard T. Krause
U.S. Food and Drug
  Administration
Washington, D.C.

James F. Lawrence, Ph.D
Food Research Division, Bureau
  of Chemical Safety
Food Directorate
Banting Research Centre
Health Protection Branch
Ottawa, Canada

Ronald G. Luchtefeld
U.S. Food and Drug
  Administration
Total Diet Research Center
Kansas City, Missouri

Colin P. Malcolm
Ruakura Agricultural Research
  Center
Ministry of Agriculture and
  Fisheries
Hamilton, New Zealand

Gregory C. Mattern, Ph.D
Miles, Inc.
Environmental Research Station
Stilwell, Kansas

Patricia A. Nugent
Formulations and Environmental
  Chemistry Laboratory
DowElanco
Midland, Michigan

Bruce E. Richter, Ph.D
Lee Scientific Division
Dionex Corporation
Salt Lake City, Utah

Joseph D. Rosen, Ph.D
Department of Food Science
Cook College
Rutgers University
New Brunswick, New Jersey

Lamaat M. Shalaby, Ph.D
E. I. du Pont de Nemours & Co.
Du Pont Agricultural Products
Experimental Station
Wilmington, Delaware

Hans-Jürgen Stan, Dr. rer. nat.
Institute of Food Chemistry
Technical University Berlin
Berlin, Germany

Harald Steinwandter, Ph.D
Department of Organic Residues
  and Environmental Chemicals
Hessian Agricultural Research
  Station
Darmstadt, Germany

Michael J. Wraith
Shell Research Ltd.
Sittingbourne Research Centre
Sittingbourne, England

Alan S. Wright, Ph.D
Shell Research Ltd.
Sittingbourne Research Centre
Sittingbourne, England

TABLE OF CONTENTS

Part III. Emerging Technologies

Part I. Extraction and Cleanup

Chapter 1

DEVELOPMENT OF MICROEXTRACTION METHODS IN RESIDUE ANALYSIS

Harald Steinwandter

TABLE OF CONTENTS

I. INTRODUCTION

In residue analysis, the used amount of chemicals and solvents released into the environment is usually a factor 10^8 larger than the quantity of the polluting analytes to be determined.

Therefore, this report draws attention to a little heeded and yet very grave problem that influences our analytical work, namely, the huge discrepancy between those quantities of solvents and chemicals that we use on the one hand and those of harmful analyte substances on the other.

Additionally, an attempt has been made to establish which precepts have guided scientists and analytical chemists in the past, and whether they still afford adequate guidance and protection in an age of permanent destruction of the environment. Do we require for the future new, extended guidelines, i.e., a timely "Categorical Imperative" (see Section II) that also takes into account conservation of the environment and related topics, a new paradigm that will enable us to reevaluate our chemical problems and take the appropriate measures. Furthermore, this report will offer some practical suggestions and results that can be derived directly from the moral compulsion of the "Extended Categorical Imperative" and which can provide the answer to our chemical problems.

Without anticipating this matter, the right answer to our chemical problems is the consistent use of micro methods by which the solvent and chemical amounts for the extraction and cleanup steps are reduced to 1/10 to 1/100 of that used by the macro methods.

Whereas any conventional cleanup method could be miniaturized very easily,[3–9] miniaturization of conventional extraction methods[10–14] was not possible without greater inconvenience, because by these methods the extraction, filtration, and partitioning steps are conducted off-line, that is, these steps are separated from each other by space and time.

This situation changed in 1985, when a new universal extraction technique was developed[9,15–17] by which the filtration, partition, and shake-out steps were eliminated.

Because all individual steps of this so-called on-line method are conducted in the same vessel either simultaneously or sequentially, it is inherent to the on-line method that this technique was for the first time perfectly suitable for miniaturization: for example, if a 1-g fruit sample is extracted on-line with 2 ml of acetone, 2 ml of petroleum ether, and ca. 0.5 g of NaCl, the Ultra-Turrax extraction is finished after ca. 15 s, so that not only is the chemical emission largely reduced, but also the results can be obtained far more quickly. This increased awareness of the chemist for the protection of both public health and the environment is in accordance with the 1989 postulated "Extended Categorical Imperative",[1,2] which involves not only the moral aspect of Immanuel Kant's "Categorical Imperative",[18] but also includes the general conservation of both life and the environment and unites the person with the environment.

Returning to residue analysis, we can say that analysis should be conducted in such a way that both the damage, i.e., the chemical emissions, and the benefits, i.e., the information of analysis, are in balance and therefore in harmony with the paradigm of the "Extended Categorical Imperative".[1,2]

II. THE "EXTENDED CATEGORICAL IMPERATIVE"

During Immanuel Kant's time (1724–1804), in which the "Categorical Imperative" (1786) was first presented in his *Grounding for the Metaphysics of Morals,*[18] relatively little or no knowledge existed about physics, chemistry, biology, astronomy, medicine, nuclear engineering, etc. compared to today. In addition, the possible destruction of the environment brought about by current human activity was also not known at that time. If Immanuel Kant had had any idea about environmental destruction by man, it is suggested that Kant would have formulated his "Categorical Imperative" in another form.

Therefore, we will try to integrate all our present knowledge into Kant's untimely "Categorical Imperative", so that this "Extended Categorical Imperative", which is now a timely imperative, is able to postulate human being and the environment as one unity. Principally, this "Extended Categorical Imperative" covers all human areas of economical, ecological, and scientific aspects, that is, of all producible and reproducible areas of human actions.

Although the "Extended Categorical Imperative" is a moral formula, it is also — because of the polarity of ethical and environmental aspects — another form, another expression of the demand to slow down the entropic increase of human actions determined by the second and fourth laws of thermodynamics.

Because of the continuously increasing and irreversible character of this entropy, which is finally the reason of any environmental destruction (acid rain, greenhouse effect, death of forests), the new paradigm of the above extended imperative requires keeping the entropy of human actions to a minimum, so that finally a balance can be reached between the mass independent entropy ($=$ information) and the mass dependent entropy ($=$ damage) of human actions and activities.

In the following section, the "Extended Categorical Imperative" is discussed exclusively in connection with specific analytical problems, and an attempt is made to establish the conditions under which environmental analytics are either beneficial or harmful.

A. THE "EXTENDED CATEGORICAL IMPERATIVE" AND ENVIRONMENTAL ANALYSIS

"Damage", in the context of pesticide analysis, is not described by the definition laid down by the European Economic Community (EEC) Standard

Commission, which recently pointed out that, according to its cautious estimates, the damage done to the national economies of all EEC countries by spurious results in trace analysis runs into billions in European Currency Units (ECU), not to mention misjudgments with possible criminal consequences.

It is clear that this type of damage is a risk that can be conserved by insurance and should be sorted out by bankers among themselves. The term "damage", however, is used in our context to denote the extremely large quantities of chemicals and solvents that we use to perform our environmental analyses. In short, the purpose of this chapter is, among other things, to highlight the fact that chemicals and solvents are being discharged from laboratories into the environment in quantities that are inconceivably large compared with the harmful substances that are being analyzed — or, to express it another way, the information yielded by the analyses is in no reasonable proportion to the attendant damage that the analyses themselves cause. These chemical emissions are becoming of more and more interest to the public, and not only for logistical reasons. Ultimately, there are economic and, especially, ecological aspects that make this a matter of top priority and public concern. They are economic because, in some cases, the cost of local disposal and waste treatment is greater than that of the pure chemicals and compounds themselves — and this also affects the national economy. They are ecological because the ultimate disposal of the chemicals and solvents and the mass flow into the environment is, in principle, one of the most serious problems facing us today and in the future. Only two catchphrases are mentioned here in this connection: the greenhouse effect and the hole in the ozone layer. The consequences, therefore, are being felt at not only national but also international levels.

Although all of this is known, nothing of significance is being done about it in the analytical world. It would not be surprising if analytical science soon found itself where it does not want to be: on the public firing line. The work of the environmental analyst today is very closely associated with pollution — and even destruction — of the environment. It is becoming more and more important to examine this other aspect of our analytical work as objectively as possible and with a critical eye. There is no room for discussion of the fundamental work and importance of the analyst; it is only a matter of answering the questions of "how" and "with what means". By what principles and standards should analytical chemists and scientists be led? The first thought that springs to mind is that of legislation. However, legislation is something that we would rightly neither agree with nor welcome since it would soon signal the demise of freedom of the sciences.

There must be other, innate, moral alternatives, in short, ethical maxims that can be applied to the sciences. In this connection, Immanuel Kant's "Categorical Imperative"[18] has been the guiding light for all actions of the enlightened western world since the 18th century. The "Categorical Imperative", which Kant formulated and postulated in his *Grounding for the*

Metaphysics of Morals,[18] is: "Act only according to that maxim by which
you can at the same time will that it shall become a universal law." Only
brief reference is made to this subject, since it is a painful experience to see
how scientists have behaved and still behave in the light of this grandiose
maxim. We need only recall, for example, the last two World Wars and the
recent use of poison gas in regional conflicts. None of these would ever have
been so horrifying if the best scientists at the time had not been involved.
Was their involvement a reflection of their morals or does the "Categorical
Imperative" lack a compulsion, with the result that its moral compass needle
failed to point out the right direction for scientists to follow?

The root of the trouble probably lies in the "Categorical Imperative"
iteself, which in the author's opinion and when viewed in retrospect, is too
individualistic and ultimately too vague. It is not without reason that not only
extreme individualism, but also socialism, identify with it. And which of
them is right? In the final analysis, this was bound to happen to the "Categorical Imperative" because Kant simply translated the logical expressions
for the most important forms of judgment in the relation directly to the
command formula of morals and, as is well known, there is nothing less
compatible than logic and morals.

Perhaps this whole dilemma could be solved by expanding Kant's "Categorical Imperative".[18] In its extended form, it would show the way as follows:
"Think and act not only according to that maxim by which you can at the
same time will that it should become a universal law but think and act also
according to that maxim that you can at the same time will, that also the
general conservation of the environment and the human being are included
in that maxim."

What does all this mean in practice — for example, in residue analysis
— and how are we to understand it? Consider the following scenario. What
should be the ultimate precept behind our actions when we as scientists stand
between the transgressions committed and the product that has to be protected.
And let us first consider what the "Categorical Imperative" and then the
"Extended Categorical Imperative" — with its environmental dimension
acting as compass needle and top priority for our actions — have to say on
the subject. We see immediately from the former that the analyst who "just"
performs his analysis well has done more than enough. His results are reproducible, correct, and the coefficients of variation are small. In other words,
the analyst could say that the maxim behind his actions is to work accurately
and precisely and that he would like this to be a general law.

However, on the basis of the new "Extended Categorical Imperative",[1,2]
this maxim would be totally inadequate, because good analytical work would
no longer be accorded exclusive top priority, but would ultimately find its
justification in combination with actions that conserve the environment or
minimize damage caused to it by analytical activity. It is no longer good

enough just to be a good analyst. There can be no doubt that one of the many consequences of continuing to do what we have always done will be that in the next 10 or 20 years we will have to analyze everything that we — in good faith, but with little consideration — are currently discharging and have discharged into the environment.

For this reason, we must now call a halt and consider by what means we are going to solve this urgent problem with solvents and chemicals, if not radically — because then we would not be able to perform any more analyses — then at least to a reasonable extent. The problem is made all the more urgent by the toxicological properties of the chemicals, an aspect not to be underestimated. Without going into too much detail, we at Darmstadt have solved this whole sorry problem of chemicals in analytical work simply by using micro methods for all of our work. This micro strategy — which ultimately is the direct consequence of the "Extended Categorical Imperative", which postulates all the time: act global, think global — has been pursued by us in Darmstadt for more than 15 years.

By this micro method, not only is the chemical emission into the environment greatly reduced, but also the information, i.e., the benefits of analyses, is in a balanced relation to the damage caused and is therefore in harmony with the paradigm of the "Extended Categorical Imperative".[1,2] The development of microextraction methods is discussed in the following section.

III. MICROEXTRACTION METHODS

A. REQUIREMENTS OF A MICROEXTRACTION METHOD

A microextraction method is considered to be a technique by which solvent and chemical consumption is reduced to 1/10 to 1/100 of a macro method and to be a procedure in which all pesticide residues are transferred from each and every sample matrix into the organic phase. Since the variety of pesticides ranges in terms of polarity from nonpolar (water-insoluble) to very polar (water-soluble), it is easy to imagine that a truly universal method by this definition never will be found. At best, the method of choice always will be "relatively" universal. Considering the possible differences in the chemistry of the sample matrix (soil, milk, meat, etc.), this will be even more so.

The problem of the sample chemistry can be eliminated, to a large extent, when the water content of all samples is adjusted to roughly 5 g, corresponding to about 60 to 95% of the sample plus added water. This usually is accomplished by presoaking the sample in water or by adding water to enhance the extractibility of pesticide residues. The goal of this procedure is to bring all samples to the same starting conditions prior to the extraction.

For the partitioning of the pesticides, it is necessary to use a nonpolar solvent, which should be miscible with the polar solvent and nonmiscible with water. Therefore, if the original sample extract, which is a binary solvent system (existing of the polar solvent and water), is treated with the nonpolar solvent, a ternary solvent system is obtained that separates into two layers: an aqueous and an organic phase. By this procedure, the pesticides are distributed from the aqueous phase into the organic phase. The extent of this distribution into the organic phase depends mainly on the solubility of each pesticide in both solvent layers, and is discussed in the following section.

B. DISTRIBUTION OF PESTICIDES

The quantitative aspect of pesticide distribution between two or more "nonmiscible" solvents are presented in the following paragraphs.

The Nernst distribution coefficient K — Suppose that an equivolume two-phase system contains two or three partially miscible solvents. Suppose also that the initial concentration of the pesticide of interest in the lower (usually polar) phase is a. Then, after mixing the solvents, the fraction x distributes into the upper (usually nonpolar) phase. The concentration remaining in the lower phase will then be $a - x$. At constant temperature

$$\frac{x}{a - x} = \text{constant } (K)$$

independent of the total amount of dissolved pesticide present. The latter distributes itself between two layers in a constant ratio. The value K is called the Nernst distribution coefficient.

The distribution value p — In residue analysis, the p value is more commonly used than the K value:

$$p = \frac{x}{a}$$

where p gives the fraction x of the total amount of a distributed into the equivolume upper phase. For example, a p value of 0.80 indicates that 80% of the pesticide is in the upper phase, while 20% is in the polar (lower) phase. This is easier to handle than the corresponding K value of 4.

From the preceding, it is evident that the higher the K or p value of the pesticides, the better the extraction efficiency. If the p values are low, a quantitative extraction can be reached by multiple extraction steps (see Section III.C.2.b).

Note: In residue analysis, only such solvents — including water — should be used for which the K or p values of all pesticides are greater than 10 or 0.9, respectively.

C. SOLVENT SYSTEMS FOR PESTICIDE EXTRACTION

In residue analysis, either two or three solvents — at least one of which must be nonmiscible with the others — are used to extract and to partition

FIGURE 1. Binary liquid system with completely miscible solvents, for example, water (A) + acetone (B).

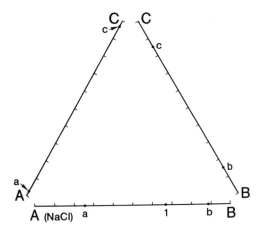

FIGURE 2. Three binary liquid systems with partially miscible solvents. For example, AC = water + petroleum ether; AB = water (NaCl) + acetone; BC = acetonitrile + petroleum ether.

the pesticides of interest. Binary and ternary systems and some theoretical and practical aspects concerning the extraction and partition of pesticides are discussed in the following sections.

1. Binary Solvent Systems
a. Solvents that are Completely Miscible
Let us consider the extraction of a 5-g sample containing 4.5 g of water. After blending the sample with 10 ml of acetone, for example, a binary solvent system is obtained. A binary solvent system can be represented by a straight line, where the two endpoints are the pure solvents (A and B in Figure 1) and the points in between give the composition of any mixtures. From here on, water is always placed at the left side of the line. Point 1 in Figure 1 shows the binary system obtained above, where the solvents are completely miscible.

b. Solvents that are Partially Miscible
Important partially miscible binary solvent systems for the partitioning of pesticides are (1) water and petroleum ether or dichloromethane; (2) acetonitrile and petroleum ether; and (3) water and acetone saturated with solid sodium chloride.

Consider, for example, that two partially miscible solvents, A (water saturated with NaCl) and B (acetone), are mixed to produce a total composition indicated by point 1 shown in Figure 2. The two conjugate solutions then

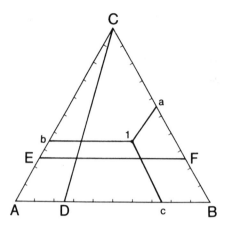

FIGURE 3. Ternary liquid system with three completely miscible solvents. (From Stein-wandter, H., in *Analytical Methods for Pesticides and Plant Growth Regulators*, Vol. 17, Sherma, J., Ed., Academic Press, San Diego, 1989, 35. With permission.)

have compositions indicated by points a and b. The relative amounts V of layers a and b are given by

$$\frac{\text{V of a}}{\text{V of b}} = \text{distance } \frac{1b}{1a}$$

The closer the point 1 is to a, the larger will be the proportion of that layer.

Three partially binary systems, AB, AC, and BC, are shown in Figure 2. They are ordered in the form of a triangle on purpose. From here it is easy to proceed to ternary systems: water always is in the left-hand corner, the polar solvent is in the right-hand corner, and the nonpolar solvent is at the top.

2. Ternary Solvent Systems
a. Ternary Systems with Three Completely Miscible Solvents

The equilateral triangle, introduced by Gibbs, is most suitable for the discussion of the ternary system with solvents A, B, and C. The advantages can be seen in Figure 3:

1. The corners of the triangle represent the pure solvents A, B, and C.
2. The distance from point 1, which represents the composition and a ternary mixture, to any side, gives the fraction of the solvent occupying the opposite corner, when measured parallel to either of the others. The distances 1a, 1b, and 1c represent the fractional volume X(A), X(B), and X(C) of A, B, and C in the ternary system indicated by point 1. The sum of these distances is always the same and is equal to one side of the triangle.

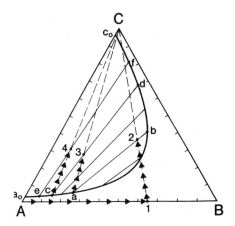

FIGURE 4. Ternary liquid system with one pair of partially miscible solvents. Demonstration for a multiple extraction with solvent C. AB = water + acetone; AC = water + dichloromethane; BC = acetone + dichloromethane.

3. Taking each side as unity and expressing the volume of the three solvents as fractions of total, it follows:

$$X(A) = \frac{V(A)}{V}; \quad X(B) = \frac{V(B)}{V}; \quad X(C) = \frac{V(C)}{V}$$

where $V(A) + V(B) + V(C)$ = total volume V.

4. Assume that a line is drawn from corner C of the triangle to point D of the opposite line AB. Then all points on line CD have the same ratios of A and B. A line parallel to one of the sides — for example, EF — represents a constant volume of C with variable amounts of the others.

b. Ternary Systems with One Pair of Partially Miscible Solvents

If the solvents in the ternary system are A, B, and C and it is assumed that the pairs AB and BC are completely miscible when A and C are only partially miscible, a system with the corresponding miscibility gap results (see Figure 4).

It is of interest to note that this system represents all acetone[10,12] and acetonitrile[14] methods, with A = water, B = acetone or acetonitrile, and C = petroleum ether and dichloromethane. If acetonitrile is the polar solvent, only dichloromethane can be used. For all further discussions, the arrangement of solvents is always the same as already mentioned in the section concerning the binary solvent system: water is in the left corner of the triangle, the polar organic solvents are in the right corner, and the nonpolar organic solvent is in the top corner. Using Figure 4, the principle of the extraction procedure of these methods can be described as follows. Assume, for example, that C is added to a binary mixture of A and B with composition 1, also containing pesticides, so that a ternary mixture with composition 2 is reached. Two conjugate ternary solutions with compositions a and b are obtained: point a corresponds to the water-rich layer and point b corresponds to the organic-rich layer into which the pesticides are distributed.

In cases when the pesticides of interest are not extracted quantitatively by a single extraction procedure, additional extraction steps are necessary. For that purpose, layers a and b are separated and layer a is further extracted with solvent C; the total composition is now that of point 3, and the conjugate solutions joined by the corresponding tieline have compositions c and d. If the extraction is repeated, the layer with composition c is extracted with another portion of solvent C, so that we obtain phases e and f. If this procedure is continued, the composition of the two layers moves along the binodal curve — which is obtained when the various points, representing the compositions of the conjugate solution, are joined — to finally reach the binary system AC with the corresponding composition of the two conjugate layers a_0 and c_0, which indicate that by a multiple extraction with solvent C we have isolated both the pesticides and polar solvent B from the original sample extract 1. Finally, extracts b, d, and f are combined and evaporated for further pesticide determination.

c. Ternary Systems with Two Pairs of Partially Miscible Solvents

If solvents A and C, as well as B and C, are partially miscible, there are two binodal curves which will overlap to form a continuous band from side AC to side BC with common tielines. Similar figures are obtained if the solvent pairs AC and AB on one hand, and AB and BC on the other hand, are partially miscible.

If the miscibility gaps of solvent pairs AC and BC are overlapping, one obtains (see Figure 5) the Mills et al. method[11] with A = water, B = acetonitrile, and C = petroleum ether, while Steinwandter's on-line method[15] is represented by the two partially miscible solvent pairs AB and AC forming a continuous band (Figure 6) with A = water (NaCl), B = acetone or acetonitrile, and C = petroleum ether, dichloromethane, or ethyl acetate.[9]

D. THE EVOLUTION OF EXTRACTION METHODS

The evolution of the off-line and on-line extraction methods is discussed with the concept of ternary liquid systems in mind. To simplify the discussion, the following premises are made:

1.   The extraction solvents are placed in the corners of the Gibbs triangle, always in the same sequence: water is in the left corner, the polar organic solvent is in the right corner, and the nonpolar solvent is in the top corner.

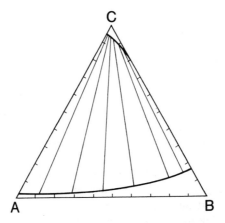

FIGURE 5. The ternary liquid system of the Mills et al. method.[11]

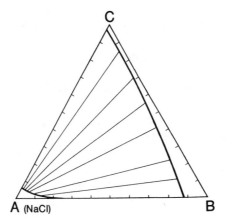

FIGURE 6. The ternary liquid system of the Steinwandter on-line method.[15]

2. Only samples with a water content adjusted to about 90% are used and discussed for extraction.

3. Only the most important solvents, such as acetone and acetonitrile, are used as the extraction agent, while for the water-removing and partitioning step, petroleum ether, dichloromethane, or sodium chloride are employed.

4. The Gibbs triangle for each method discussed has more qualitative than quantitative character.

5. Off-line extraction methods are techniques by which all steps (extraction, filtration, taking aliquots, partitioning) are separated each from another by space and time.

6. On-line extraction methods are techniques by which all steps are conducted in one vessel. Therefore, no filtration step and no separatory funnel are necessary. In principle, all steps are conducted simultaneously; that is, all steps occur in the same space and at the same time.

Although the evolution of extraction methods has already been described,[9,16,17] some remarks are necessary for better understanding. If we consider all the arrangements that are possible in the Gibbs triangle and that do not give a new figure using water, acetone, acetonitrile, petroleum ether and dichloromethane and NaCl as electrolytes, theoretically eight different Gibbs triangles (see Figure 7) for the off-line methods are obtained. From these figures, one is a three-phase system and cannot be used. Therefore, only seven basic methods could be developed in the past, so that one can easily understand that the extreme disproportion of real, existing extraction methods, compared to published methods comes from the fact that minor variations have led to ''new'' published methods. One can see that most methods are only variations of a few known procedures. Moreover, if we introduce only one step into all existing off-line methods, that is, if NaCl is added to each method using acetone or acetonitrile as the extracting agent, all steps before and after the partitioning procedure can be standardized to one basic extraction method,[9,15–17] as shown in Figure 8. In addition to petroleum ether and dichloromethane, ethyl acetate can also be used[9] as the nonpolar solvent. However, by combining all fundamental off-line extraction methods[10–14] into one single standard off-line procedure, one problem still has to be eliminated: in order to determine the sample weight in the analyzed extract, the water in the sample extracts, and therefore in the original sample, must be known. This problem is overcome by the so-called on-line technique, by which all working steps are combined into one single extraction step; that is, by saturating the sample extract with NaCl and simultaneously driving away the water with petroleum ether, dichloromethane, or ethyl acetate in the same reaction vessel. Although quite different in its procedures, this on-line method has the same Gibbs triangle as the standard off-line method. Therefore, we must find a way to distinguish between these two working techniques.

E. TECHNIQUES OF OFF-LINE AND ON-LINE EXTRACTION METHODS

In the following sections, the different working techniques and working steps of the off-line and the on-line methods — demonstrated with the corresponding micro method — are discussed in order to learn more about the underlying principles. In the following, acetone is used instead of acetonitrile because acetone is more volatile, less expensive, less toxic, and is highly efficient.

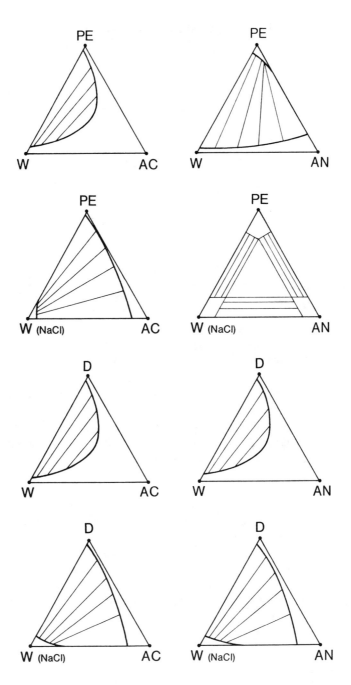

FIGURE 7. The eight basic extraction methods with the solvents water (W), acetone (AC), acetonitrile (AN), petroleum ether (PE), and dichloromethane (D).

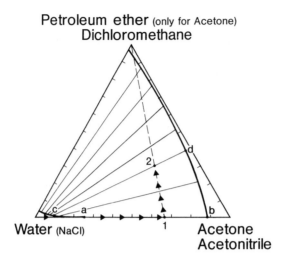

FIGURE 8. Common principle of all extraction methods represented by the Gibbs triangle. (From Steinwandter, H., *Fresenius J. Anal. Chem.*, 342, 150, 1992. With permission.)

Partition of pesticides into phase b with sodium chloride — The basis of the presented extraction procedure is the use of sodium chloride (salting-out effect) for the partitioning of pesticides from the acetone-water extract with composition 1 into the organic phase b (see Figure 2 and Figure 8). The extent of this partition with sodium chloride can be analyzed if the pesticides are dissolved in the system acetone-water (2:1 v/v) having point 1 and are then partitioned with sodium chloride. This procedure gives >90% recoveries for nonpolar pesticides, while the rates for polar pesticides are somewhat lower. The value of dimethoate, for example, is >80%.

The off-line extraction method — Assume that a 5-g fruit or vegetable sample, which contains 4.5 g of water and is represented by the left corner of the Gibbs triangle, is extracted with 10 ml of acetone. By this extraction step, the sample composition moves from the left water corner of Figure 8 to point 1, which is now a binary solvent system existing of water plus acetone. The sample extract with composition 1 is now filtered and a measured aliquot of this sample extract is poured into a separatory funnel. After addition of ca. 2 g of NaCl, a partially miscible solvent system is reached with phases a and b. Into the latter, which is the acetone-rich phase, the analytes are partitioned. Ten milliliters of the above-mentioned nonpolar solvents are now added for water removal from the sample extract. By this procedure, the binary system is transferred into a ternary system, so that the total composition of the system moves from point 1 to point 2. The conjugate solutions joined by the corresponding tieline have compositions c and d. The organic phase d with the pesticides of interest is dried and analyzed. Therefore, the following six working steps are characteristic for the standard off-line method:

1. The extraction step (E)
2. The filtration step (F)
3. The measurement of an aliquot (A)
4. The transfer of the aliquot into a separatory funnel (S)
5. The partitioning step (P)
6. The water removal step (R)

In this connection, it is important to note that it is inherent to each off-line method that steps 1 to 4 are separated from each other by space and time: therefore, the name off-line method. That is, for example, the extraction (E) of the sample is conducted in a separate extraction vessel. The filtration (F) step requires that the sample extract must be transferred from the extraction vessel through a filter (F) into a separate beaker. From this beaker, an aliquot (A) is then taken with a separate measuring cylinder. This aliquot is then poured into a separatory funnel (S), which is again separated from the previous step (measuring cylinder) by space and time. In contrast to the above four off-line steps E, F, A, and S, the two following P and R steps are conducted in the same vessel (separatory funnel), so that both steps can be conducted, therefore, either simultaneously (the same space and time) or sequentially (the same space but separated by time).

The on-line extraction method — Although the on-line method has the same Gibbs triangle as the off-line extraction method (see Figure 8), the working of the on-line technique is fundamentally different, as can be seen below. Let us assume that the same 5-g fruit or vegetable sample, which was extracted by the above-mentioned off-line method, is used. Then, using the on-line method, 10 ml of acetone, 10 ml of the above-mentioned nonpolar solvent, and ca. 2 g of NaCl are added directly to the 5-g sample followed by a 1-min extraction. Therefore, by this technique, the extraction (E) step, the partitioning (P) step, and the water removal (R) step are conducted simultaneously in one step, so that by this on-line technique no filtration (F) step, no measurement of an aliquot (A), and no separatory (S) funnel are necessary.

Therefore, the on-line extraction can be described as follows: the original composition of the sample — represented by the left water corner in Figure 8 — moves directly over point 1 to point 2 in one step, yielding the two conjugate layers c and d. In the latter, the pesticides can be determined. The following steps are therefore — in contrast to the off-line method — characteristic for the on-line method:

1. The extraction step (E)
2. The partition step (P)
3. The water removal step (R)

In addition, because all steps are conducted in the same extraction vessel (the same space), all three working steps can be conducted either simultaneously (the same space and time) or sequentially (the same space but separated by time). It is inherent to the on-line method that no filtration step (F), no measuring of an aliquot (A), and no separatory (S) funnel are necessary.

F. THEORETICAL ASPECTS OF THE OFF-LINE AND ON-LINE EXTRACTION METHODS

In the previous sections, it was shown that there are fundamental differences between the off-line and on-line extraction methods; however, these cannot be seen from the corresponding Gibbs triangles, which are the same for both techniques; the Gibbs triangle also has its limitations. The reason for this is that the Gibbs triangle only represents the composition of binary and ternary mixtures by a point in the system and does not describe individual steps by which the composition of that point in the system is reached. From that, it follows that the composition of one point in the Gibbs triangle is independent, whether it was reached in one or several stages. Now one can see why all different extraction methods cannot be differentiated from each other using the Gibbs triangle.

Therefore, we must find a way to show how different working and extraction procedures, which are separated by space and time, or how simultaneously and sequentially conducted steps can be distinguished using the Gibbs triangle.

This problem is overcome if the sequence of the working steps is completely described, which leads to one of the two organic phases b and d (see Figure 8) of the off-line and on-line methods. The resulting sequence of the working steps is then indicated in brackets behind the points b and d. Now it is possible to describe precisely the various extraction sequences of both the off-line and the on-line extraction principles. Their working can be summarized as follows:

1. For the above-mentioned individual working steps on extraction (E), filtration (F), taking of an aliquot (A), use of a separatory (S) funnel, partitioning (P) of pesticides, and removal (R) of water, the symbols E, F, A, S, P, and R are used.

2. The above symbols are always in the same sequence: E, F, A, S, P, R for the off-line methods, and E, P, R for the on-line methods. The extraction (E) is always the first step and R is always the last step.
3. If the working steps are separated by space and time, the symbols are separated from each other by a semicolon, for example, E; F, which represents the two first steps of the off-line extraction method.
4. If the working steps are sequentially conducted in the same vessel, meaning that they are separated only by time but not by space, the symbols are separated by a comma, for example, E, P, which represents the two first steps of the on-line method: at first the extraction step E takes place, followed by the partition step P.
5. If the working steps are simultaneously conducted in the same vessel, i.e., the steps are not separated by space and time, the semicolon or the comma between the symbols of the working steps is omitted, for example, EP, which represents the first two steps of the on-line method, but in another extraction mode.
6. All working steps that are conducted to obtain the organic phases b or d are written within brackets behind the organic phases b or d, for example, b (EP) or d (E, PR), so that we know phase b is reached by the simultaneously conducted steps E and P, while phase d is reached at first by the single extraction step E followed by the simultaneously conducted steps P and R.

G. THE EXTENDED GIBBS TRIANGLE FOR THE REPRESENTATION OF THE OFF-LINE AND ON-LINE METHODS

If the history of individual steps of each technique is described, which leads to one of the organic phases b and d and then is set into brackets after the two points b and d, then each method can be exactly characterized.

1. The Extended Gibbs Triangle of the Off-Line Method

In the off-line method, the working steps E, F, A, and S are each separated by space and time, so that these working steps are separated by a semicolon while steps P and R are conducted in the same vessel (separatory funnel), so that these two working steps can be done either sequentially or simultaneously (see Sections III.F.4 and III.F.5).

Suppose we ask, for example, what are the working sequences to obtain the organic phase b? It follows from the above points in Sections III.F.1 through 6, that this phase is reached — as can be seen in Figure 9 — only by the working sequence b (E; F; A; S; P).

On the other hand, if we ask what steps are necessary to reach phase d, the following two working sequences are possible: d (E; F; A; S; P, R) and d (E; F; A; S; PR), as shown in Figure 9. This means that only the partitioning (P) step and the water removal (R) step can be conducted either sequentially

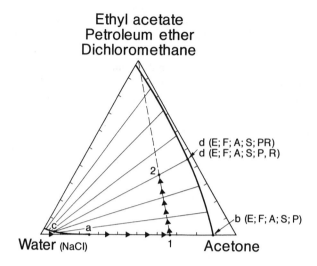

FIGURE 9. The three extraction sequences for the characterization of the standard off-line method to reach the organic phases b and d using the ternary liquid system.

d (E; F; A; S; P, R) or simultaneously d (E; F; A; S; PR), while all the other steps are separated by space and time, and, therefore, each symbol is separated by a semicolon. The reason is that — in contrast to the E, F, A, and S steps — the P and R steps are performed in the same vessel (separatory funnel).

2. The Extended Gibbs Triangle of the On-Line Method

Because all working steps are conducted in the same extraction vessel, it is inherent to the on-line method that all working steps can be done both simultaneously and sequentially, leading to a total of six possible sequences, as discussed below.

If we ask, for example, what are the working sequences to obtain the organic phase b, we have to write — as can be seen in Figure 10 — the two sequences b (EP) and b (E, P), which means that the sample can be extracted (E) and partitioned (P) either sequentially or simultaneously (see points III.F.4 and III.F.5). To reach the organic phase d, the four working sequences d (EPR), d (E, PR), d (EP, R), and d (E, P, R) are possible, as shown in Figure 10. This means that in addition to the straight sequential or simultaneous working steps, any combination of both can be used.

H. ADVANTAGES OF THE ON-LINE METHOD

From the previous Sections III.G.1 and III.G.2, one can see that the advantages of the on-line extraction technique over the off-line method are:

1. No filtration step is necessary.
2. No aliquot of the sample extract is necessary.

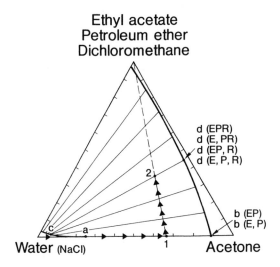

FIGURE 10. The six extraction sequences for the characterization of the on-line method to reach the organic phases b and d using the ternary liquid system. (From Steinwandter, H., *Fresenius J. Anal. Chem.*, 342, 150, 1992. With permission.)

3. No separatory funnel for the partitioning step is necessary.
4. All working steps are conducted in the same extraction vessel.
5. The volume of the organic phase obtained is independent of the sample water content.
6. The volume of the measured organic phase is the sum of the added volume of the polar and the nonpolar solvent.
7. In addition to the straight simultaneous and sequential working steps, any combination of both can be used.
8. A minimum of extraction steps results in a minimum of analytical error.
9. A minimum of glass equipment is necessary.
10. The extraction is finished after 15 to 30 s.
11. Instant phase separation is obtained in most cases.

These points 1 through 11 are inherent to the on-line technique and make it, therefore, perfectly suitable for the development of micro extraction methods.

I. THE MICRO ON-LINE EXTRACTION METHOD

Microextraction methods are such that the amount of solvent and chemicals is reduced to 1/10 to 1/100 of the macroextraction methods.

1. The Principle

Because all working steps of the on-line method are conducted in the same reaction vessel, any variation of simultaneous and sequential procedures is practicable. Therefore, various strategies are offered by the on-line method

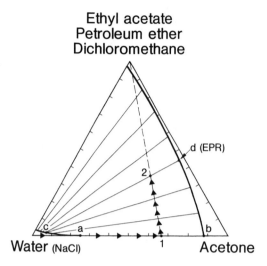

FIGURE 11. The extraction sequence d (EPR) of the on-line method for the determination of pesticides in samples with water contents >70%.

to decide what extraction sequence for what kind of sample is the most suitable at any time.

One remark about the micro on-line method seems to be necessary. Generally, each analyzed sample can contain varying amounts of water, so that it is necessary to bring all samples into the same starting condition prior to the extraction, as described earlier.[9,15–17] This is achieved by adjusting the acetone/water ratio (we use acetone instead of toxic acetonitrile) in all samples to a value of about 2. In addition, to use the micro on-line extraction method, the preparation of representative samples is also important. Useful guidelines of composition and subsampling are described elsewhere.[19–24]

For reanalyses, the sample should be stored in 100- or 200-g portions in a deep-freezer. If the samples are subsampled and stored in small pieces, homogenize or blend the sample with a suitable apparatus prior to the microextraction. For the extraction, take an aliquot of the homogenized and, therefore, representative sample. On this condition the microextraction method can be described as a universal extraction method.

To demonstrate the practicability, flexibility, and simplicity of the on-line method, the microextraction of some samples with different water contents is described in the following section using the Gibbs triangle.

2. Solvent Selection

The use of dichloromethane in the following illustration of the Gibbs triangle (see Figures 11 and 12) has only historic value and does not reflect the present practice in the author's laboratory.

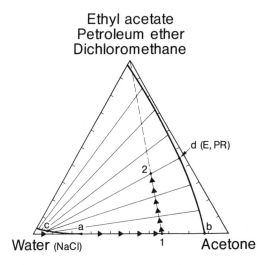

Ethyl acetate
Petroleum ether
Dichloromethane

FIGURE 12. The extraction sequence d (E, PR) of the on-line method for the determination of pesticides in samples with water contents <15%.

Besides its carcinogenic potential, dichloromethane is also problematic in terms of environmental questions. Therefore, petroleum ether and ethyl acetate should be used as the nonpolar solvents. Most of the earlier conducted investigations with dichloromethane described in this paper were restudied with petroleum ether or ethyl acetate, and it was found that similar results were obtained.

3. The Micro On-Line Extraction of Fruits and Vegetables (Water Content >70%)

Weigh 5 g of a homogenous and representative subsample into the extraction vessel (i.d. = 25 mm, h = 100 mm) and add 10 ml of acetone, 10 ml of the above-mentioned nonpolar solvents, and ca. 2 g of NaCl. Extract with the Ultra-Turrax for 15 to 30 s. Let stand for 5 min. Pour the organic extract containing the pesticides into a 50-ml Erlenmeyer flask containing some Na_2SO_4. Stopper the flask and dry the extract on a magnetic stirrer using a stirrer bar. Transfer an aliquot, for example, 10 ml of the organic phase (= 2.5-g sample), into a 50-ml pear-shaped flask and reduce the volume to ca. 0.5 ml. Reconcentrate with a solvent of your choice. Repeat the evaporation twice. Dry the extract with ca. 50 mg of Na_2SO_4. Transfer the extract quantitatively to a 1-ml graduated flask.

As can be seen from the above extraction procedure, all individual steps are conducted simultaneously, so that the organic phase d is directly obtained, and the history (see Figure 11) to reach the organic phase d is d (EPR). This means that E, P, and R are carried out at the same time (simultaneously) in the same extraction vessel.

TABLE 1

**Extraction of Pesticide Residues from 13 Samples with Unknown
Spray History by the Micro- and Macroextraction Methods**

			Residue (µg/kg) analyzed by		
Laboratory sample no.	Analyzed sample	Identified compound	Micro method using D	Micro method using P	Macro method using D
P 1	Brussels sprouts	Bromophos-ethyl	71	ND	78
P 2	Tomatoes	Pyrazophos	9	ND	10
P 3	Clementines	Methidathion	680	ND	650
P 4	Cabbage	Chlorpyriphos	20	ND	18
P 5	Apples	Chlorpyriphos	24	ND	22
		Parathion	9	ND	10
P 6	Green peppers	Ethion	13	ND	12
P 7	Grapes	Chlorpyriphos	4	5	5
P 8	Tomatoes	Pyrazophos	42	43	35
P 9	Carrots	Bromophos-methyl	110	110	100
		Parathion	17	15	13
P10	Lettuce	Parathion	18	20	15
P11	Cucumbers	Dimethoate	320	360	360
P12	Brussels sprouts	Bromophos-ethyl	120	170	120
P13	Pears	Chlorpyriphos	16	19	18

Note: D = dichloromethane; P = petroleum ether; ND = not determined.

From Steinwandter, H., *Fresenius J. Anal. Chem.*, 336, 8, 1990. With permission.

To show the capability of the micro on-line method,[9,25,26] 13 contaminated fruit and vegetable samples with unknown spray history were analyzed both with the described microextraction method and the macro method described earlier.[15] Because dichloromethane is a problematic solvent in terms of its toxic properties, PE was included in these investigations as the partition solvent. The results are listed in Table 1. One can see a good correlation of the analyzed values.

For recovery studies of those pesticides that were analyzed in the samples of unknown spray history, an untreated sample was fortified with the pesticides bromophos-ethyl, chlorpyriphos, dimethoate, ethion, methidathion, parathion, and pyrazophos at levels of 100 to 500 µg/kg. The results are listed in Table 2, indicating that the on-line extraction method works quantitatively.

Note: The same weight g in the aliquot volume V of the organic extract is

$$g = g_0 \frac{V}{V_0}$$

where g_0 = total weight of sample taken, V = aliquot volume of organic phase, and V_0 = total volume of organic phase. As mentioned above, the

TABLE 2
Recovery Studies by the Micro On-Line Extraction Method

		Residue (µg/kg) analyzed by		
Compounds	Fortification level (µg/kg)	Micro method using D	Micro method using P	Macro method using D
Bromophos-ethyl	100	110	120	100
Chlorpyriphos	100	100	100	110
Dimethoate	200	170	170	180
Ethion	100	105	110	100
Methidathion	100	110	110	95
Parathion-ethyl	100	100	110	110
Pyrazophos	500	450	470	470

Note: D = dichloromethane; P = petroleum ether.

From Steinwandter, H., *Fresenius J. Anal. Chem.*, 336, 8, 1990. With permission.

total volume of the organic phase is the sum of the added polar and nonpolar solvent. In our case, therefore, V_0 = 20 ml.

4. The Microextraction of Animal Feedstuff and Dried Fruit and Vegetable Samples (Water Content <15%)

Weigh 1 g of a representative subsample into the extraction vessel (i.d. = 25 mm, h = 100 mm) and add 5 ml of water and 10 ml of acetone. Mix with the Ultra-Turrax for ca. 15 s. Let stand for ca. 10 min. Add ca. 2 g of NaCl and 10 ml of the above-mentioned nonpolar solvents and mix again for ca. 30 s. Pour the organic phase into a 100-ml Erlenmeyer flask and continue as described in Section III.I.3.

If we now describe how the organic phase d is produced, the following procedure is applicable:[23] at first the extraction step E is carried out and then, in sequence, the two simultaneous steps P and R are performed, so that we can write d (E, PR), as shown in Figure 12.

Results of a collaborative study conducted by the GDCh Arbeitsgruppe "Pesticide"[27] in 1989, in which any method could be used, are listed in Table 3. Each of the participating laboratories was sent a dried and pulverized carrot sample fortified with seven pesticides. One can see a good correlation of the micro on-line method with the theoretical values.

5. Micro On-Line Extraction of Soils (Water Content <15%)

Weigh 5 g of a 2-mm sieved and representative soil sample into a 100-ml flask and add 10 ml of water and 20 ml of acetone. Close the flask and extract the pesticide by shaking it on a mechanical shaker (ca. 120 cycles/min) overnight. Open the flask and add ca. 3 g of NaCl and 20 ml of the above-mentioned nonpolar solvents. Close the flask again and mix vigorously by hand for 2 min. Pour the organic phase into a 100-ml Erlenmeyer flask and continue as described in Section III.I.3.

TABLE 3
Extraction of Pesticide Residues from a Fortified Carrot Sample by the Microextraction Method[a]

Pesticides	Fortification level (μg/kg)	Concentration (μg/kg) analyzed by GDCh[27] after fortification	Residue (μg/kg) analyzed by the micro on-line method[26] using P
Diazinon	800	700	620
Dichlofluanid	800	450	500
Dichloran	400	350	390
Lindan	400	360	370
Parathion	500	430	490
Phosalone	1000	1020	980
Vinclozolin	500	440	480

Note: P = petroleum ether.

[a] A collaborative study of the GDCh-Fachgruppe "Pesticide".

From Steinwandter, H., *Fresenius J. Anal. Chem.*, 336, 8, 1990. With permission.

TABLE 4
Recovery Studies of the Micro On-Line Method

Compounds	Micro method (%)
Atrazine	84 ± 6.1 (n = 9)
Simazine	89 ± 5.4 (n = 9)
Terbutylazine	87 ± 5.5 (n = 9)
Deethylatrazine	80 ± 3.8 (n = 9)
Deisopropylatrazine	81 ± 3.0 (n = 9)
Deethylterbutylazine	90 ± 5.0 (n = 9)
Hydroxyterbutylazine	88 ± 5.7 (n = 9)

From Steinwandter, H., *Fresenius J. Anal. Chem.*, 339, 30, 1991. With permission.

We can see that the organic phase d is reached by d (E, PR), because the extraction (E) step with acetone is carried out overnight, followed by the two simultaneously conducted steps P and R, as shown in Figure 12.

Recovery studies from untreated soil samples fortified with triazine herbicides, including some metabolites, show[28] (see Table 4) that the micro on-line extraction method is quantitative. The recoveries ranged from 80 to 90%.

6. Alkaline Hydrolysis of Cereals Prior to the Micro On-Line Extraction

Weight 0.5 to 1 g of a powdered and representative cereal sample into the extraction vessel (i.d. = 35 mm, h = 100 mm) and add ca. 10 ml of a 0.1 M NaOH. Heat on a sand bath for 30 to 60 min. Cool the reaction mixture.

Add 1 ml of 10% H_2SO_4 to the alkaline solution so that the mixture is acidic. Add 20 ml of actone and mix with the Ultra-Turrax for 15 to 30 s. Add ca. 3 g of NaCl and 20 ml of the above-mentioned nonpolar solvents and mix again for ca. 30 s. Pour the organic phase into a 100-ml Erlenmeyer flask and continue as described in Section III.I.3.

The described procedure for the determination of chlorophenoxyalkane carboxylic acids is also an example of the flexibility and simplicity of the on-line extraction technique. The hydrolysis reaction and the micro on-line extraction of the chlorophenoxy acids are conducted in the same reaction vessel, so that a minimum of different operations and glass equipment is necessary.

Although an alkaline hydrolysis step is performed prior to the micro on-line extraction procedure, this has no influence on the extraction scheme. Phase d is obtained after two sequentially conducted steps. First (as shown in Figure 12), the E step is carried out, followed by the simultaneous sequence PR, so that we can also write d (E, PR). Recovery studies from untreated control samples fortified with 6 chlorophenoxy acids ranged, after acid-base extraction and methylation, between 80 and 90% and are comparable with the results obtained by the macro on-line method.[29]

7. BF₃ Treatment of Soil Samples Prior to the Micro On-Line Extraction

Weigh 5 g of a 2-mm sieved and representative soil sample into a 100-ml flask and add 10 ml of BF_3-methanol solution. Close the flask tightly with a screw cap. Shake overnight (ca. 15 h) on a mechanical shaker (ca. 60 cycles/min). Now the following three extraction procedures are practicable to isolate the pesticides:

1. Open the flask and add 10 ml of water and 20 ml of acetone. Extract with the Ultra-Turrax for 1 min. Then add ca. 3 g of NaCl and 20 ml of the above-mentioned nonpolar solvents and extract again for 1 min.
2. Open the flask and add 10 ml of water and 20 ml of acetone. Close the flask and shake for at least 5 h on a mechanical shaker (ca. 120 cycles/min). Open the flask and add ca. 3 g of NaCl and 20 ml of the above-mentioned nonpolar solvents. Close the flask and shake by hand vigorously for 2 min.
3. Open the flask and add 10 ml of water and 20 ml of acetone. Close the flask and shake for 1 min. Open the flask and pour the extract into a blender jar. Add ca. 3 g of NaCl and 20 ml of the above-mentioned nonpolar solvents and extract for 1 min.

In each case, pour the organic phase into a 100-ml Erlenmeyer flask and continue as described in Section III.I.3.

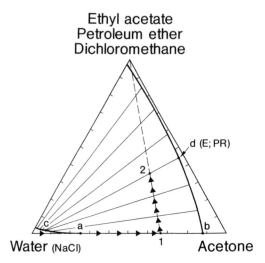

FIGURE 13.   The extraction sequence d (E; PR) of the on-line method for the determination of pesticides in samples with water contents <15%.

Although each soil sample is pretreated with BF_3, the working sequence of the first two described extraction procedures (Section III.I.7) to obtain the organic phase d is d (E, RP) (as shown in Figure 12), because the individual steps E, R, and P are conducted in the same reaction vessel. However, E is the first step — only separated by time — followed by the simultaneous steps RP.

In contrast to the above two d (E, RP) modes, it is of interest to note that if the above BF_3 sample extract is first extracted in the original flask with 10 ml of water and 20 ml of acetone and then poured into a separate blender jar (see Section III.I.7) for the partition of analytes with ca. 3 g of NaCl and 20 ml of the nonpolar solvents, we obtain the working sequence d (E; RP), as shown in Figure 13. That means the extraction step E, conducted in the original extraction flask, is separated by space and time from the next two simultaneously conducted R and P steps in the blender jar. The variability of the above three described extraction modes is again an indication for the great flexibility of the micro on-line extraction technique.

Results of two weathered soil samples, which were contaminated about 20 years ago with technical HCH and were analyzed by the microextraction method, showed that the recovery of these analyzed compounds is increased up to 300% by this procedure, and agree with the results obtained by the macro on-line method.[30]

8. Super Micro On-Line Extraction Method

Upon trying to reduce solvent and chemical amounts as far as possible, it was found that reduction to 1/100 of the original method is probably the limit. The procedure is as follows: weigh 1 g of representative and homogenized fruit or vegetable sample into the extraction vessel (i.d. = 25 mm, h = 100 mm) and add 2 ml of acetone, 2 ml of the nonpolar solvents, and ca. 0.3 g of NaCl. Extract with the Ultra-Turrax for 15 s. Pour the organic phase into a 25-ml Erlenmeyer flask containing some Na_2SO_4. Close the flask and dry on a magnetic stirrer using a stirrer bar. Transfer an aliquot, e.g., 2 ml, of the organic phase (= 0.5 g of sample) into a 10-ml pear-shaped flask and reduce the volume to ca. 0.5 ml. Concentrate down with a solvent of your choice. Repeat the evaporation twice. Dry the extract with ca. 10 mg of Na_2SO_4. Transfer the extract quantitatively into a 1-ml graduated flask. Fill up with the appropriate solvent.

The results obtained by the super micro on-line method are comparable with that of the micro and the macro on-line methods. The extraction procedure is d (EPR), as shown in Figure 11.

Note: In some cases we found that coextractives of some samples have a depressing influence on the recovery of compounds containing NH groups. We immediately solved this problem by adding some sodium carbonate to the sodium sulfate in the Erlenmeyer flask during the drying process, so that acidic sample compounds are eliminated and fixed on the surface area of the Na_2CO_3 particles. The recovery of compounds such as dimethoate from citrus samples could be increased if Na_2CO_3 was used.

9. Influence of Temperature on the Extraction of Atrazine from Soils

Weigh 5 g of soil into a 100-ml flask and add 10 ml of water and 20 ml of acetone. Close the flask and extract atrazine both by shaking on a mechanical shaker (ca. 120 cycles/min) for 16 h overnight at 20°C and by mixing for 16 h at 20 and 60°C in a water bath or in a drying oven using a magnetic stirrer and a stirrer bar. If a water bath is used, seal the magnetic stirrer watertight with polyethylene. Open the flask, cool it down, and add ca. 3 g of NaCl and 20 ml of dichloromethane or petroleum ether. Close the flask again and mix vigorously by hand for 2 min. Pour the upper organic phase into a 50-ml Erlenmeyer flask and continue as described in Section III.I.3. The organic phase b is reached by the following steps: first (as shown in Figure 12) the E step is carried out, followed by the simultaneous sequence PR, so that we can write: d (E, PR).

We found[31] that the fraction of the extracted atrazine is increased up to 70% if the temperature is raised from 20 to 60°C. This increased atrazine recovery is difficult to explain on the basis of "bound atrazine residue", which is defined as that additional residue[32,33] that can only be extracted by such drastic extraction techniques like the extraction with supercritical methanol[34] and the high temperature distillation.[35-37] From that point of view, the term "bound residue" has to be rediscussed.

Water (NaCl)   1   Acetone

FIGURE 14.   Binary liquid system with the two partially miscible solvents water (NaCl) +
acetone.

J. THE MICRO ON-LINE EXTRACTION SYSTEM USING THE BINARY SOLVENT SYSTEM ACETONE–WATER

A further reduction of the solvent consumption and, therefore, of the
solvent emission into the environment is reached if anhydrous $MgSO_4$, instead
of the above-mentioned nonpolar solvents, is used to remove water from the
organic phase b (see Figures 8 to 13) containing the pesticides.

Because the relative amounts of the two layers a and b are proportional
(as shown in Figure 14) to the segments of the tieline, the volume of phase
b depends at the composition of point 1. That is, if the water amount of the
sample is not known, the volume of phase b cannot be calculated, so that if
an aliquot of phase b is treated with anhydrous $MgSO_4$, the volume of the
remaining acetone phase for the determination of the sample weight is not
known. Therefore, it is necessary to standardize this sample extraction; in
order to always have the same amount of water in phase b, the composition
of point 1 in Figure 14 must be kept constant. This is realized by adjusting
a constant ratio of acetone to water = 2:1 for each extraction. For example,
if a sample is extracted with 10 ml of acetone, the total water amount of the
sample must be adjusted to 5 ml. This procedure is described in the following
section.

1. Standardization of the Sample Extraction

To bring all samples to the same starting condition, each sample is adjusted
to a total of 5 g of water prior to the 10 ml of acetone extraction. This is
done by the use of the corresponding tables in which the average water contents
of the original samples are listed.[38,39]

Suppose, for example, a potato and a cereal sample with an average water
content of 80 and 10%, respectively, should be extracted. If we add to a 5-g
potato subsample 1 ml of water and to a 1-g cereal subsample 4.90 ml of
water, both samples now have exactly 5 g of water.

The samples can now be extracted with 10 ml of acetone and ca. 2 g of
NaCl to obtain the two layers a and b (see Figure 14). Phase b is dried with
anhydrous $MgSO_4$, so that a pure acetone phase is obtained containing the
pesticides. To determine the sample weight in an aliquot acetone volume, it
is necessary to analyze the solvent composition of phase b.

2. The Solvent Composition of Phase b

The water content of phase b is determined by mixing 5 g of water, 10
ml of acetone, and ca. 2 g of NaCl. For this mixture, the total volume of

FIGURE 15. The two extraction sequences of the on-line method to reach the organic phase b using the binary liquid system. (From Steinwandter, H., *Fresenius J. Anal. Chem.*, 342, 150, 1992. With permission.)

phase b is 9.1 ml. The water amount of this phase, measured by the Karl-Fischer titration, is 1.1 ml, so that the acetone volume of phase b = 8.0 ml.

3. Determination of the Sample Weight in the Aliquot Acetone Volume

After standardization, the acetone-rich phase b consists, as mentioned above, of 8.0 ml of acetone and 1.1 ml of water. If this 1.1 ml of water is removed by anhydrous $MgSO_4$, a pure acetone phase with a total volume of 8.0 ml remains. Usually an aliquot of this latter phase is reduced to 1 ml for GC determination. With $g_0 = 5$ g and $V_0 = 8$ ml, the sample weight g in the aliquot volume V of the acetone extract is

$$g = g_0 \frac{V}{V_0}$$

with g_0 = total weight of sample taken, V = aliquot volume of acetone phase, and V_0 = total volume of acetone phase.

4. Operation of the On-Line Method Using the Binary Solvent System

Let us assume a 5-g representative fruit or vegetable subsample adjusted to 5 g of water. Add to this sample 10 ml of acetone and ca. 2 g of NaCl and extract for ca. 30 s. Because the extraction step is conducted in the same vessel, the extraction step E and the partition step P take place simultaneously in one step. Therefore, the on-line extraction of this technique can be described as follows: the original composition of the sample represented by the left water point of the straight line (see Figure 15) moves after the addition of 10 ml of acetone and ca. 2 g of NaCl to point 1, yielding the two conjugate layers a and b.

The following two steps are, therefore, characteristic for this on-line extraction technique: (1) the extraction step E and (2) the partition step P. Therefore, if we ask, as discussed earlier,[25] what are the working sequences to reach the organic phase b, we can write the two sequences b (EP) and b (E, P), as shown in Figure 15. That means that the samples can be extracted (E) and partitioned (P), either simultaneously or sequentially, depending on the water content of the original sample. This strategy is described in the following sections.

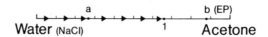

FIGURE 16. The extraction sequence b (EP) of the on-line method for the determination of pesticides in samples with water contents >70%. (From Steinwandter, H., *Fresenius J. Anal. Chem.*, 342, 150, 1992. With permission.)

5. The Microextraction Procedure

The use of these two different extraction modes b (EP) and b (E, P) depends, as mentioned above, on the original water content of the samples to be extracted.

Products with water contents >70% — Weigh, for example, 5 g of a homogeneous and representative potato subsample into the extraction vessel, add 1 ml of water (see Section III.J.1), 10 ml of acetone, and ca. 2 g of NaCl. Extract with the Ultra-Turrax for 15 to 20 s. Let it stand for 5 min, or centrifuge directly after extraction. Pour the upper organic phase into a 25-ml Erlenmeyer flask containing ca. 2 g of anhydrous $MgSO_4$. Stopper the flask and dry for 15 min on a magnetic stirrer using a stirrer bar. Take an aliquot, e.g., 4.0 ml, of the acetone phase (= 2.5 g of sample) and reduce the volume to ca. 0.5 ml. Transfer the solution quantitatively into a 1-ml graduated flask and fill up with acetone. Because the extraction (E) and the partitioning (P) steps are conducted simultaneously, the corresponding working sequence of this extraction procedure to reach the organic phase b is b (EP), as shown in Figure 16.

Note: Recovery studies from untreated fruit and vegetable samples fortified with the pesticides bromophos-ethyl, chlorpyriphos, dimethoate, ethion, methidathion, parathion, and pyrazophos ranged between 90 and 110%. The results obtained from samples of unknown spray history are comparable with those obtained by the micro on-line method using the ternary solvent system listed in Table 5.

Products with water contents <15% — Weigh, for example, 1 g of a representative cereal subsample into the extraction vessel, add 4.90 ml of water (see Section III.J.1) and 10 ml of acetone. Mix with the Ultra-Turrax for 10 s. Let stand for 10 min. Add ca. 2 g of NaCl and mix again for 15 to 20 s. Continue as described in Section III.J.3. Phase b is reached by b (E, P), as shown in Figure 17, because first the extraction step E is carried out, followed by the separate partitioning step P.

K. THE MICRO ON-LINE EXTRACTION SYSTEM USING THE BINARY SOLVENT SYSTEM ETHYL ACETATE–WATER

For several years, we have been searching for another solvent that can be used in addition to acetone as a universal solvent. Finally we decided after some screening experiments with different matrix samples and analytes to use ethyl acetate as the second universal extraction solvent in our laboratory.

TABLE 5

**Extraction of Pesticide Residues from Nine Samples with
Unknown Spray History by the Microextraction Methods
Using the Ternary (D) and Binary Solvent (MgSO₄) Systems**

Laboratory sample no.	Analyzed samples	Identified compound	Residue (µg/kg) analyzed by	
			Micro method using D	Micro method using MgSO₄
L 1	Peaches	Phosalone	230	240
		Azinphos-ethyl	53	61
		Chlorpyriphos	70	63
L 2	Apples	Phosalone	220	260
L 3	Pears	Fenitrothion	76	75
L 4	Cherries	Phosalone	70	60
		Azinphos-ethyl	40	50
L 5	Apples	Phosalone	80	74
L 6	Peaches	Pyrazophos	75	66
L 7	Apples	Phosalone	110	110
L 8	Apples	Phosalone	500	530
L 9	Apples	Phosalone	1110	1230

Note: D = Dichloromethane.

FIGURE 17. The extraction sequence b (E, P) of the on-line method for the determination of pesticides in samples with water contents <15%. (From Steinwandter, H., *Fresenius J. Anal. Chem.*, 342, 150, 1992. With permission.)

However, if ethyl acetate is used as a universal extraction solvent, the same principles must be considered as discussed in Section III.A.

In this connection it is of interest to note that if the polar organic solvent acetone in the above-mentioned ternary solvent system water-acetone-ethyl acetate of the on-line method (see Section III.I.2) is eliminated, we obtain directly the binary solvent system water-ethyl acetate for the extraction discussed earlier.[40–43] So, we can see that both extraction systems are related methods.

The extraction of atrazine from weathered soils — Weigh 5 g of soil into a 100-ml flask and add 10 ml of water and 20 ml of ethyl acetate. Close the flask and extract atrazine by shaking on a mechanical shaker (ca. 120 cycles/min) for 16 h overnight at 20°C. Pour the organic extract into a 50-ml Erlenmeyer flask and continue as described in Section III.I.3. The proposed micro method is equally as effective as the described micro on-line method using acetone (see Section III.I.5) and more effective than the soxhlet extraction using acetone or methanol.

The extraction of pesticides from fruits and vegetables — Weigh 5 g of a homogeneous and representative subsample into the extraction vessel (i.d. = 35 mm, h = 100 mm) and add 20 ml of ethyl acetate and then 10 g of Na_2SO_4. Extract with the Ultra-Turrax for 2 min at low speed and then for 1 min at high speed. Pour the organic extract into a 50-ml Erlenmeyer flask and continue as described in Section III.I.3. The proposed micro method is equally effective as the described micro on-line method using acetone.

The extraction of pesticides from products which are dried and pulverized — Weigh 1 g of a representative subsample into the extraction vessel (i.d. = 35 mm, h = 100 mm) and add 5 ml of water. Let stand for 10 to 15 min. Add 20 ml of ethyl acetate and then 10 g of Na_2SO_4. Continue as described above. The proposed micro method is equally effective as the described micro on-line method using acetone (see Section III.I.4).

IV. INTERNAL LABORATORY CONTROL

In spite of the limited space for this section, some general remarks are necessary to this important matter.

If microextraction methods are developed, laboratories must ensure that once an analysis is performed, it must not be repeated in any way, so that the amount of chemicals released into the environment is kept to a minimum. The amount of these released chemicals is usually a factor 10^8 larger than the quantity of the polluting analytes to be determined and, if the "Extended Categorical Imperative" is considered (see Section II), one can easily understand that the contamination aspect of the analytical procedure, and not the quality of the corresponding result, is of primary concern. However, both aspects can be satisfied, if self-audited lab operations, called Internal Laboratory Controls, are implemented. In principle, each responsible and ethical scientist would like to find for his laboratory a universal control system, which by formal logic can be expressed as a universal implication. This can be written symbolically by the relation:

$$S:(Y \rightarrow Z)$$

which explains that for all analyzed samples S, if the controls Y are conducted (e.g., Y = blanks, recovery studies, and sample preparation), the true value Z is obtained.

If the "truth" Z cannot be reached with these controls, we have to correct the starting point condition Y to an extended Y, which is now for example, $Y \wedge Y_1$ (with Y = blanks, recovery studies, and sample preparation and, for example, Y_1 = calibration curves), so that we now obtain:

$$S:[(Y \wedge Y_1) \rightarrow Z]$$

which means that for all analyzed samples S, if the controls Y and Y_1 are conducted, then the "true" value Z is obtained.

For all these iterative procedures (controls) beginning with Y, until Y $\wedge$ Y_1 $\wedge$ Y_2 $\wedge$ . . . Y_n, the "truth" Z cannot be reached. So the analyst is living between doubt and investigation aspiring to the "truth".

Nevertheless, if the analytical chemist is acting with the above-mentioned analytical conscience and responsibility, he or she will automatically find the right network, which is the right Internal Laboratory Control, to minimize the contamination of the environment when performing an analysis and to achieve, for discussion of the corresponding findings, analytical results as true as possible at the same time.

REFERENCES

1. **Steinwandter, H.,** Environmental Analysis — Benefits or Damage, lecture presented at the 101st VDLUFA Congress in Bayreuth, 1989.
2. **Steinwandter, H.,** *VDLUFA-Schriftenreihe,* 30, 543, 1990.
3. **Holden, A. V. and Marsden, K.,** *J. Chromatogr.,* 44, 481, 1969.
4. **Stijve, T. and Brand, E.,** *Dtsch. Lebensm. Rundsch.,* 73, 41, 1977.
5. **Greve, P. A. and Grevenstuck, W. B. F.,** *Meded. Fac. Landbouwwet. Rijksuniv. Gent.,* 40, 1115, 1975.
6. **Steinwandter, H. and Schlüter, H.,** *Dtsch. Lebensm. Rundsch.,* 74, 139, 1978.
7. **Steinwandter, H.,** *Fresenius Z. Anal. Chem.,* 316, 493, 1983.
8. **Steinwandter, H.,** *Fresenius Z. Anal. Chem.,* 321, 600, 1985.
9. **Steinwandter, H.,** in *Analytical Methods for Pesticides and Plant Growth Regulators,* Vol. 17, Sherma, J., Ed., Academic Press, San Diego, 1989, 35.
10. **Goodwin, E. S., Goulden, R., and Reynolds, I. G.,** *Analyst,* 86, 697, 1961.
11. **Mills, P. A., Onley, J. H., and Gaither, R. A.,** *J. Assoc. Off. Anal. Chem.,* 46, 186, 1963.
12. **Becker, G.,** *Dtsch. Lebensm. Rundsch.,* 67, 125, 1971.
13. **Luke, M. A., Froberg, J. E., and Masumoto, H. T.,** *J. Assoc. Off. Anal. Chem.,* 58, 1020, 1975.
14. **Storherr, R. W., Ott, P., and Watts, R. R.,** *J. Assoc. Off. Anal. Chem.,* 51, 513, 1971.
15. **Steinwandter, H.,** *Fresenius Z. Anal. Chem.,* 322, 752, 1985.
16. **Steinwandter, H.,** Universal extraction methods for the determination of organic residues — beginning and evolution, lecture presented at the 100th VDLUFA Congress in Bonn, 1988.
17. **Steinwandter, H.,** *VDLUFA-Schriftenreihe,* 28, 1069, 1988.
18. **Kant, I.,** in *Grounding for the Metaphysics of Morals,* Hartknoch, J. F., Ed., Riga, 1786.
19. **Lykken, L.,** *Residue Rev.,* 3, 19, 1963.
20. **Zweig, G. and Sherma, J., Eds.,** in *Analytical Methods for Pesticides and Plant Growth Regulators,* Vol. 6, 1972, 1.
21. **Arbeitsgruppe "Pesticide",** 3. Empfehlung. Mitteilungsblatt GDCh-Fachgruppe Lebensmittelchem. gerichtl., *Chemistry,* 28, 219, 1974.

22. Rückstandsanalytik von Pflanzenschutzmitteln, Mitteilung VI der Senatskommission für Pflanzenschutz-, Pflanzenbehandlungs- und Vorratsschutzmittel der Deutschen Forschungsgemeinschaft, Methodensammlung der Arbeitsgruppe "Analytik", Lieferung, Section VIII-1, 1982, 6.
23. **Thier, H. P. and Zeumer, H., Eds.,** in *Manual of Pesticide Residue Analysis*, Vol. 1, Verlag Chemie, Weinheim, Germany, 1987, 17.
24. **Horwitz, W.,** in *Advances in Pesticide Science,* Vol. 3, Geissbühler, H., Ed., Pergamon Press, Elmsford, NY, 1979, 649.
25. **Steinwandter, H.,** *Fresenius Z. Anal. Chem.,* 335, 475, 1989.
26. **Steinwandter, H.,** *Fresenius J. Anal. Chem.,* 336, 8, 1990.
27. **Thier, H. P., Stijve, T., and Diserens, H.,** *Lebensmittelchem. Gerichtl. Chem.,* 43, 121, 1989.
28. **Steinwandter, H.,** *Fresenius J. Anal. Chem.,* 339, 30, 1991.
29. **Steinwandter, H.,** *Fresenius Z. Anal. Chem.,* 334, 133, 1989.
30. **Steinwandter, H.,** *Fresenius Z. Anal. Chem.,* 327, 309, 1987.
31. **Steinwandter, H.,** *Fresenius J. Anal. Chem.,* 340, 389, 1991.
32. **Kahn, S. U.,** *Residue Rev.,* 84, 1, 1982.
33. **Calderbank, A.,** *Residue Rev.,* 108, 71, 1989.
34. **Capriel, P., Haisch, A., and Kahn, S. U.,** *J. Agric. Food Chem.,* 34, 70, 1986.
35. **Führ, F. and Mittelstaedt, W.,** *J. Agric. Food Chem.,* 28, 122, 1980.
36. **Khan, S. U. and Hamilton, H. A.,** *J. Agric. Food Chem.,* 28, 126, 1980.
37. **Khan, S. U.,** *J. Agric. Food Chem.,* 28, 1096, 1980.
38. **Thier, H. P. and Zeumer, H., Eds.,** in *Manual of Pesticide Residue Analysis*, Vol. 1, Verlag Chemie, Weinheim, Germany, 1987, 386.
39. **Luke, M. A. and Masumoto, H. T.,** in *Analytical Methods for Pesticides and Plant Growth Regulators,* Vol. 15, Zweig, G. and Sherma, J., Eds., Academic Press, San Diego, 1989, 161.
40. **Watts, R. R. and Storherr, R. W.,** *J. Assoc. Off. Anal. Chem.,* 48, 1158, 1965.
41. **Watts, R. R., Storherr, R. W., Pardue, J. R., and Osgood, T.,** *J. Assoc. Off. Anal. Chem.,* 52, 522, 1969.
42. **Kadoum, A. M. and Mock, D. E.,** *J. Agric. Food Chem.,* 26, 45, 1978.
43. **Gorder, G. W. and Dahm, P. A.,** *J. Agric. Food Chem.,* 29, 629, 1981.

Chapter 2

SOLID-PHASE PARTITION COLUMN TECHNOLOGY

Marvin L. Hopper

TABLE OF CONTENTS

I. INTRODUCTION

Pesticide methods are continually being revised and improved with new and conventional technologies. These methods revolve around liquid-liquid partition chromatography, which is the heart of most pesticide methods. Pesticide procedures are being improved with solid-phase extraction (SPE) columns. This technology incorporates bonded-phase and solid-phase sorbents into solid-phase partition extraction columns.

The bonded-phase technology offers advantages for improving existing sample preparation techniques. This technique offers batch processing, time savings, solvent reduction, high selectivity, trace enrichment, elimination of emulsions, and ease of automation. The potential advantages of bonded-phase extraction columns have not yet been fully realized in the isolation of pesticide residues from food matrices, but they have been adapted to environmental water analysis with good results. Future applications will reveal the full potential of this technique.

Solid-phase sorbent technology, using diatomaceous earth, has not significantly changed over the years, but continues to serve a vital role in new sample preparation techniques. Conventional liquid-liquid partition separations have been replaced by procedures utilizing solid-phase partition columns containing diatomaceous earth. This technique offers batch processing, elimination of emulsions, and time savings and streamlines the partition step by eliminating the many steps associated with separatory funnel partition.

II. BASIC PRINCIPLE

A. THEORY

Liquid-liquid chromatography (LLC), also recognized as liquid-liquid partition chromatography, was developed by Martin and Synge in 1941 using an organic-aqueous two-phase system and a silica support.[1] This process, described by Snyder and Kirkland,[2] involves a carrier liquid moving across a finely dispersed stationary phase. The stationary phase is coated with a liquid phase immiscible in the mobile phase or coated with a chemically bonded phase. The analytes and extractants rapidly reach equilibrium between the two phases because of the large interface. Separation occurs as the various solutes are distributed between the different phases according to their distribution coefficients. The separation power of the liquid-liquid system is dependent on the polarity difference between the two phases and the solubility parameter of each phase.[3] The liquid-liquid partition chromatography columns can be used in the simple extraction of analytes from a sample extract or in the separation of individual extracted analytes.

The traditional separatory funnel partitioning is similar to LLC in that solutes are distributed between two immiscible solvents after equilibrium is

TABLE 1
Bonded-Phase Column Sorbents

Nonpolar		Polar		Ion-exchange	
C-18	Octadecyl	CN	Cyanopropyl	SCX	Benzenesulfonylpropyl
C-8	Octyl	2OH	Diol	PRS	Sulfonylpropyl
C-2	Ethyl	SI	Silica	CBA	Carboxymethyl
C-H	Cyclohexyl	NH₂	Aminopropyl	DEA	Diethylaminopropyl
P-H	Phenyl	PSA	*N*-Propylethylenediamine	SAX	Trimethylaminopropyl

achieved with shaking. This type of extraction results in a simple one-stage batch extraction of analytes and extractants with limited separation of individual analytes.

The supports used in LLC fall into two general categories: porous and superficially porous materials, as described by Kirkland.[4] Porous silica beads, silica gel, and diatomaceous earth materials are considered to be porous supports. Superficially porous supports are beads constructed from an impervious center with surface etching or a thin porous shell.

III. GENERAL APPLICATIONS

Solid-phase extraction is generally accepted to involve columns containing a chemically bonded silica-based sorbent. Other materials, such as diatomaceous earth, cellulose, and silica gel, can also be used as a sorbent for solid-phase partition columns.

A. SOLID BONDED-PHASE PARTITION SORBENT

Bonded silica sorbents are normally constructed from irregular particles with a porosity of approximately 6 nm (60 Å). The functional group is attached to the substrate through a silyl ether linkage that is stable within a pH range of approximately 2 to 7.5. Interactions with unreacted silica substrate silanol groups are minimized through endcapping reactions.

A silica-based SPE column normally contains from 100 to 1000 mg of sorbent with a 40- to 90-μm particle size range. The SPE column can be purchased with a wide variety of silica bonded-phase sorbents. The sorbents fall into three general classes: nonpolar, polar, and ion-exchange. The activity of the column is dependent on the properties of the bonded phase and any active sites not endcapped on the sorbent. The most common SPE column sorbents[5] are shown in Table 1.

There are four steps normally associated with SPE methods.[6] The column is initially conditioned with different solvents to improve reproducibility and reduce any background contaminants. Second, the isolates and matrix extractants are adsorbed on the conditioned column. Third, the column is rinsed with a weak solvent that elutes matrix extractants and leaves the isolates.

Finally, a stronger solvent is used to elute the isolates while leaving as many extractants as possible on the column.

The column is initially rinsed with the solvent used to elute the isolates. Then it is eluted with methanol followed by water. The column is usually not allowed to go dry before the sample solution is added. The sample is forced through the column at a specified flow rate, usually between 5 to 10 mL/min, but not to exceed 25 mL/min. Gravity feed is used in special applications. Many samples have to be prefiltered before they are put through the column. The column, in most cases, is dried with air after the addition of the sample. The isolates can then be eluted with organic solvents such as ethyl acetate and dichloromethane, which are favorite elution solvents for SPE columns.

A single elution system usually will not give optimum recoveries for all compounds of interest. Different solvent elution systems can be developed by using the solvent classification scheme proposed by Snyder,[7] which separates solvents into eight groups according to their selectivities.

Octadecyl and octyl, as shown in Table 2, are the most common SPE sorbents used in pesticide analysis. These applications show that SPE sorbents can be used to extract herbicides, carbamates, organophosphorus pesticides, organochlorine pesticides, polycyclic aromatic hydrocarbons (PAHs), polycyclic organic materials (POMs), and fungicides from water and other matrices. The recoveries obtained with SPE extraction columns[8,9] are comparable to recoveries obtained with conventional liquid-liquid extraction techniques. The SPE columns offer sufficient cleanup for many eluants determined by liquid chromatography without further cleanup. Other cleanup steps are still generally needed for eluants analyzed by gas chromatography and/or mass spectrometry. The solid phase sorbents offer the capability of producing mixed bed columns that allow selectivity to be built into screening procedures. This allows specific classes of compounds to be separated from other classes of compounds or matrix components. The SPE columns are multifunctional in that samples can be collected in the field and transported to the laboratory, where they are stored until they can be analyzed.

B. SOLID-PHASE PARTITION SORBENT

Diatomaceous earth and associated products are the most common supports used for solid-phase partition sorbents. Diatomaceous earth or diatomite is made up of siliceous skeletons of minute aquatic plants known as diatoms. Diatomaceous earth is also called kieselguhr or infusorial earth. Diatomaceous earth contains 65 to 87% SiO_2, 2.3 to 11.7% Al_2O_3, and up to 3% Fe_2O_3.[10] Small proportions of the oxides of magnesium, potassium, calcium, and sodium along with 5 to 14% water are also in diatomaceous earth. Diatomaceous earth is often calcinated in order to improve color and remove organic matter to improve the internal porosity. Other common products, such as Celite, Hyflo Super-Cel, Chromosorb W, and Chromosorb P, are derived from diatomaceous earth.

TABLE 2
Bonded-Phase Partition Column Extraction Applications

Column support	Sample matrix	Pesticides analyzed	Ref.
C-18	Milk	Chlorsulfuron	20
C-18	Rice paddy water	Carbofuran, 7-phenol, 3-keto carbofuran, 3-keto-7-phenol, 3-hydroxy carbofuran, and 3-hydroxy-7-phenol	21
C-18	Well water	Carbofuran, atrazine, simazine alachlor, and cyanazine	22
C-18	Pond water	Trifluralin, methyl paraoxon, methyl parathion, fenvalerate, and 2,4-D dimethylamine	23
C-18	Surface and ground water	Alachlor, atrazine, cyanazine, metolachlor, metribuzin, propachlor, trifluralin, carbaryl, carbofuran, chlordane, p,p-DDE, endrin, lindane, heptachlor epoxide, chlorpyrifos, fonofos, ethyl parathion, and 8 PAHs	24
C-8, C-18	Deionized water	TEPT, dichlorovos, ethoprop, phorate, diazinon, dimethoate, methyl parathion, parathion, tokuthion, famphur, EPN, and azinphos methyl	25
XAD-7, XAD-4	Environmental water	Methamidophos, acephate, trichlorfon, phorate, dimethoate, fonofos, diazinon, disulfoton, fenitrothion, malathion, ethyl parathion, tetrachlorvinphos, fensulfothion, and phosmet	8
C-18	Seafood	HCB, DDE, DDT, mirex, 8-ketone, 9-kepone, and kepone	9
C-18, XAD-2	Surface and ground water	Alachlor, atrazine, cyanazine, metolachlor, metribuzin, propachlor, trifluralin, carbaryl, carbofuran, chlordane, p,p-DDE, endrin, lindane, heptachlor epoxide, chlorpyrifos, fonofos, ethyl parathion, 6 PAHs, 4 nitro-PAHs, and 6 POMs	26
C-18	Serum	α-Hexachlorocyclohexane, heptachlor, heptachlor epoxide, γ hexachlorocyclohexane, oxychlordane, cis-chlordane, trans-chlordane, trans-nonachlor, dieldrin, p,p-DDE, p,p-DDD, and p,p-DDT	27
C-18	Groundwater	Chlorpyrifos, isofenphos, carbaryl, iprodione, and triadimefon	28
C-18	Water	Propoxur, carbofuran, carbaryl, propham, captan, chloropropham, barban, and butylate	29

TABLE 2 (continued)
Bonded-Phase Partition Column Extraction Applications

Column support	Sample matrix	Pesticides analyzed	Ref.
C-18	Water	EPTC, butilate, molinate, triflura-lin, α-BHC, simazine, atrazine-d5, atrazine, lindane, propazine, trieta-zine, diazinon, heptachlor, alach-lor, simetryne, ametryne, prome-tryne, terbutryne, aldrin, tiocarbazil, endosulfan I, and en-dosulfan II	30
C-18	Water	Alachlor, atrazine, chlordane, cyan-azine, *p,p*-DDE, endrin, fonofos, heptachlor epoxide, lindane, meto-lachlor, metribuzin, and trifluralin	31
C-8, C-18 enmeshed in a membrane	Ground water	Vernam, atrazine, diazinon, dyfon-ate, metribuzin, alachlor, sulpro-fos, heptachlor, aldrin, endosulfan and 4 phthalates	32

Diatomaceous earth (Celite) is usually preconditioned before it is used in a specific analytical method. The Celite quality and method determine the degree of preconditioning. Celite is usually preconditioned with an acid wash-ing followed by rinsing with water and methanol, respectively. The Celite is then air dried to remove the solvent. Further conditioning may require heating in an oven or muffle furnace.

The sample and column size dictate the amount of Celite needed for a particular solid-phase partition column procedure. The optimum Celite to immobile aqueous phase ratio is usually 1 g to slightly less than 1 mL liquid,[11] but can be as high as 1 g to 1.4 mL liquid. The immobile phase can be aqueous or nonaqueous, and both are used in conjunction with immiscible organic solvents. Celite is usually inert in most applications, but any adsorp-tion introduced by the siliceous earth can usually be neutralized by modifying either the immobile or the mobile phase. The Celite and immobile phase are then premixed and added to the column and compressed into a uniform bed by tamping with a glass rod. While this is the most common packing technique, special applications may require alternate packing techniques.

The isolates are eluted from the column with a suitable solvent at a controlled flow rate, which generally is <10 mL/min. A stopcock or a screw-clamp on tubing attached to the column outlet is used to meter the mobile phase with the aid of gravity, air pressure, or vacuum. Reproducible flow rates are achieved with these techniques.

Extraction methods have been developed for organochlorine and organo-phosphate pesticides using solid-phase sorbents for holding the immobile phase. Rogers[12] developed a method for the extraction of chlorinated pesticides from large quantities of extracted fats and oils. This method suspends the fats or oils on Micro-Cel-E and Celite, which is then blended with 5% acetone-acetonitrile. The mobile phase is separated from the suspended fat oil by filtration. The filtrate is then mixed with water and the pesticides are extracted with petroleum ether. The extraction procedure gives good recoveries for lindane, aldrin, heptachlor epoxide, *p,p*-DDE, and dieldrin, with less than 2 g of fat carryover from 30 g of extracted fat.

A Celite column method for purifying aldrin, dieldrin, and endrin from visceral material is described by Kurhekar et al.[13] This method involves adding visceral material extracts in hexane to a column containing 2 g of Celite mixed with 1.2 mL of 0.5 N H_2SO_4. Hexane is used to elute the pesticides from the column with good recoveries. The column gives adequate cleanup for sample extracts equivalent to 20 g of visceral material.

A ready-to-use disposable minicolumn of diatomaceous earth was used by Muccio et al.[14] to efficiently partition organophosphorus pesticides into the mobile acetonitrile phase from the immobile *n*-hexane fat solution adsorbed on the column. Recoveries between 80 and 107% were obtained for diazinon, etrimfos, chlorpyrifos-methyl, pyrimiphos-methyl, chlorpyrifos, bromophos, bromophos-ethyl, malathion, and fenithrothion spiked in 2 g of lipidic material at levels ranging from 0.1 to 5 ppm. This method was evaluated on lipids from stored cereals, oil seeds, and legumes.

A method for analyzing organophosphorus and organochlorine pesticides in lanolin was developed by Diserens.[15] This method involved adsorbing a light petroleum solution of lanolin onto a disposable minicolumn of diatomaceous earth. The pesticides were eluted with acetonitrile saturated with light petroleum. Recoveries ranged from 80 to 90% for HCB, α-BHC, γ-BHC, dieldrin, *p,p'*-DDE, *p,p'*-TDE, *p,p'*-DDT, diazinon, dichlofenthion, chlorfenvinphos, bromophos-ethyl, ethion, and carbophenothion pesticides frequently found in lanolin.

The above extraction methods demonstrate that conventional solid-phase sorbents can be efficiently used to separate organochlorine and organophosphorus pesticides from fats with good recoveries.

IV. UNIVERSAL SOLID-PHASE PARTITION EXTRACTION COLUMN

Multiresidue methods (MRMs) are important procedures used in screening table-ready foods and raw produce for pesticide residues. The more common MRMs used for screening pesticide residues include the Storherr et al.,[16] Mills et al.,[17] and Luke et al.[18] procedures. The Luke procedure is widely used for the analysis of organochlorine and polar and nonpolar organophosphorus pesticides in nonfatty food items because it screens for the largest number of pesticides.

The Luke procedure involves blending the sample with acetone, filtering it, and shaking it with petroleum ether and methylene chloride, which partitions the pesticides into the organic phase from the aqueous filtrate. The partitioning step in this method uses the traditional separatory funnel shake out, which is efficient but time consuming and labor intensive.

A solid-phase partition column as described by Hopper[19] is a direct substitute for the separatory funnel partition step used in the Luke procedure. The solid-phase partition column in this procedure is reusable and gives quantitative and reproducible recoveries for organochlorine and organophosphorus pesticides fortified in nonfatty food items as shown in Table 3. This solid-phase partition column streamlines the partition step and offers batch processing, elimination of emulsions, time savings, and the elimination of the manipulations associated with the separatory funnel partition. The following discussion describes the solid-phase partition column procedure in detail.

A. TECHNIQUE

Liquid-liquid partition chromatography, as described above, is used with an organic solvent to extract analytes and matrix extractants from a water-acetone mixture. A liquid-liquid solid-phase partition column is prepared from a specially modified form of diatomaceous earth. The diatomaceous earth is in the form of pellets ranging from 10 to 30 mesh. The column is pretreated with a buffered solution to neutralize any active sites that may be present.

The column is ready for sample analysis after removing the buffering solution with acetone. The column is first presaturated with methylene chloride. All solvents are collected as a sample extract consisting of an acetone-water mixture and are added to the column for extraction. The methylene chloride mixes with the acetone, forcing the water to be absorbed on the large surface area of the diatomaceous earth support. The analytes and matrix extractants are extracted with methylene chloride as it trickles down through the column and comes in contact with the film of aqueous sample. The column is reusable after removing the absorbed water with acetone followed by a methylene chloride rinse.

B. SOLID-PHASE EXTRACTION COLUMN PROCEDURE
1. Equipment and Reagents

This technique requires equipment usually associated with a laboratory equipped for pesticide analysis, including a glass chromatography tube with a Teflon stopcock (28 mm O.D., 25 mm I.D. × 500 mm, Kontes, Cat. No. K-420540-0245), a Kuderna-Danish (K-D) 500-mL concentrator, a three-ball Snyder column, a graduated receiving tube, a microevaporative concentrator (No. 6709, Ace Glass Inc., Vineland, NJ), and assorted sizes of flasks and graduated cylinders. A gas chromatograph (GC), equipped with a flame photometric detector (phosphorous mode), an electron capture detector, and an

TABLE 3
Recovery Data for Universal Solid-Phase Partition Extraction Column

Pesticide	Spike level (ppm)	Number of samples fortified	Average % recovery	% Relative standard deviation (RSD)
Aldrin	0.01	13	90	12
Dicloran	0.01	9	86	20
Vinclozolin	0.01	11	86	18
Diclofop-methyl	0.02	11	82	17
Fonofos oxygen analog	0.02	5	96	9
Pirimiphos-ethyl	0.05	12	102	14
Dichlorvos	0.05	15	105	10
Fenthion oxygen analog	0.05	10	97	10
Phorate oxygen analog	0.05	9	100	6
Tris(chloropropyl) phosphate	0.08	7	98	13
Aspon	0.10	9	101	7
Chlorfenvinphos, α	0.10	9	98	9
Dichlofenthion	0.10	8	96	5
Fenthion	0.10	6	97	7
Chlorfenvinphos, β	0.10	27	97	10
Monocrotophos	0.10	11	105	12
Parathion-methyl oxygen analog	0.10	11	100	10
Phenthoate	0.10	11	98	8
Phorate sulfone	0.10	11	109	13
Malathion	0.10	15	106	11
Propetamphos	0.10	8	97	17
Metasystox thiono isomer	0.15	10	80	21
Quinalphos	0.20	6	94	6
Diazinon oxygen analog	0.20	10	91	12
Chlorpyrifos	0.20	3	102	7
Pyrazophos (afugan)	0.25	12	106	7
Chlorpyrifos oxygen analog	0.25	16	105	10
Carbophenothion	0.30	8	99	6
Fenthion sulfone	0.40	15	97	8
Phoxim oxygen analog	0.40	5	87	12
Carbophenothion sulfoxide	0.50	11	106	7
Propargite	0.50	3	111	14
Thiabendazole	0.50	3	114	17
Carbophenothion sulfone	0.60	18	106	9
Phosalone oxygen analog	0.75	9	98	7

electrolytic conductivity detector, is normally used in the quantification of the analytes extracted using this technique.

A unique pelletized diatomaceous earth (Hydromatrix material, No. CT0001, Analytichem International, Harbor City, CA) is the solid-phase support used in this solid-phase partition column procedure. Monobasic potassium phosphate, carborundum chips, and pesticide-grade solvents are the other reagents used with this extraction technique.

2. Column Preparation

The fines are removed from 50 g of the pelletized Hydromatrix material by sieving with a No. 30 sieve. A small plug of glass wool is loosely placed in the bottom of a glass tube with a Teflon stopcock. The column is filled with 40 g of the sieved support. A large plug of glass wool is added to the top of the column after it is lightly tapped on a hard surface to settle the support.

The column is then washed with 150 mL of 0.1 M KH$_2$PO$_4$. The column is eluted with 300 mL of acetone after the buffer solution flow rate has slowed to 3 to 5 mL/min. The flow rate of the acetone is adjusted to 50 to 60 mL/min after 100 mL of acetone has eluted. The column is washed with 300 mL of methylene chloride, with the flow readjusted to 50 to 60 mL/min after the first 100 mL. The column is now ready for use. The established flow rate should not be changed.

3. Sample Extraction

The column is washed with 200 mL of acetone to remove any water that may be absorbed on the pelletized diatomaceous earth. The column is pre-washed with 200 mL of methylene chloride before each sample extraction. A K-D concentrator (500 mL) and attached 4-mL graduated tip are placed under the column for solvent collection. A 40-mL aliquot of the acetone-aqueous sample filtrate is transferred to the column and allowed to elute through the column. A 50-mL aliquot of methylene chloride is added to the column when the initial sample effluent reaches a flow rate of 3 to 5 mL/min. A second 50-mL aliquot of methylene chloride is added to the column after the first fraction has drained to the top of the column support. The column is then eluted with an additional 200 mL of methylene chloride after the second 50-mL aliquot has disappeared into the column. The column effluent is collected until the flow rate has slowed to a drip (ca. 1 mL/min). Boiling chips and a Snyder column are added to the K-D receiver, and the effluent is concentrated to <5 mL with a steam bath. After adding 50 mL of acetone through the Snyder column, the eluate is reconcentrated to <5 mL without cooling. The eluate is concentrated to <1 mL with a microevaporative concentrator after cooling. The final volume is adjusted to 1 mL with acetone and the sample is ready for GC analysis.

4. Restoration of the Column

The column sorbent becomes discolored after many samples, and over time the color elutes from the column in subsequent samples. The column can be restored by washing with 200 mL of acetone and enough 0.1 M KH$_2$PO$_4$ buffer (200 to 300 mL) until no additional color elutes from the column. The column is then eluted with 300 mL of acetone followed by 200 mL of methylene chloride, and it is again ready for sample analysis.

V. CONCLUSION

Applications have shown that bonded-phase and solid-phase partition columns can be substituted for conventional separatory funnel liquid-liquid extractions. These columns have streamlined existing pesticide residue procedures, which saves resources and time. The saving of resources and time is an important goal in the development of procedures, and future research should reflect efficient use of the available technologies.

REFERENCES

1. **Martin, A. J. P. and Synge, R. L. M.**, *Biochem. J.*, 35, 1358, 1941.
2. **Snyder, L. R. and Kirkland, J. J.**, in *Introduction to Modern Liquid Chromatography*, John Wiley & Sons, New York, 1974, 197.
3. **Unger, K. K.**, in *Packings and Stationary Phases in Chromatographic Techniques*, Unger, K. K., Ed., Marcel Dekker, New York, 1990, 235.
4. **Kirkland, J. J.**, in *Modern Practice of Liquid Chromatography*, Kirkland, J. J., Ed., John Wiley & Sons, New York, 1971, 161.
5. **Van Horne, K. C.**, in *Sorbent Extraction Technology*, Analytichem International, Harbor City, CA, 1985, 35.
6. **Furton, K. G. and Rein, J.**, *Anal. Chim. Acta*, 236, 99, 1990.
7. **Snyder, L. S.**, *J. Chromatogr.*, 92, 223, 1974.
8. **Mallet, C. and Mallet, V. N.**, *J. Chromatogr.*, 481, 37, 1989.
9. **Kohler, P. W. and Su, S. Y.**, *Chromatographia*, 21, 531, 1986.
10. **Scinta, J.**, *U.S. Patent* 4,448,669, 1984.
11. **Levine, J.**, *J. Pharm. Sci.*, 52, 1015, 1963.
12. **Rogers, W. M.**, *J. Assoc. Off. Anal. Chem.*, 55, 1053, 1972.
13. **Kurhekar, M. P., D'Souza, F. C., and Meghal, S. K.**, *J. Assoc. Off. Anal. Chem.*, 58, 548, 1975.
14. **Muccio, A. D., Cicero, A. M., Camoni, I., Pontecorvo, D., and Dommarco, R.**, *J. Assoc. Off. Anal. Chem.*, 70, 106, 1987.
15. **Diserens, H.**, *J. Assoc. Off. Anal. Chem.*, 72, 991, 1989.
16. **Storherr, R. W., Ott, P., and Watts, R. R.**, *J. Assoc. Off. Anal. Chem.*, 54, 513, 1971.
17. **Mills, P. A., Onley, J. J., and Gaither, R. A.**, *J. Assoc. Off. Anal. Chem.*, 46, 186, 1963.
18. **Luke, M. A., Froberg, J. E., Doose, G. M., and Masumoto, H. T.**, *J. Assoc. Off. Anal. Chem.*, 64, 1187, 1981.
19. **Hopper, M. L.**, *J. Assoc. Off. Anal. Chem.*, 71, 731, 1988.
20. **Long, A. R., Hsieh, L. C., Malbrough, M. S., Short, C. R., and Barker, S. A.**, *J. Assoc. Off. Anal. Chem.*, 72, 813, 1989.
21. **Beauchamp, K. W., Jr., Liu, D. W., and Kikta, E. J., Jr.**, *J. Assoc. Off. Anal. Chem.*, 72, 845, 1989.
22. **Nash, R. G.**, *J. Assoc. Off. Anal. Chem.*, 73, 438, 1990.
23. **Swineford, D. M. and Belisle, A. A.**, *Environ. Toxicol. Chem.*, 8, 465, 1990.
24. **Junk, G. A. and Richard, J. J.**, *J. Res. Natl. Bur. Stand.*, 93, 274, 1988.
25. **Loconto, P. R. and Gaind, A. K.**, *J. Chromatogr. Sci.*, 27, 569, 1989.

26. **Junk, G. A. and Richard, J. J.,** *Anal. Chem.,* 60, 451, 1988.
27. **Saady, J. J. and Poklis, A.,** *J. Anal. Toxicol.,* 14, 301, 1990.
28. **Brooks, M. W., Tessier, D., Soderstrom, D., Jenkins, J., and Clark, J. M.,** *J. Chromatogr. Sci.,* 28, 487, 1990.
29. **Marvin, C. H., Brindle, I. D., Hall, C. D., and Chiba, M.,** *Anal. Chem.,* 62, 1495, 1990.
30. **Bagnati, R., Benfentai, E., Davoli, E., and Fanelli, R.,** *Chemosphere,* 17, 59, 1988.
31. **Richard, J. J. and Junk, G. A.,** *Mikrochim. Acta,* 1, 387, 1986.
32. **Hagen, D. F., Markell, C. G., and Schmitt, G. A.,** *Anal. Chim. Acta,* 236, 157, 1990.

Chapter 3

SUPERCRITICAL FLUID EXTRACTION METHODS

Bruce E. Richter

TABLE OF CONTENTS

I. INTRODUCTION

In the last several years, many advances have been made in the area of sample analysis, including improvements to chromatographic and spectroscopic instrumentation. However, similar improvements in sample preparation have not kept pace. For most laboratories analyzing samples containing pesticides and pesticide residues, sample preparation is one of the major rate-limiting steps in the entire sample throughput equation. Separating or extracting contaminants from a matrix may be accomplished in several ways. Purging, heating, or digesting samples as well as liquid extraction are some standard means of separating analytes from their matrix. Each of these techniques has corresponding advantages and disadvantages. For example, standard liquid sample extraction techniques, such as Soxhlet or pulsed sonication, are time consuming and use copious amounts of organic solvents, making them costly (because of initial purchase and disposal costs of solvents) and hazardous to laboratory workers. The use of solid phase extraction schemes has reduced the volume of solvents used for some extractions, but the techniques are not directly applicable to all solutes and matrices. There are new, high-speed Soxhlet-type extraction apparatus commercially available, but they still use approximately 100 mL of organic solvents for each sample prepared. Clearly, a different approach to the problem of sample preparation needs to be examined.

Supercritical fluid extraction (SFE) has been shown to be a viable alternative to more conventional sample preparation techniques. One of the biggest advantages of SFE over the use of standard liquid extractions is the fact that many sample preparations can be done with nonpolluting, nontoxic fluids such as CO_2. The use of SFE on an industrial scale has occurred for many years, but SFE has not been applied to analytical scale sample preparation until recently. This chapter will discuss the physicochemical properties of supercritical fluids, the instrumentation used in SFE, and finally, many of the practical aspects of SFE, including applications of SFE to pesticides and metabolites.

II. PHYSICOCHEMICAL PROPERTIES OF SUPERCRITICAL FLUIDS

A material becomes a supercritical fluid when it is maintained at temperatures and pressures above the critical point. The critical point is defined by a critical temperature and a critical pressure above which the substance is neither a gas nor a liquid, but possesses properties of both. At temperatures and pressures above the critical point the substance cannot be liquified regardless of the pressure exerted on it, and it is called a supercritical fluid. Supercritical fluids have also been called dense gases, and this term is helpful

TABLE 1
Comparison of Physical Properties of Supercritical CO_2
and Liquid Solvents at 25°C

	CO_2[a]	C_6H_{14}	$MeCl_2$	$MeOH$
Density (g/mL)	0.746	0.660	1.326	0.791
Viscosity $\times$ 10^4 (PaS)	0.80	2.94	4.11	5.47
Diffusivity of dilute benzoic acid $\times$ 10^9 (m²/s)	6.0	4.0	2.9	1.8

[a] At 200 atm and 55°C.

in explaining the supercritical fluid phenomenon. For carbon dioxide, the most widely used fluid for analytical scale extractions, the critical temperature is 31°C and the critical pressure is 73 atm (1073 psi).

Supercritical fluids exhibit several properties that make them desirable as extraction solvents. One of the most interesting properties is the relationship of solvent strength to density. There have been many complex mathematic relations derived to try and predict the solvent strength of supercritical fluids such as cubic equations of state like the Redlich-Kwong or Peng-Robinson equations or the modified Hildebrand solubility parameters. It is not the purpose of this chapter to discuss the physical chemistry relations in supercritical fluids. Suffice it to say that the solvent strength of pure supercritical fluids is generally directly related to the density of the fluid. Since the density of the fluid is a function of its pressure and temperature, precise control of the pressure and temperature can be used to obtain a solvent with a more narrow window of solvating strength than is the case with liquid solvents. It is possible, therefore, to perform selective extractions using supercritical fluids, something that is often not achievable with organic solvents.

As the pressure of the fluid increases, higher molecular weight compounds become more soluble, and as the pressure is reduced, the fluid loses some of its solvent strength. If the pressure is reduced all the way to atmospheric pressure, the fluid loses essentially all of its solvating ability, and the extracted compounds fall out of solution. The process of condensing solutes from supercritical fluids by depressurization was first discovered in 1879 by Hannay and Hogarth[1-3] and is a fundamental part of SFE as a sample preparation technique.

Although the density and solvent strength of supercritical fluids are comparable to many organic liquids, the diffusivities are higher and the viscosities are lower. Density, viscosity, and diffusivity are compared for three commonly used organic solvents and supercritical carbon dioxide in Table 1.

Based on this comparison to liquids, supercritical fluids would be expected to give rise to more rapid mass transfer, and therefore faster extractions, based on the following reasons:

1. The high diffusivities observed in supercritical fluids promote rapid molecular diffusion. This means that the solubilized solutes will diffuse into the bulk fluid faster to be transported away from the matrix.
2. The low kinematic viscosities (viscosity divided by density) enhance free-convection mass transfer. Free-convection mass transfer includes contributions from buoyancy effects, such as differences in density between the fluid and the solutes, and gravity effects, as well as concentration gradients.
3. Low kinematic viscosities also allow increased turbulence in an extraction system, which contributes to more rapid forced-convection mass transfer.
4. The lower viscosities of fluids allow them to penetrate the pores and interstices of the matrix better, which promotes rapid mass transfer and solute extraction.

Another advantage of using supercritical fluids as extraction solvents can be realized if fluids that are gases at atmospheric pressure are used. After extractions using liquid solvents are complete, there is the necessity of concentrating the extract before analyses can be performed. This can be done using rotary evaporation, a Kuderna-Danish apparatus, or simple solvent evaporation under a stream of a clean compressed gas. Often several hundred milliliters of solvent remain after the extraction and must be evaporated, causing more environmental concerns. However, if CO_2 or N_2O are used as supercritical fluids for extraction, they can be allowed to escape into the air after decompression, and a clean extract in little or no organic solvent remains. Since extractions can be done with SFE in minutes and not hours, there are considerable savings in time as well as cost from the reduced use of organic solvents, which are costly to purchase and dispose of properly.

Another common practice in SFE that should be mentioned in connection with the physicochemical properties of supercritical fluids is the use of co-solvents, entrainers, or modifiers. These are compounds that are added to the primary fluid to enhance the extraction efficiency of the compounds of interest. For example, methanol is a common cosolvent added to CO_2 to improve the recovery of many analytes. There are two basic reasons for this. The first is to overcome ''matrix effects'' caused by the solutes being strongly bound to the matrix via chemisorption or physisorption mechanisms. An example of this is the extraction of pesticides from soil. If chlorinated pesticides, such as aldrin, dieldrin, or DDT, are extracted from a clean matrix onto which they have been spiked, quantitative recovery can be achieved using pure CO_2.

However, if an attempt is made to recover these same compounds from an incurred or natural soil that has been weathered, then the use of methanol or acetone in the CO_2 is necessary to get quantitative recovery; pure CO_2 cannot extract the bound pesticides by itself. The second reason is to increase the polarity of the primary fluid and, hence, its solvent strength. An example of this would be the extraction of ionic surfactants from sewage sludge. Pure CO_2 does not solubilize these compounds, but CO_2 with 30 to 40% methanol can extract these polar compounds. Many organic cosolvents in concentrations from 0.5 to 40% (wt/wt) have been used in SFE with good success.

III. INSTRUMENTATION FOR SUPERCRITICAL FLUID EXTRACTION

In principle, the instrumentation for SFE is quite simple. It is not the purpose of this chapter to discuss in depth all instrumental aspects of SFE, but to give the reader an understanding sufficient to comprehend the operation of SFE.

An extraction system consists of four basic components: (1) a pressure controlled pump, (2) a heated region for temperature control of the extraction cells, (3) a region for the decompression of the supercritical fluid, and (4) a region for trapping the analytes as they precipitate during the decompression step. Various levels of complexity exist in the instrumentation that has been reported in the literature and is available currently from commercial vendors.

The pump delivers the pressurized material to the extraction oven. Syringe pumps or reciprocating pumps have been used for this purpose. Generally the pressures are in the range of 100 to 680 atm or 1500 to 10,000 psi. The pumping system must be able to provide an uninterrupted supply of fluid during the entire length of the extraction.

The heated region must be capable of uniformly heating the extraction cells as well as any associated connecting tubing. It is especially important to preheat the fluid before it enters the extraction cells, particularly if high flow rates or large extraction cells are used. Of course, the operating temperature is determined by the fluid used and the nature of the analytes being extracted; typical temperatures reported in the literature range from about 30 to 130°C. The cells typically used in analytical SFE are similar in construction to empty liquid chromatography (LC) columns. They are made of stainless steel and are available in various internal diameters and lengths. Volumes range from 0.5 to 50 mL with pressure ratings from 4000 to 20,000 psi. One major difference between commercially available extraction cells and empty LC columns is that the extraction cells have been designed with polymeric seals, allowing them to be opened and closed many times, which is not the case with LC columns.

The region where decompression occurs is important for the complete recovery of the extracted analytes. The most common items used for this region are valves, back-pressure regulators, small I.D. orifices, or pieces of small I.D. capillary tubing, usually fused silica. During the decompression step, a great deal of cooling occurs if CO_2 or N_2O are used because of the Joule-Thompson effect upon expansion. This cooling can cause precipitation of the analytes and plugging of the decompression region or restrictor. It is common, therefore, to heat this decompression region to overcome the cooling effect and help prevent the loss of analytes and the plugging of the restrictors. Temperatures range from slightly above ambient to 250°C. Residence times in these restriction zones are so short (4 to 50 msec) that decomposition of temperature-sensitive analytes does not take place.

As equally important as the decompression region, the analyte trapping region must be functioning properly to ensure good recovery of the extracted compounds. The trapping region may be as simple as a vial containing an organic solvent through which the extractor effluent passes, or it may be much more complex. Cryogenically cooled packed-bed traps and gas chromatograph ovens have been used. The configuration of the trapping region is often determined by the mode of extraction, either on-line or off-line. A comparison of these two modes of operation will be discussed next.

A. ON-LINE SUPERCRITICAL FLUID EXTRACTION

With on-line SFE, the extraction process replaces the normal sample injection process into the chromatograph, and the extraction apparatus is directly coupled with the chromatographic instrument used for the analysis subsequent to the extraction. The extracted analytes are transferred to and collected in a chromatographic injection loop, a thermal or sorbent trap, prior to the chromatographic column, or in the stationary phase at the head of the analytical column itself. SFE has been coupled on-line with gas chromatography (GC), high performance liquid chromatography (HPLC), and supercritical fluid chromatography (SFC).

On-line SFE has the advantage of being inherently sensitive because all of the extracted analytes can be loaded directly on the chromatographic column. There is also less sampling with on-line extraction, and the entire system is easily automated. The trapping of analytes is usually better controlled because it is an instrument-controlled parameter.

There are some drawbacks with on-line SFE that must be considered. When performing on-line or coupled SFE, the SFE parameters, the analyte trapping conditions, and the chromatographic separation all must be understood and optimized for the analysis to be successfully completed. A sample extracted on-line is dedicated to the coupled chromatographic system. Once the on-line SFE analysis is completed, the extract is no longer available for evaluation using different techniques or chromatographic parameters.

B. OFF-LINE SUPERCRITICAL FLUID EXTRACTION

In contrast to on-line SFE, off-line SFE is performed with the analysis and extraction systems completely separate and decoupled. This is inherently simpler because only the extraction and analyte trapping need be considered. Once collected, the analytes can be subjected to several analytical techniques depending on the information desired from the sample. This technique allows the extraction and the chromatographic systems to do what they do best, and that is sample preparation and separations, respectively. This gives rise to higher throughput and better utilization than with on-line techniques in which either the chromatograph or extractor sit idle.

With the advantages of off-line SFE, one must not forget that disadvantages do exist. Off-line techniques may not be as sensitive as on-line procedures. For example, on-line SFE/GC using a 1-mg sample in the extraction cell yields the same sensitivity (in terms of analyte concentration in the bulk sample and the amount of analyte delivered to the column) as off-line SFE of a 1-g sample when the analytes are collected in 1 mL of solvent followed by analysis using a 1-µL injection on-column for GC. There is also more sample handling with off-line extraction than with on-line, and increased sample handling can mean more chances of sample loss or contamination. Although simple in principle, the decompression area and analyte trapping regions can be more problematic with off-line than with on-line SFE. This is probably due to the fact that larger samples are often used, and with larger sample sizes, more analytes are present and higher flow rates are used, causing trapping and decompression to be more difficult to control.

C. DYNAMIC AND STATIC SUPERCRITICAL FLUID
EXTRACTION

Analytical SFE has been performed using three different modes: (1) dynamic (in which the supercritical fluid is continuously flowing through the cell), (2) static (in which the cell is pressurized with supercritical fluid and the extraction is allowed to proceed without any outflow of the supercritical fluid until extraction is finished), and (3) a combination of static followed by dynamic extraction. As far as the hardware or instrumentation is concerned, the only difference between these methods is the inclusion and use of additional valves near the inlet and outlet of the extraction cells for performing static extraction. In practice for static extraction, the cells are pressurized with the supercritical fluid, the inlet and outlet valves are closed for a given period, both valves are opened, and the fluid flows dynamically through the cell carrying the extracted analytes away.

Both dynamic and static SFE have been used to achieve quantitative results, but dynamic extraction might be expected to yield more rapid recoveries by continuously providing pure extraction fluid to the sample. Dynamic extractions can also be performed without any values between the extraction cell and the collection medium. The elimination of the valve between the extraction and collection media is attractive, particularly for the extraction of trace compounds, since the chances for analyte loss or contamination are reduced. Static extractions, however, have the advantage that modifiers can be added directly to the extraction cell prior to the pressurization step. This static step can cause the swelling of the matrix in some cases, which can make the solutes more readily accessible for extraction. Another potential advantage of static extraction is that it is possible to do extractions with less fluid than with dynamic extraction.

IV. APPLICATIONS OF SUPERCRITICAL FLUID EXTRACTION FOR THE EXTRACTION OF PESTICIDES

SFE has been applied to the extraction of pesticides from many matrices prior to quantitative analysis. Some of the first data on the use of SFE to remove pesticide residues were reported by Stahl and Rau in 1984.[4] In this work, the authors showed the successful extraction of HCH, aldrin, DDT, and α-endosulfan from senna leaves. These results were preliminary, but they demonstrated the potential of SFE for the removal of pesticide contaminants while leaving much of the matrix interference behind. This section will discuss the work that has been reported in the literature to date. The data will be presented according to the matrix from which the pesticides were extracted: soils and sediments, animal lipids and tissues, plant lipids and tissues, aqueous materials, and sorbent materials.

A. SOILS AND SEDIMENTS

Table 2 summarizes the results reported for the extraction of pesticides from soils and sediments. After initial work demonstrated the potential of SFE for the selective extraction of pesticides, attention was focused on several specific applications and more in-depth studies to investigate the efficacy of SFE for extracting pesticides.

Capriel[5] and co-workers investigated the extraction of bound residues in soil. They used supercritical methanol at 250°C and 150 bar as the extraction fluid. The extraction efficiency of SFE was compared to high temperature distillation (HTD) for the recovery of ^{14}C labeled pesticides and metabolites. The compounds studied were atrazine, prometryn, deltamethrin, diuron, 2,4-D, methylparathion, dieldrin, carbofuran, and some of their associated metabolites. Identification of the compounds was done by GC/MS, and quantification was performed using combustion followed by liquid scintillation or GC with either an electron capture detector (ECD) or nitrogen phosphorus detector (NPD). Several different soils, all with varying composition, were investigated. Under the conditions used, the soil type seemed to have no

TABLE 2
Pesticides Extracted from Soil

Pesticides	Extraction conditions	Ref.
Prometryn, atrazine, deltamethrin, dieldrin, carbofuran, diuron, 2,4-D and methylparathion	Methanol, 250°C, 150 bar, 1 mL/min, 2 h	5
DDT	CO_2, 40°C, 100 atm, 0.7 g/sec, 10 min	6
DDT	CO_2 with 5 wt% methanol or toluene, 40 or 100°C, 0.7 g/sec, 5 min	7
Lindane, aldrin, DDT	CO_2, 138 bar, 15 min	8
Linuron, diuron	CO_2 with ethanol, methanol, or acetonitrile, 75 to 120°C, 100 to 400 bar, 2.5 to 8.5 mL/min, 15 to 180 min	9, 10
Sulfonylureas	CO_2 with 2% methanol, 40°C, 223 bar, 6.0 mL/min, 2 to 15 min	11
Simazine, atrazine, propazine, terbutylazine, cyanazine	CO_2 at 48°C, 230 bar, 1.7 mL/min, 30 min	12
Organochlorine, organophosphorus	100% CO_2 and with 10% methanol, 50 to 70°C, 150 or 300 atm, 25 to 60 min	13
DDT, DDE, DDD, lindane, aldrin	CO_2 with 5% acetone or methanol, 75°C, 400 atm, 60 min	14

influence on recovery of the analytes. In all cases, the recovery of the ^{14}C-labeled compounds was better with SFE than with HTD. However, in some cases, possible reaction products were seen, probably due to reactions that may have taken place between methanol and the pesticides or their metabolites. Even with these possible chemical alterations, the results were still significantly better than with standard procedures.

The extraction of DDT was studied by Brady et al.[6] using much more mild conditions. In this study, CO_2 at 40°C and 100 atm pressure and a flow rate of 0.7 g/s (65 mL/min) was used as the extraction fluid. Quantification was performed using GC with an ECD. The moisture content and pore size of the soils were measured to determine their effect on extraction efficiency. It was found that DDT was not completely removed from soils with high organic content. This was presumably due to the interaction of the DDT with humic acid components in the soil. The addition of water to the soils did not reduce the amount of DDT extracted, but it did slow down the extraction rate, so the extraction had to proceed for a longer period of time. The highest recovery achieved with these mild conditions was 70%. In the same study, the recovery of a more nonpolar material, Arochlor 1254, was investigated. The polychlorinated biphenyls (PCBs) were more easily extracted (90% under the same conditions). In order to study the effects of long-term exposure of soil to these compounds, the extraction rate of a laboratory-spiked soil and a soil from a spill site were compared. The laboratory-spiked soil had a more rapid decrease of concentration vs. time than the incurred sample, but in both cases, the maximum recovered was 70% as compared to standard Soxhlet extraction procedures.

Follow-up work from this same research group[7] discussed the effect of methanol and toluene as cosolvents or entrainers. The addition of toluene (5 wt%) showed very little improvement over pure CO_2. However, adding methanol (5 wt%) did have a profound impact on recovery. With this modified fluid, 95% of the DDT (from both laboratory-spiked and incurred soil samples) was extracted in 5 min compared to 70% in 10 min with toluene-modified fluid or pure CO_2. The explanation for this observation is that the methanol is able to hydrogen bond with active sites in the soil and displace the DDT molecules, allowing them to become soluble in the fluid and be swept out of the cell.

Engelhardt and Groß[8] showed the feasibility of extracting lindane, aldrin, and DDT from soil. They extracted spiked soil samples with CO_2 at 138 bar for 15 min, and the analysis was done by packed column SFC coupled on-line with the SFE system. No recovery data were reported in this work.

McNally and Wheeler[9–11] have reported studies on the extraction of urea and sulfonylurea herbicides from soil and other matrices. In these studies, the extraction temperature, flow-rate, pressure, and extraction fluid modifier type and concentration were examined for their effect on the extraction efficiency of the compounds of interest from soils. The two urea herbicides studied were diuron and linuron. Analyses of the extracts were performed using liquid scintillation counting of the ^{14}C-labeled compounds. In some cases, further analyses were performed by HPLC or GC.

Increases in temperature yielded increased extraction efficiencies of diuron and linuron. Higher extraction efficiencies over shorter time periods resulted from increases in extraction pressures. Increases in flow rates yielded higher extraction efficiencies until the pressure drop across the extraction cell caused by the high flow rate became significant. Higher concentrations of modifiers gave better recovery of the compounds, but the identity of the entrainers also influenced the results. Methanol gave the best results with diuron, and ethanol was best with linuron.

In the second study, static and dynamic extraction were investigated.[10] For this system, static extraction seemed to give quantitative recovery in a shorter period of time than did dynamic extraction. This may be because swelling of the sample matrix may occur during static extraction and not during dynamic extraction.

Studies of the sulfonylurea herbicides[11] yielded many of the same conclusions as the work with the urea herbicides. Few recovery data were reported in this work, but the ability to extract and analyze sulfonylurea herbicides from soils in less than 45 min was shown with on-line SFE/SFC instrumentation.

The extraction efficiency of *s*-triazine herbicides from sediment was reported by Janda et al.[12] Simazine, atrazine, propazine, terbutylazine, and cyanazine were spiked onto dried sediment samples. SFE was performed using CO_2 at 48°C and 230 bar for 30 min. In some cases, methanol was added to the extraction cell prior to extraction with the CO_2. The extracted analytes were trapped in methanol. Analyses of the extracts were done using GC or HPLC. At the 150- to 400-ppb level, essentially 100% recovery of the five compounds was achieved. It should be kept in mind that although it is a good place to start for method development, data collected on laboratory-spiked samples is not always representative of the recovery that will be obtained with aged or incurred samples.

Lopez-Avila et al.[13] recently reported on a preliminary study investigating the use of SFE for the extraction of pesticides and other pollutant materials as part of a proposed U.S. Environment Protection Agency protocol. The effects of extraction pressure, temperature, moisture content, time, static or dynamic mode of extraction, cell volume, and sample size were studied. There were 41 organochlorine and 47 organophosphorus pesticides spiked on clean sand or soil. Conditions used were CO_2 with 10% methanol, 70°C, and 250 atm for 30 min. Analyses were done by GC with an ECD.

Excellent recoveries were achieved for 38 of the 41 organochlorine pesticides. Only the recoveries for hexachlorocyclopentadiene, chlorbenzilate, and DBCP were low. All but five of the organophosphorus pesticides were recovered almost quantitatively. Diazinon, phorate, demeton-S, demeton-O, and TEPP gave low recoveries, possibly due to hydrolysis caused by the methanol or decomposition during the GC analyses.

Recently, the results of studies comparing the extraction efficiency of SFE to pulsed sonication for the recovery of DDT, DDE, DDD, aldrin, and lindane from soil were reported.[14] Extractions were done at 400 atm, 75°C, for 45 min with CO_2 containing either 5% acetone or 5% methanol. In these studies, separate 1-g samples were extracted by SFE and pulsed sonication. These were incurred samples and not laboratory-spiked samples. The analyses were performed using GC with an ECD. In essentially all cases, quantitative recovery of the pesticides was achieved as compared to the pulsed sonication results. In some cases, more material was determined after SFE than after pulsed sonication. Clearly, with optimized conditions, SFE can achieve quantitative extraction of pesticides from soils and sediments.

B. ANIMAL LIPIDS AND TISSUES

Table 3 summarizes the data that have been reported on the extraction of pesticides from animal lipids and tissues. Work performed as part of a National Cancer Institute contract reported some of the earliest data on the use of SFE for the extraction of pesticides and metabolites from animal tissues.[15] On-line SFE coupled with capillary SFC was used in this work. These data were

TABLE 3
Pesticides Extracted from Animal Tissues

Pesticides	Extraction conditions	Ref.
Aldicarb, carbaryl, captan, malathion	CO_2 at 50°C, 250 atm, 0.5 mL/min, 15 min	15
Organochlorine	CO_2 at 50°C, 140 atm, 60 min	16
DDT, DDE	None given	17
Organochlorine, thiophosphates, phenoxyesters, triazines, carbamates	CO_2 at 50°C, 137 to 170 atm, 60 min	18
Diuron, alachlor, bendiocarb, carbaryl	CO_2 at 100°C, 75 to 300 atm, 40 to 70 mL/min as expanded gas, 30 min	19
Organochlorine	CO_2 at 40°C, 190 to 270 bar, 1 mL/min, 20 min	20
Organochlorine, organophosphorus	CO_2 at 80°C, 10,000 psi (680 atm), 5 L/min as expanded gas, 20 min	21

compared to standard extraction techniques and either HPLC or liquid scintillation counting for quantification. The SFE conditions were typically CO_2 at 50°C and 250 atm for 15 min. Four pesticides were extracted from rat hepatocyte isolates: aldicarb, carbaryl, captan, and malathion. Carbaryl and malathion were also extracted from urine, feces, kidney, and liver samples as part of an *in vivo* rat metabolism study. The recoveries of the parent pesticide and the major metabolites as compared to the standard procedures were 75 to 100%. Of course, the major advantage of SFE is the reduction in time: 15 min compared to 12 h for liquid extraction.

Nam et al. demonstrated good recovery of chlorinated pesticides from fish tissue.[16] The conditions used in this off-line extraction system were CO_2 at 50°C and 140 atm for 60 min. The compounds extracted were lindane, heptachlor epoxide, chlordane, dieldrin, endrin, DDD, DDE, DDT, heptachlor, and aldrin. Analyses were done by GC equipped with an ECD. Quantitative extraction was achieved for all of these compounds spiked into fish tissue. No change in extraction efficiency was seen over an analyte concentration range of two orders of magnitude. A 40-fold difference in lipid concentration appeared to have no effect on the recovery of these spiked compounds.

Pipkin has also shown good recovery of DDD, DDT, and DDE from fish tissue.[17] No details of extraction conditions were given, but the recoveries averaged 112% and the precision was greatly improved as compared to solvent extraction procedures.

In follow-up work, Nam et al.[18] used a multiple-cell off-line system capable of extracting four samples simultaneously. Organochlorine pesticides were spiked onto pig whole blood and milk, and fish tissue was spiked with thiophosphates, phenoxyesters, triazines, and carbamates. The samples were subsequently extracted with CO_2 at 50°C and at pressures ranging from 137 to 170 atm. After a 30 min equilibration time (static extraction), the contents of the cells were purged through hexane in the collection vessel. A second extraction was carried out in an analogous manner. Any coextracted lipids were removed by using gel permeation chromatography (GPC). Analyses of the extracts were done by GC with an ECD. The recoveries of the chlorinated pesticides from blood and milk were essentially the same as with liquid solvent extractions. The recoveries of the pesticides from the fish tissue were not as good because of the increased polarity of some of the compounds. The thiophosphates and phenoxyesters were recovered quantitatively, but the recoveries of simazine (48%), atrazine (46%), metribuzine (60%), butylate (68%), and carbofuran (64%) were low. The use of a modified fluid would improve these recovery numbers.

Murugaverl and Voorhees[19] have combined SFE with the cleanup step that is often needed following the extraction of lipid-containing materials with supercritical fluids. In this work, an on-line apparatus was constructed which incorporated a second stainless steel column packed with sorbent material after the extraction cell to retard migration of the coextracted lipids. In addition, the adsorbent material was added directly to the cell prior to extraction. Lard, beef fat, and bacon fat were spiked with known levels of diuron, alachlor, bendiocarb, and carbaryl (levels ranged from 0.7 to 110 ppm). A switching valve and photodiode array UV detector were placed in the system to monitor the migration of the lipid fraction. Just before the lipid material entered the SFC column, the switching valve was actuated, and the lipid components were vented to waste. This prevented the heavy triglycerides from fouling the SFC column. Extraction conditions were CO_2 at 100°C, the pressure was increased from 75 to 300 atm at 15 atm/min, and the final pressure was maintained for 30 min. The flow as 40 to 70 mL/min as a gas. The analyses were done with a capillary SFC instrument with a flame ionization detector (FID). Recoveries for these compounds were in the 80 to 85% range.

France et al.[20] have described a similar off-line approach to cleaning up lipids that are contaminated with pesticides. In this work, incurred poultry fat samples that had been examined previously and found to contain low ppm levels of heptachlor epoxide, dieldrin, and endrin were used along with lard spiked with lindane, heptachlor, heptachlor epoxide, dieldrin, endrin, and DDT at 0.5 to 2 ppm. Neat fat or lard was injected into the SFE cleanup system using an injection valve. A column containing either alumina or silica was used to retard the lipids. Since the pesticides were soluble under the conditions used in the experiments (CO_2 at 40°C and 190 to 270 bar pressure) they were swept out of the extraction cell and collected in hexane. The analysis of the hexane solution was done by GC with an ECD. Recoveries of all pesticides were 90 to 100%. Based on the f test and the t test, precisions and recoveries for the SFE cleanup technique and the conventional method were not statistically different at the 95% confidence limit.

Hopper and King[21] have shown very impressive results using SFE to remove organochlorine and organophosphorus pesticides from butter fat and hamburger. The pesticides were spiked into the butter fat at levels ranging from 6 ppb to 2 ppm. The hamburger sample was an incurred or unspiked sample that was also extracted by conventional organic solvent methods. The levels of pesticides ranged from 0.5 to 5 ppb. In this work, the authors discuss the use of an extraction enhancer that was added to the sample to adsorb water and make the wet or soft samples robust enough to remain in the extraction cell during extraction. The material was a granular diatomaceous earth. In addition to adsorbing water, the enhancer creates a homogeneously permeable extraction bed that can be efficiently extracted without channeling.

After being mixed with the enhancer, the samples were extracted with CO_2 for 20 min at 10,000 psi and 80°C. Flow rates were 5 L/min as expanded gas. Cleanup of the samples was done with GPC, and the assays were done by GC and an ECD, flame photometric detector (FPD), or Hall electrolytic conductivity detector (HECD). The recovery from the butter fat ranged from 70 to 100%, while the recovery from the hamburger averaged 88%. Clearly, from these data SFE is a viable alternative to liquid extraction procedures that use copious amounts of organic solvents.

C. PLANT LIPIDS AND TISSUES

Table 4 summarizes the data reported on the extraction of pesticides from plant tissues or lipids. The first indication that supercritical fluids have good potential for extracting pesticides from plant tissues was shown by McRae and Wheldon,[22] who investigated the solubility of hops α- and β-acids in liquid carbon dioxide. They also reported that many of the pesticides used by hops growers (such as endosulfan, mephosfolan, dicofol, dinocap, and triadimefon) are soluble in liquid carbon dioxide. No quantitative data were given.

After this initial report, little was discussed in the scientific literature concerning the supercritical fluid extraction of pesticides from plant tissues. The first study showing the quantitative removal of pesticides was given by Stahl et al.[4,23] These workers showed the extraction of HCH isomers (lindane), HCB (hexachlorobenzene), heptachlor, aldrin, DDT, endosulfan, and PCBs from senna leaves and pods without affecting the glycoside content. The conditions used were CO_2 at 60°C and 113 to 380 bar pressure. The extraction of urea herbicides from thorn apple leaves[23] without decreasing their tropa alkaloid content was also discussed. These examples clearly showed the enhanced selectivity of SFE over extraction with liquid organic solvents.

TABLE 4
Pesticides Extracted from Plant Tissues

Pesticides	Extraction conditions	Ref.
Endosulfan, mephosfolan, dicofol, dinocap, triadimeton	CO_2 at 10°C, 1000 psi	22
HCH, HCB, heptachlor, aldrin, DDT, endosulfan	CO_2 at 60°C, 113 to 380 bar, 20 min to 2 h	4
Prometryn, atrazine, deltamethrin, dieldrin, carbofuran, diuron, 2,4-D	Methanol, 250°C, 150 bar, 1 mL/min, 2 h	5
Linuron and diuron	CO_2 with ethanol, methanol, or acetonitrile, 75 to 120°C, 110 to 400 bar, 2.5 to 8.5 mL/min, 15 to 180 min	9, 10
Sulfonylureas	CO_2 with 2% methanol, 40°C, 223 bar, 6.0 mL/min, 2 to 15 min	11
Diazinon, chlorpyrifos, malathion, aldrin, diuron	None given	17
Diuron, alachlor, bendiocarb, carbaryl	CO_2 at 100°C, 75 to 300 atm, 40 to 70 mL/min as expanded gas, 30 min	19
Organochlorine, organophosphorus	CO_2 at 80°C, 10,000 psi (680 atm), 5 L/min as expanded gas, 20 min	21
Aldrin	CO_2 at 75°C, 150 atm, 150 mL/min as gas, 15 min	24

An interesting study of SFE was the work of Capriel and co-workers,[5] already discussed in the section on the extraction of soils. In addition to studying the ability of SFE to extract bound residues from soils, these workers also investigated the potential of SFE to extract bound pesticides and metabolites from radishes and corn. The extraction of the ^{14}C-labeled prometryn, atrazine, deltamethrin, dieldrin, carbofuran, diuron, 2,4-D, and methylparathion with supercritical methanol was compared to HTD. Gas chromatography with an ECD or NPD, or GC/MS, was used for quantitation and identification. Combustion of the plant materials followed by scintillation counting was done to determine mass balance. In all cases, the recovery of the bound residues was 94% or greater. It must be remembered that the conditions for supercritical methanol, 250°C and 150 bar pressure, may cause the decomposition of many sensitive compounds. Methanol can also form explosive mixtures at 250°C. There may be mixed fluid systems that could give the same recoveries with less safety concerns.

In addition to studying the recovery of urea and sulfonylurea herbicides from soil, McNally and Wheeler[9-11] demonstrated the ability of SFE to quantitatively remove these compounds from wheat grain, wheat flour, and wheat straw. In these experiments, the effect of methanol, ethanol, and acetonitrile as cosolvents in CO_2 was determined. Temperatures between 40 and 120°C and pressures of 200 to 400 atm were used. Both static and dynamic extraction were investigated. Extraction times were 6.5 to 90 min, and assays were done by HPLC, SFC, and liquid scintillation counting. With optimized conditions, greater than 90% recoveries of the components were obtained.

Pipkin[17] showed that diazinon, chlorpyrifos, malathion, aldrin, and diuron could be extracted from alfalfa, onions, almonds, oranges, strawberries, and lettuce. No actual recoveries were reported, and no experimental conditions were given.

In addition to extracting pesticides from pork lard and bacon fat, Murugaverl and Voorhees[19] investigated the extraction of diuron, alachlor, bendiocarb, and carbaryl from soybean oil. An on-line SFE-SFC system was used for the extraction and analysis. The conditions for extraction were CO_2 at 100°C, and pressures were programmed from 75 to 350 atm during the extraction. Capillary SFC with an FID was used to assay the extracts. The use of the FID limited the detection level to about 2 ng for the various pesticides; with the small sample sizes used (6 to 20 mg) this corresponds to the ppm range. These workers used a C_{18} sorbent material mixed with the lipid materials to help retain them in the column and to help fractionate the triglycerides from the pesticides.

The use of an adsorbent or enhancer was first reported by Hopper and King,[21] as mentioned previously. In addition to animal tissues, pelletized diatomaceous earth has been used in conjunction with the extraction of organochlorine and organophosphorus pesticides from lettuce, carrots, potatoes, and peanut butter. Extraction conditions were CO_2 at 80°C and 10,000 psi for 20 min. After the SFE, GPC was used for the separation of the triglyceride fraction from the pesticide fraction. After appropriate concentration steps, analyses were done by GC with an HECD or FPD. Quantitative recovery of the pesticides was achieved at the ppm down to the sub-ppb levels.

Murphy and Richter[24] used both on-line and off-line SFE to determine the recovery, linearity, and reproducibility for the extraction of aldrin from rice hulls, mill rice, soybean meal, refined soybean oil, lecithin, and crude soybean oil. Extraction conditions were CO_2 at 75°C and 150 atm for 15 min. Analyses were done with GC with either an FID or ECD. Samples of 1 mg were used for the on-line work, while 1 g samples were used for the off-line experiments. Recovery at 10 ppb was 96% with 3.1% relative standard deviation (RSD) for the off-line work with the ECD. The calculated minimum detection level at a signal to noise level of three was reported to be 0.23 ppb.

D. AQUEOUS MATERIALS

There has been only one example reported in the literature of the direct extraction of pesticides from water. Hedrick and Taylor[25] demonstrated the extraction of diisopropyl methylphosphonate (DIMP) from water using an extraction cell held in a vertical position that allowed the supercritical fluid to sparge up through the water. The fluid was recycled through the extraction cell containing water until equilibrium had been obtained. An aliquot of the fluid was analyzed by SFC with either an FID or Fourier transform infrared (FTIR) detection. Extractions were done at 350 atm and 50°C with pure CO_2. The addition of NaCl shortened the extraction time from 1.5 h to 45 min, probably because it increased the ionic strength of the water and reduced the solubility of the DIMP, making it easier for solubilization in the supercritical fluid. Response was linear over the range studied (800 ppb to 800 ppm) and the RSD ranged between 1.5 and 15%.

E. SORBENT MATERIALS

Raymer and co-workers[26,27] have investigated the extraction of radio-labeled parathion and lindane from Tenax-GC and four additional polyimide sorbent resins. Extractions were done at 3000 psi and 40°C for about 15 min. Analyses were performed by thin layer chromatography (TLC), scintillation counting, and mass spectrometry. Results from the SFE experiments were compared to those obtained with thermal desorption. In all cases, supercritical fluid extraction (desorption) was superior to thermal desorption for the recovery of these pesticides.

V. SUMMARY

Using SFE, sample preparation can be done faster, cheaper, and more completely than can be achieved by conventional extraction procedures. SFE is not necessarily a panacea for all sample preparation problems, but it seems to be ideally suited for the extraction of pesticides from many matrices. Several manufacturers of instrumentation can provide equipment that is reliable and can be used routinely. As more workers investigate the use of SFE and our understanding expands, it will become an even more powerful sample preparation tool that will be utilized by many analytical chemists.

REFERENCES

1. **Hannay, J. B. and Hogarth, J.,** *Proc. R. Soc. London,* 29, 324, 1879.
2. **Hannay, J. B. and Hogarth, J.,** *Proc. R. Soc. London,* 30, 178, 1880.
3. **Hannay, J. B.,** *Proc. R. Soc. London,* 30, 484, 1880.
4. **Stahl, E. and Rau, G.,** *Planta Med.,* 2, 171, 1984.
5. **Capriel, P., Haisch, A., and Khan, S. U.,** *J. Agric. Food Chem.,* 34, 70, 1986.
6. **Brady, B. O., Kao, C.-P. C., Dooley, K. M., Knopf, F. C., and Gambrell, R. P.,** *Ind. Eng. Chem. Res.,* 26, 26, 1987.
7. **Dooley, K. M., Kao, C.-P. C., Gambrell, R. P., and Knopf, F. C.,** *Ind. Eng. Chem. Res.,* 26, 2058, 1987.

8. **Engelhardt, H. and Groß, A.**, *J. High Res. Chromatogr.*, 11, 726, 1988.

9. **McNally, M. E. P. and Wheeler, J. R.**, *J. Chromatogr.*, 447, 53, 1988.

10. **Wheeler, J. R. and McNally, M. E. P.**, *J. Chromatogr. Sci.*, 27, 534, 1989.

11. **McNally, M. E. P. and Wheeler, J. R.**, *J. Chromatogr.*, 435, 63, 1988.

12. **Jandra, V., Steenbeke, G., and Sandra, P.**, *J. Chromatogr.*, 479, 200, 1989.

13. **Lopez-Avila, V., Dodhiwala, N. S., and Beckert, W. F.**, *J. Chromatogr. Sci.*, 28, 468, 1990.

14. **Richter, B. E., Campbell, E. R., Rynaski, A. F., Murphy, B. J., Nielsen, R. B., Porter, N. L., and Craig, C. A.**, Proc. 2nd Int. Symp. Supercritical Fluid Chromatography and Extraction, BYU Press, Provo, Utah, 1991, 121.

15. Small Business Innovative Research (SBIR) Phase II Final Report, Biochemical monitoring of pesticides, solvents and their metabolites by supercritical fluid chromatography, the National Cancer Institute, Bethesda, MD, Contract No. N44-CP-71086, 1989.

16. **Nam, K. S., Kapila, S., Viswanath, D. S., Clevenger, T. E., Johansson, J., and Yanders, A. F.**, *Chemosphere*, 19, 33, 1989.

17. **Pipkin, W.**, *Am. Lab.*, November, 40D, 1990.

18. **Nam, K. S., Kapila, S., Yanders, A. F., and Puri, R. K.**, *Chemosphere*, 20, 873, 1990.

19. **Murugaverl, B. and Voorhees, K. J.**, *J. Microcol. Sep.*, 3, 11, 1991.

20. **France, J. E., King, J. W., and Snyder, J. M.**, *J. Agric. Food Chem.*, 39, 1871, 1991.

21. **Hopper, M. L. and King, J. W.**, *J. Assoc. Off. Anal. Chem.*, 74, 661, 1991.

22. **McRae, J. B. and Wheldon, A. G.**, *J. Inst. Brew. London*, 86, 296, 1980.

23. **Stahl, E., Quirin, K. W., Glatz, A., Gerard, D., and Rau, G.**, *Ber. Bunsenges. Phys. Chem.*, 88, 900, 1984.

24. **Murphy, B. J. and Richter, B. E.**, *J. Microcol. Sep.*, 3, 59, 1991.

25. **Hedrick, J. and Taylor, L. T.**, *Anal. Chem.*, 61, 1986, 1989.

26. **Raymer, J. H. and Pellizzari, E. D.**, *Anal. Chem.*, 59, 1043, 1987.

27. **Raymer, J. H., Pellizzari, E. D., and Cooper, S. D.**, *Anal. Chem.*, 59, 2069, 1987.

Part II. Multiresidue Approaches
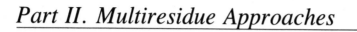

Chapter 4

MULTIRESIDUE ANALYSIS OF FRUITS AND VEGETABLES

Patrick T. Holland and Colin P. Malcolm

TABLE OF CONTENTS

I. INTRODUCTION

Multiresidue analysis for pesticides has its main application in the monitoring of foodstuffs destined for human consumption. Fruits and vegetables are the staples that are most likely to receive direct applications of pesticides in the field and to retain a proportion as residues in or on the edible portion at harvest. Multiresidue (MR) methods have, therefore, been required for the analysis of samples to determine residues of a wide range of pesticides that could have been used in production of the crops or in postharvest protection.

The continuing development of new pesticides provides a constant challenge to MR methodology. Compound classifications, such as insecticides in the organochlorine (OC), organophosphorus (OP), or carbamate structural classes, have had utility in defining analytical strategies. However, the ever-broadening range of chemical classes and increasing diversity of structures/properties represented within a class have made it difficult to develop methods that can adequately determine many classes or even all pesticides within a class. The gradual deregistration of older pesticides has not always led to their removal from analytical lists. Production and use may continue in some countries or these pesticides may have become persistent and ubiquitous environmental contaminants, such as DDT.

The high potency of many newer pesticides, such as synthetic pyrethroid insecticides or sulfonylurea herbicides, has led to use rates in the 1- to 10-g/ha range. Detecting residues, even soon after application at these rates, may be beyond the means of standard monitoring methods. Achieving low detection limits and evaluating their significance in scientific and regulatory senses are major challenges for monitoring programs.

Although biological/bioassay and simplified chemical methods such as thin-layer-chromatography continue to have a place in pesticide residue analysis, MR methodology has increasingly become dependent on instrumental analysis techniques. The advances in gas chromatography (GC) and high performance liquid chromatography (HPLC), both in separation power and in sensitive and selective detection systems including mass spectrometry (MS), have given the residue analyst the necessary tools to determine a wide range of pesticides with adequate sensitivity. Integrators or computers can aid data capture, processing, and interpretation of complex chromatograms. The increased capital costs and more complex and demanding techniques make it difficult for developing nations to set up and maintain adequate laboratories for residue analysis. Even in developed nations, the high cost of residue testing on a large scale is an incentive for research into more cost-effective methods.

Developments in MR analysis of food up to the mid 1980s have been given thorough coverage in several reviews.[1-3] There have been few recent changes in the basic technology of sample extraction, cleanup, and pesticide determination. However, a variety of monitoring strategies and methods have been used in different countries. This review examines some of the regulatory rationales for residue monitoring such as the scope of assays that interact with laboratory requirements. Current approaches to residue monitoring are compared. Developments toward more cost-effectiveness in MR analysis are examined, including miniaturization and automation.

II. THE REGULATORY ENVIRONMENT

The principal functions of residue monitoring on food commodities in most countries are the enforcement of legislated national maximum residue limits (MRLs) and the estimation of overall exposure to pesticides from the diet. The assumption is that enforcement of MRLs reduces exposure of the population to possibly harmful residues in the diet. Monitoring of crops also exerts pressure on growers to avoid the use of unregistered pesticides and take due care with ones that may affect soil or water quality or the health of farm workers. These perceptions have been given more political weight in recent years as the public in affluent countries has become increasingly concerned about possible health risks from residues of xenobiotics in the diet. Questions about the dependence of modern agriculture on agrochemicals and the negative effects on various environmental parameters have also impinged on policy making for pesticide use on food and for monitoring programs. Wessel and Yess[4] have reviewed the role of the U.S. Food and Drug Administration (FDA) in monitoring pesticide residues in food imported into the U.S., including the complex issues of which pesticides are routinely monitored and how the lower levels of residue for triggering regulatory action are set.

Ekstrom and Akerblom[5] have examined similar issues in an international context.

A. TYPES OF RESIDUE MONITORING PROGRAMS

There are five broad categories of monitoring where multiresidue analysis of fruits and vegetables is an important component. Each has particular objectives that influence the choice of analytical methodology.

1. Surveillance Monitoring

Food commodities entering domestic commerce are tested to enforce national residue limits. The sampling is generally random over all food types, with weighting of numbers based on relative importance to the diet. There may be some biasing toward commodities or sources that have tended to have residue problems in the past. MR methods are used to detect as many pesticides as possible, but single compound methods may also be required to achieve the desired scope. Detection limits are set so that illegal residues can be routinely found and confirmed under laboratory conditions of high sample throughput. The surveys are usually planned to be on-going with quarterly or annual reporting. Special subsurveys may also be organized to study particular pesticides or commodities.

2. Compliance Monitoring

Consignments from particular countries or growers are targeted because of previous residue problems. This targeting is an efficient way to reduce the opportunity for commodities with illegal residues to reach market and exerts pressure on food producers to correct unacceptable pesticide use practices. Compliance testing is generally an adjunct to surveillance monitoring. In FDA monitoring during 1989 of a total of over 18,000 samples, 685 were analyzed for compliance.[6] In Swedish monitoring during 1985 to 1989 of over 23,000 samples, 3200 were analyzed for compliance.[7]

3. Total Diet Studies

Composite samples of commodities from typical "market baskets" are tested in a form ready for consumption to determine the average exposure of individuals to pesticide residues from common diets.[8] The overall dietary intakes estimated for each pesticide are compared to acceptable daily intake (ADI) figures. These studies often also determine other dietary hazards such as heavy metals. Although surveillance residue data can be used to calculate dietary intakes, there may be overestimation because losses from food preparation and cooking were not taken into account, or underestimation for low level residues that may have fallen below surveillance reporting limits.

The generally low and uneven distribution of residues among raw commodities and the losses upon food preparation lead to very low residues in total diet samples. Detection limits of 0.001 mg/kg are required if residues of the principal agrochemicals used in production of foods are to be represented in the final dietary consumption estimates.[6,8] Low detection limits are also required to account for bioaccumulative residues of deregistered OC pesticides, PCBs, PCDDs, and PCDFs that may reach food from the environment. Due to the analytical difficulties, many studies have been limited to OC and OP compounds. Recent FDA studies[6] have evaluated a wider range of pesticides and used compositing at the individual commodity level rather than at a crop grouping level or total market basket level to reduce some of the problems arising from infrequent residues.

4. Pre-Export Testing

Food consignments are tested individually or by means of surveys to establish their acceptabililty to destination markets. Complex issues can arise, particularly for new or uncommon crops where MRLs may not be set in many countries yet sprays are necessary to meet quarantine standards.[9] Furthermore, political pressures in countries such as the U.S. are for a high degree of surveillance of imports.[4] Until MRLs in large markets such as the EEC and the U.S. are harmonized and set on a logical basis of crop groupings, exporting countries must ensure access by careful consideration of spray programs and the resultant patterns of residues on the harvested crops. Responsibility for this testing lies primarily with the exporter, but government agencies may be involved where crops are of high national economic importance. A subset of pesticides can generally be targeted for testing from local knowledge of the types of agrochemicals available and likely to be used. Indeed, such testing is often an adjunct to recommended spray schedules and other extension programs.

5. Special Interest Testing

Groups such as consumer organizations, food processing companies, supermarket food companies, or "organic" grower groups may carry out residue surveys. A variety of objectives can be involved. The testing has generally been carried out in private laboratories rather than in the large government laboratories where most method development has taken place. The criteria for residue acceptability may be different from those set as national standards.

B. RESULTS OF PREVIOUS MONITORING

The large amount of data that has been gathered in previous monitoring programs should be taken into account when planning future work.

1. Distribution of Residue Levels

Most surveillance programs have found that a high proportion of fruit or vegetable samples has no detectable residues and that a low proportion has residues that exceed MRLs (including nil MRL). FDA random sampling of imports during 1989 found no residues on 60% of fruit samples and 68% of vegetable samples,[6] while 2.3 and 4.5%, respectively, had illegal residues. These results were similar to those found over the previous 7 years. State of California monitoring of fresh food crops in 1989 found no residues in 78% of samples, while 0.71% contained illegal residues.[10] Swedish testing of fruit and vegetable samples for 1985 to 1989 found 79% were without detectable residues and 2.1% exceeded MRLs.[7]

The most common violations reported in most domestic or import surveillance monitoring programs were for modest levels of pesticides that had no registration or MRL for particular crops, but were acceptable on other crops. High residues in excess of set MRLs were less common, as were residues of completely banned pesticides.[6,7,10,11]

2. Frequency of Pesticide Occurrence

Surveillance monitoring and total diet studies have also found that residues are commonly found for only a small proportion of the possible individual pesticides. For example, in the 1989 FDA Total Diet Study involving 288 food items and 200 pesticides, 22 pesticides had a frequency of occurrence of greater than 1%, and only 3 (malathion, DDT, and chlorpyrifos methyl) occurred in more than 10% of samples.[6] In the more targeted sampling of FDA surveillance and compliance programs from 1982 to 1986 involving 19,851 samples of raw agricultural commodities, 20 pesticides had frequencies of detection greater than 1% and only 4 (methamidophos, chlorpyrifos, omethoate, and endosulfan) were found in more than 7% of samples.[11] Similar low occurrence frequencies have been reported for monitoring in Sweden.[7] These findings are consistent with residues generally only being expected for pesticides that are directly applied to the commodity either as field sprays or as postharvest treatments. This largely excludes herbicides. The list of insecticides or fungicides in common use on fresh foods is a relatively restricted subset of the 600 or so pesticides that have been registered for some use in crops.

3. Implications

The low proportion of samples with illegal residues, the high proportion with negligible residues, and the relatively restricted range of pesticides found in most monitoring programs have important implications for sampling strategies. Elementary statistical considerations show that if accurate measures are required of the types and levels of residues, then large numbers of samples must be tested.

Marketing of export commodities depends on avoidance of residue problems. If a 95% probability is required that the incidence of illegal residues be less than 1%, then 400 samples must test as being lower than MRLs. One over-limit sample in the 400 decreases the assurance probability to 80%. Conversely, there is an 82% probability that a true 1% incidence of illegal residues would be undetected if only 20 samples were tested.

Surveillance programs must also be large if they wish to influence as well as monitor the level of residues in the diet. Enforcing MRLs and regulating the use of agrochemicals through random sampling appear to be quite inefficient processes due to the low rate of violations detected. However, the presence of a monitoring program can influence grower practices, particularly when aligned with targeted sampling in the field, as, for example, in California.[10]

C. PRIORITY PESTICIDES

Any monitoring program will be subject to limitations arising from available analytical methods and laboratory resources. The priority for detectability of individual pesticides will mainly be determined by their toxicological and environmental properties and their agricultural use patterns. The FDA has established a Surveillance Index system for this type of information that is used to guide monitoring programs.[12] The pesticides of major economic importance internationally have been summarized together with their hazard classifications and ADIs.[5] The reassessment and review of particular pesticides for their toxicological or environmental acceptability can often lead to restrictions on usage or outright bans. These regulatory actions can have major effects on monitoring programs by causing large and possibly rapid changes in relative priorities for analysis of various pesticides.

A National Academy of Science study[13] has examined U.S. policy regarding regulation of pesticide residues in food. The report listed over 50 potentially oncogenic pesticides identified by the U.S. Environmental Protection Agency. These pesticides, along with others such as organochlorine insecticides and captafol, which have already been subject to regulatory actions in many countries, should have a high priority for inclusion in monitoring programs.

D. SETTING DETECTION LIMITS FOR SURVEILLANCE

Most surveillance monitoring programs have used nominal reporting limits of about 0.05 mg/kg for each pesticide. When reporting limits are lowered to the best detection limits routinely achievable by current technology, a higher proportion of samples can be expected to have detectable residues. For example, a small survey of fruit from Ontario, Canada, showed that 28% of samples had "negligible" residues (<0.1 mg/kg) of 55 insecticides and fungicides, while only 14% had no residues at detection limits of 0.005 to 0.02 mg/kg.[15] This demonstrates that nondetectable residues in crop samples at the detection limits used in many surveillance programs do not necessarily indicate that no agrochemicals had been used on the crops in question.

The furor in the U.S. in 1987 over residues of the plant growth regulator daminozide (Alar) in apples was partly due to the use by consumer groups of survey data gathered by methodology capable of detecting daminozide residues below 0.1 mg/kg. This data was interpreted as revealing much more widespread use of Alar on apples than had been indicated by FDA surveillance monitoring. The FDA method had a detection limit of 0.2 mg/kg, adequate to enforce the relatively large MRLs set for daminozide.[6]

Difficult situations also arise when dealing with low-level residues of pesticides that have no MRL on the crop in question. In some countries, a nominal "insignificant" MRL of 0.1 mg/kg may be set for pesticides of no particular toxicological concern. Codex similarly sets low values that are based on estimated limits of determination.[5] In the U.S., the statutes state nondetectable residues.[4] In this case, the levels above which regulatory action will be taken are largely decided by the detection limits of the primary and confirmatory methods. These decisions are under administrative rather than legislative control. The action levels are therefore likely to be lowered as analytical technologies improve.

E. CURRENT ISSUES IN MULTIRESIDUE MONITORING

Concern that U.S. monitoring programs may be inadequate in some respects led to a recent reassessment of technology for detection of pesticide residues in food.[16] The conclusions were that federal pesticide residue regulatory programs needed to put more effort into development of multiresidue methodologies. They were considered to offer the most cost-effective approach to overcoming constraints in current monitoring in the following areas:

1.   Coverage — Scope in regard to proportion of possible pesticides detectable, including new pesticides, metabolites, and coformulants
2.   Resources — Trained personnel, instrumentation, and laboratories
3.   Confirmation — Provision of alternative test results confirming violative residues so legal sanctions can be applied
4.   Timely analyses — Rapid testing to enable detection of consignments with violative residues

These problems are common to monitoring programs in all countries. They are also problems that are not amenable to absolute solutions, as methodology tends to lag behind new pesticide developments, legislative requirements change, and resources are always limited.

Rapid provision of results is always desirable. However, surveillance programs do not always rely on detention of shipments for enforcement. Consignments of fruits or vegetables with illegal residues that are direct, objective health risks are rare. Most countries use trace-backs, compliance sampling, and other sanctions to force changes from food suppliers. These responses can operate with test results coming available within a few days from sampling, rather than the 24 hours that may be required for detentions. This can have major implications for the laboratory systems used for multi-residue analysis.

III. ANALYTICAL METHODOLOGY

The methods developed and thoroughly tested by various national laboratories are an invaluable resource when published in full detail. Residue analysis manuals are available in English from U.S.,[16] Canadian,[17] German,[18] and The Netherlands[19] federal laboratories. MR methods are presented, as are many single compound or subclass methods. The manuals also contain a wealth of useful information on sampling, reagents, instrumentation, and general residue techniques. The Official Methods Manual of the Association of Official Analytical Chemists[20] also contains detailed residue methods that have met rigorous standards, including collaborative study.

A degree of commonality has been reached in the MR methods in wide-spread current use for fruits and vegetables.[1] The basic strategy involves solvent extraction of a subsample, concentration/cleanup including partitioning or gel permeation chromatography, and optional partition column chromatography for additional fractionation. The pesticides are determined by GC using various column/detector combinations to achieve the necessary selectivity and sensitivity. These standard MR methods have been validated for a large proportion of the pesticides in common use or of high regulatory concern.[3] However, the following widely used pesticides are not directly determinable due to losses in extraction, partitioning, or GC:

1. Very polar herbicides — amitrole, diquat, paraquat, glyfosate, glufosinate-ammonium
2. Acidic herbicides — chloro-phenoxy acids, chloro-benzoic acids, triclopyr, fluazifop, diclofop, oryzalin
3. Phenylurea herbicides (some) — diuron, monuron, chlortoluron, neburon, metoxuron
4. Sulfonylurea herbicides — chlorsulfuron, metsulfuron, triasulfuron
5. Growth regulators — chlormequat, daminozide, chlorethepon
6. Benzimidazole fungicides — benomyl, carbendazim, thiophanate-methyl
7. Dithiocarbamate fungicides — maneb, mancozeb, zineb, thiram, metiram
8. Very polar fungicides — fosethyl-Al, phosphorus acid, dodine
9. Oxime *N*-methyl carbamate insecticides — methomyl, oxamyl, aldicarb
10. Miticides — tin compounds, clofentazine, hexythiazox

Integration of all these classes into standard MR residue methods is highly desirable. At present, some can be accommodated by subclass methods based on derivatization or HPLC. Quite a number can only be analyzed by time-consuming single compound methods.

The following sections look in more detail at some of the significant steps in current standard MR methods and at recent developments that may improve their scope or efficiency.

A. LABORATORY SUBSAMPLING

The primary sample size (weight or number) and portion of various crops to be taken for analysis are given in national guidelines[16-18] and as Codex recommendations.[21] These recommendations are relatively fixed due to the need to obtain a sample that is representative of the commodity.

The subsampling of a primary fruit or vegetable sample in the laboratory to obtain the analytical portion is generally preceded by fine chopping or blending to achieve homogeneity. The analytical portion specified in most standard MR methods is 50 to 100 g wet weight.

Reducing the weight of the analytical portion directly reduces the volumes of solvent required in residue analyses. The principal constraint is the increased variability in residue levels due to sample inhomogeneity. In analysis of field-derived cypermethrin residues on cabbage, the relative standard deviation (RSD) increased only slightly when the analytical portion was progressively decreased from 50 g down to 2 g (3–7 to 7–12%).[22] However, the increases in RSD were greater with apples (8 to 25%). This was attributed to the residues being mainly in the skin, which was difficult to homogenize.

In another study on the degree of sample homogeneity, variances were established for several residue and crop combinations using sample homogenates prepared by mechanical cutters.[23] It was concluded that 2-g analytical portions of well-mixed crop homogenates were adequate to produce a subsampling variance of less than 0.01. This criteria was set on the basis that the variances of the analytical determination step in residue work are of this order and, therefore, the overall variance should not exceed 0.02 (RSD 14%).

Thus, it appears there is considerable scope to reduce the size of the analytical portion in routine MR screening methods for crop samples provided very fine chopping/mixing can be achieved. Larger portions could be retained for the confirmatory analyses required when illegal residues are suspected.

B. EXTRACTION AND PARTITION PROCEDURES

A wide variety of solvent extraction and partitioning systems have been advocated for residue analysis of crop samples.[24] However, only a few are now in routine use for MR screening. The advantages of water miscible solvent systems, such as acetone, methanol, and acetonitrile, in extracting polar pesticides are now widely recognized.[2,24] Ethyl acetate with added anhydrous sodium sulfate was originally developed as an alternative extractant for polar OP pesticides in crops. It does not require a partition step. This system has recently been validated for a wider range of pesticides and incorporated into MR methods. Each system has certain advantages, some of which depend on the way in which the extraction/partition steps are integrated into the cleanup/ determination steps of the MR method. Solvent consumption is also an important issue for laboratories because of costs, health hazards, disposal problems, and other environmental concerns.

Table 1 summarizes the initial phases of seven MR screening procedures used in various countries. The total solvent consumption and the amount of solvent requiring evaporation (total and on a sample size adjusted basis) are given in Table 1 as partial measures of cost and time effectiveness of each method. Obviously, there are many other factors that must be considered in comparing MR method effectiveness, including overall scope, accuracy, complexity, and equipment requirements. The following sections summarize the approaches used in MR residue methods using the four most important extraction solvents.

1. Acetone

MR extraction procedures based on acetone extraction followed by dichloromethane partitioning have gained wide acceptance because of their proven effectiveness for a very wide range of pesticide residues and crops. Over 200 pesticides can be determined by the standard Luke procedure (FDA, Table 1) using GC on the concentrated extract obtained after a three-stage partitioning.[1,21] The Luke procedure has been extended to include benzamidazoles, phenylureas, and N-methyl carbamates using HPLC determination. Acidic herbicides and organotins were derivatized for GC.[26] In addition to its high extraction efficiency, acetone has the advantages of low cost and relatively low toxicity. Disadvantages of the Luke procedure are (1) the relatively high solvent usage and inclusion of dichloromethane, a suspected carcinogen; (2) the complex partitioning step; and (3) the need to remove dichloromethane by solvent exchange before GC using selective detectors.

The methods of Ambrus et al.,[27] Steinwandter,[24] and Specht and Tilkes[28] (see also Reference 18) also use dichloromethane partitioning. The German procedure has the significant advantages of a single stage partition with a low volume of dichloromethane.

Use of saline to dilute the acetone extract and petroleum ether rather than dichloromethane in the partition limits recoveries to medium to low polarity compounds.[17] Provided lack of data for polar pesticides is acceptable, the method can be useful because of lower crop coextractives. For example, aqueous acetone extraction followed by partitioning into iso-octane formed the basis for a comprehensive MR method for halogenated fumigants and industrial solvents in food.[29]

A solid-phase partitioning column using treated diatomaceous earth has been tested as an alternative to the separating funnels in the Luke procedure

TABLE 1
Multiresidue Extraction Procedures for Crops

	Country						
	U.S.-FDA	U.S.-CSDA	Germany	Hungary	Netherlands	Sweden	New Zealand/Japan
Ref.	1, 25	10, 34	18, 28	27	40	41	47, 48
Sample weight (g)	100	50	100	50	50	75	100
Extraction	Acetone	Acetonitrile	Acetone	Acetone	Ethyl acetate[c]	Ethyl acetate[c]	Methanol
Volume (ml)	200	100	200	150 + 50	100	200	100 + 100
Extract used (ml)	80	70	200	All	15	100	20
Partition	CH_2Cl_2	—	CH_2Cl_2	CH_2Cl_2	n/a	n/a	Toluene
Volume (ml)	200[a], 2 × 100	—	100	100, 2 × 50	—	—	60
Added water (ml)	—	10[b]	—	450	—	—	6
Added salts (g)	7	—	20	18	—	—	
Primary cleanup	—	—	GPC	C/MgO/Celite	GPC	GPC	C/cellulose/Florisil
Extract concentration (g crop/ml)	4.1	5	4.8	2–8	0.4–2.5	1.5	0.74–3.7
Solvent use							
Total volume (ml)	775	325	530	580	140	210	220
Volume evaporated (ml)	550	255	220	580	30	110	8
ml/g cleaned up crop	19	10	9	15	6	15	0–2

[a] 50:50 CH_2Cl_2 — petroleum ether.
[b] pH 7 Phosphate buffer-brine.
[c] + 40–50 g Anhydrous sodium sulfate.

and gave equivalent recoveries for both polar and nonpolar OPs.[30] Rather large volumes of solvent were required for column equilibration. Another approach to solid-phase partitioning used nitrogen gas to evaporate the bulk of the acetone off the column prior to elution with petroleum ether — dichloromethane. Recoveries of 85 to 100% for 17 low-medium polarity OPs, 75 to 78% for dimethoate, and a low level of crop coextractives were reported.[31]

2. Acetonitrile

The standard Mills procedure[1] uses acetonitrile extraction followed by petroleum ether plus saline partitioning. The concentrated petroleum ether phase may be cleaned up directly by Florisil column chromatography. It is, therefore, an efficient procedure for study of low level residues of relatively nonpolar OC and OP pesticides. The partitioning and column cleanup have been fully automated using a purpose-built apparatus that gave comparable recoveries and reproducibility to manual procedures.[32] In the Storherr method dichloromethane was used in the partition to recover more polar pesticides.[33] This method has largely been superceded by the Luke procedure.

The California State Department of Agriculture method (Table 1) does not use a second solvent in the partition but relies on added salt to force separation of water from the aqueous acetonitrile crop extract.[10] This method efficiently recovers pesticides over a wide polarity range with scope similar to the Luke procedure. In recent modifications,[34] the initial aqueous acetonitrile extract was passed through a C18-SPE column and pH 7 phosphate buffer was included at the partition. These steps were reported to give significantly cleaner extracts by removing some pigments, waxes, and organic acids. This method has the advantages of lower solvent consumption, no use of dichloromethane, and a single partition. However, acetonitrile is an expensive, high hazard solvent, the coextractive levels from crops may be higher than with other procedures, and solvent exchange is required to remove traces of acetonitrile before GC.

3. Ethyl Acetate

The Watts et al.[35] method for MR analysis of residues of OPs and their oxidation products on crops was based on ethyl acetate extraction in the presence of anhydrous sodium sulfate. Watts[36] showed that blending with ethyl acetate or acetonitrile was as efficient as exhaustive Soxhlet extraction with 10% methanol in chloroform in removing ^{14}C-labeled residues of four OPs or carbaryl from glasshouse- or field-treated crops. The extraction method has been quite widely used for determination of OP residues in fruits and vegetables[37,38] and more recently for *N*-methyl carbamates.[39] However, it is only recently that wide-ranging MR methods have been developed using ethyl acetate as an extractant.

An MR method based on ethyl acetate extraction followed by gel permeation cleanup has been developed in The Netherlands and tested on a wide range of foodstuffs (Table 1).[40] This extraction method has recently been adopted by the Swedish National Food Administration (Table 1),[41] supplanting the Luke procedure previously used in their MR monitoring.[42]

Comparisons have been carried out of ethyl acetate extraction and the Luke procedure using spiked samples or field-incurred residues in cereals[40] and fruits or vegetables.[43] These studies concluded that the methods gave comparable results for a wide range of pesticides. Recoveries of methamidaphos, acephate, and omethoate were substantially higher with ethyl acetate.[43]

The advantages of ethyl acetate are the elimination of partitioning, reduced solvent usage, and an extract that can be used directly in gel permeation chromatography cleanup. A disadvantage is the higher boiling point.

4. Methanol

Methanol is an inexpensive and very efficient extractant for polar residues from crops.[44] Krause's HPLC method for *N*-methylcarbamate insecticide residues was based on methanol extraction followed by a complex partitioning procedure for cleanup.[45] Methanol was also the solvent of choice for extraction of six phenylurea herbicides from crops prior to partition cleanup and HPLC determination.[46]

A multiresidue procedure based on methanol extraction followed by toluene partitioning was developed by Holland and McGhie[47] and subsequently extended by Ishii et al.[48] (see Table 1). The methanol/toluene/saline partitioning system followed by carbon column cleanup gave good recoveries of about 60 common pesticides. However, polar insecticides such as acephate, dimethoate, or methamidaphos were not recovered in the partition. The advantage was a simple procedure that gave clean extracts and required little or no evaporation of solvents. This method has been successfully miniaturized in our laboratory by use of a smaller analytical portion and partitioning in a 15-ml culture tube (with centrifugation, which improved phase separation and recoveries).

C. CLEANUP TECHNIQUES

Walters has provided comprehensive reviews of cleanup procedures.[2,49] The extraction and partition steps of standard MR methods provide a degree of cleanup in that highly water soluble plant components, such as sugars, are eliminated. However, fruits and vegetables contain a wide variety of other compounds that are extractable and may give rise to interferences or other problems. Crucifer crops and onions produce particularly strong interferences upon GC with N or S detection. More insidious effects arise when coextractives cause adsorption or decomposition of pesticides. Lipids, chlorophylls, or other high molecular materials accumulating in the injection region of gas chromatographs are particularly problematic. Fruits and vegetables are generally low in lipids,[1] but can have relatively high levels of pigments and phenolics.

Cleanup procedures developed for MR screening can have disadvantages:

1. Losses — Some losses of residues are inevitable and there may be unpredictable effects with different matrices.
2. Lack of generality — Many cleanups lack the scope to recover all the pesticides of interest.
3. Time and cost — Cleanups take time that can detract from rapid provision of test results and lead to increased costs. Some cleanups split pesticides into various fractions, which then require individual determination.

Because of these problems, several MR methods, including the Luke procedure,[1,25] have eliminated cleanup and rely on very selective GC detectors and on frequent injector and column maintenance to minimize adsorption/decomposition effects.[50] Chromatographic cleanups on Florisil, carbon, or C18-SPE columns are reserved for special samples or confirmation of particular pesticides.[1]

Other MR methods have retained cleanup steps so less specific detectors may be used, such as electron capture for GC or ultraviolet (UV) absorption for HPLC. Cleanup also allows the full power of modern capillary GC to be applied by reducing the losses in injector and column performance that high levels of coextractives rapidly induce for many pesticides. Detection limits are lowered and reproducibility may be improved.

Although a diversity of cleanup procedures have been investigated, only the following three basic techniques have been adopted in standard MR methods for fruits and vegetables.

1. Gel Permeation Chromatography

Gel permeation chromatography (GPC) or size exclusion chromatography uses a macroreticular resin matrix that progressively retards the elution of compounds based on the inverse of molecular size. Thus, large lipids, pigments, and polymeric coextractives elute earlier than most pesticides. The technique was first applied to the cleanup of OC pesticides and PCBs in fatty foods. Specht and Tillkes developed a GPC system using polystyrene resin Biobeads S-X3 (25 mm I.D. × 40 cm long) with cyclohexane-ethyl acetate (1:1) mobile phase, which was suitable for cleanup of a wide range of pesticides in various foods.[28] GPC on Biobeads S-X3 with dichloromethane-hexane (7:3) was used for cleanup in a comprehensive MR method for cereals.[51] Steinwandter has advocated cyclohexane-acetone (3:1) or acetone-petroleum ether (1:1) as GPC eluents.[24]

The GPC column diameter can be reduced from 25 mm to 14 mm[52] or 10 mm[40] with consequent savings in solvent consumption, but a corresponding reduction in column capacity. Fruit and vegetable extracts have a relatively low lipid content, and further miniaturization should be feasible.

The advantages of GPC are:

1. Reproducibility — All of each applied sample is completely eluted with virtually no irreversible adsorption. Column properties remain constant over long periods and loss of residues by adsorption or decomposition is negligible.
2. Automation — The very reproducible cycle of inject, dump (lipids and other high mol wt compounds to waste), collect (pesticide fraction), and purge (final column regeneration) is readily automated. An apparatus for 25 mm I.D. columns is available commercially. Column systems of 10 mm I.D. (1 ml injection, 1 ml/min eluent flow rate) can be built from standard HPLC components.
3. Compatibility — Ethyl acetate extracts of samples are easily prepared for GPC. The eluent is compatible with all GC detectors.

Disadvantages are:

1. Incomplete cleanup — Lower mol wt coextractives often elute in the pesticide fraction.
2. Incomplete separation of large pesticides — There is a trade-off between cleanup and recovery for some pesticides. For example, interferences from crucifer crops can be reduced by increasing the dump fraction, but large pyrethroids, which elute early, are also progressively removed.
3. Low volatility solvents — There is slow concentration of the pesticide fraction.

Overall, GPC has proved to be a very robust and versatile cleanup technique. It provides adequate cleanup for a majority of fruits and vegetables so that capillary GC (thermionic flame or electron capture detectors) can be used without further cleanup. GPC is used in standard MR methods in Germany, The Netherlands, Sweden, and other countries.

2. Adsorption Chromatography

Florisil or alumina have formed the basis for cleanup of low to medium polarity pesticides in many older standard methods. However, many pesticides or metabolites require polar solvents for elution or are lost.[1,3] The adsorbents require careful calibration, which can be affected by the nature of the coextractives. This makes them more difficult to incorporate into MR testing of a variety of crops. However, such systems are essential to achieve low detection limits for OC insecticides.

The German and Swedish standard MR methods use minicolumns of silica gel or alumina/silver nitrate to further purify and separate pesticide fractions following GPC.[18,28,42] These are particularly useful for "difficult" crops such as cabbages or onions.

Carbon columns form the basis for several standard MR methods. Carbon has a high affinity for plant pigments, which makes it very useful in cleanup of high chlorophyll crops such as leafy vegetables. A variety of forms have been used, generally dispersed on Celite or cellulose powder to aid flow.[33,35] Once a batch is calibrated, reproducible results can be obtained, and simple one-step procedures have been developed that clean up a wide range of pesticides.[27,47,48] However, not all pesticides can be readily recovered.

3. Reversed-Phase Chromatography

HPLC technology using C18 reversed-phase columns and solvent systems has been used to provide cleanup of fatty food extracts.[49,53] Polar through to nonpolar pesticides were separated from the bulk of lipids which eluted last. These systems are not as effective for MR analysis of crop extracts because pigments and other coextractives tend to overlap with the pesticide fraction, and poorly eluted plant components can alter the column characteristics over time. However, an automated HPLC procedure has been developed to determine three carbamates in total diet samples.[55] An on-line C18 cartridge with column switching gave adequate cleanup for detection limits of 1 μg/kg.

Solid-phase extraction (SPE) cartridges or minicolumns have been successfully used for partial cleanup of plant extracts. The irreversible adsorption of high mol wt coextractives is a useful part of the cleanup when using a disposable column. The capacity of these columns is adequate for extracts of fruits and vegetables with low lipid content. C18-SPE cartridges were used for cleanup of 10-g crop equivalent ethyl acetate extracts and allowed GC detection limits of 0.01 mg/kg for 14 *N*-methylcarbamtes.[54]

Newsome and Collins[56] used C18-SPE columns (0.5 g) to clean up acetone extracts of fruit and vegetable samples. Aliquots of extract were diluted with water to achieve a 20% acetone content and applied to the columns that were then washed with 40% methanol in water and the pesticides were eluted with 100% methanol. High recoveries were obtained for 12 low to medium polarity fungicides.

Normal-phase amino-propyl SPE columns have been found more effective than reversed-phase systems for cleanup of crop extracts in determination of *N*-methylcarbamates by HPLC.[57]

Some problems with SPE technology are:

1. Irreproducible separation characteristics between batches and suppliers
2. Contaminants in the polypropylene columns or frits and instability of bonded phases that can give rise to elevated blanks
3. Limited capacity that can lead to break-through of crude extracts

D. DETERMINATION

Gas chromatography remains the basis for determination of residues in most standard MR methods. This is largely because GC detectors have the high sensitivity and selectivity required for residue work. The advent of fused silica open tubular (FSOT) columns with bonded polymeric liquids phases has greatly aided residue determination by providing stable, reproducible, highly inert columns with much higher resolving power than previous packed column technology. The fundamentals of detector technology have undergone few changes in the past decade[58] with only incremental improvements.

HPLC has been increasingly applied to determination of thermally unstable pesticides that are difficult or impossible to determine directly by GC. The lack of element-selective detectors has been the major obstacle to wider application of HPLC to residue determination. However, the development of high sensitivity multiple wavelength UV detectors, HPLC/MS interfaces, and the photoconductivity detector have significantly increased the scope of HPLC.

Thin-layer chromatographic determination has the advantage or low cost and the ability to run many samples in parallel, which make it very suitable for routine screening of the compound classes that can be detected with high sensitivity. Quantitative analysis requires a densitometer.

An additional requirement for residue determination is the availability of confirmatory techniques. This may include repeat analysis for confirmation of quantitation, a repeat chromatographic run under conditions of altered selectivity (chromatographic and/or detection), or full scan mass spectrometry.

1. Gas Chromatography

a. Low Resolution Gas Chromatography

Most of the standard MR methods were developed using packed column GC.[27,47,50] In some cases, wide-bore FSOT columns (0.53 mm I.D.) have been substituted to upgrade performance with the advantage that these columns can be used in older instruments.[26,48]

In the Luke method used in FDA screening, the emphasis has been on short, rugged columns coupled to highly selective detectors to allow rapid analysis of crude extracts. Up to 16 instruments are set up for parallel analysis of extracts on columns of three different polarities with Hall electrolytic conductivity detectors in halogen/S and N modes and flame photometric detectors in P and S modes.[26,50] These systems allow identification of a very wide range of pesticides despite the lack of resolution of close eluting compounds on any one instrument. Similar approaches have been used in Canadian federal and Californian state laboratories.[10,17]

Disadvantages are:

1. The large number of instruments required to cover the various column-detector combinations leads to high capital and maintenance costs, particularly if automation is included.
2. The crude extracts can lead to rapid loss of column or detector performance.

b. High Resolution Gas Chromatography

It is a decade since high resolution gas chromatography (HRGC), using long, narrow-bore capillary columns (25 to 30 m, 0.20 to 0.32 mm I.D.), began to be used for routine residue screening in fruits and vegetables.[59,60] Initial developments concentrated on class analyses. More recently, full MR procedures based on HRGC determination have been published[42,61,62] that use FSOT columns connected to electron capture (ECD), thermionic flame (NPD), or flame photometric (FPD) detectors. Use of two instruments, each with a polar and a nonpolar column connected to a split-splitless injector, gave four column/detector combinations that allowed identification of residues of over 130 pesticides on a variety of crops.[42]

In our own laboratory, two instruments set up with single columns and effluent splitting to two detectors gave the same combinations in an arrangement that allowed better optimization of column/detector flow characteristics and more flexible injector options. Figures 1 and 2 demonstrate the performance of this system for two standard mixes run on a low polarity column split 1:4 to the ECD and NPD. Figures 3 and 4 show the chromatograms for an ethyl acetate extract of tomatoes (cleaned up by GPC) on the above system and confirmation on a more polar column with ECD and FPD. An internal standard, carbophenothion, was used.

The advantages of the HRGC approach are:

1. The high resolution and reproducibility of retention times (less than 0.05% RSD for relative retention time (RRT)) require only two columns for accurate identification and separation of pesticides from most coextractives.
2. Internal standards may be used[22] which improve quality control and aid miniaturization.
3. The inertness of FSOT columns enables reproducible chromatography at low levels for many "difficult" pesticides.
4. Automation and overnight operation are readily achieved.

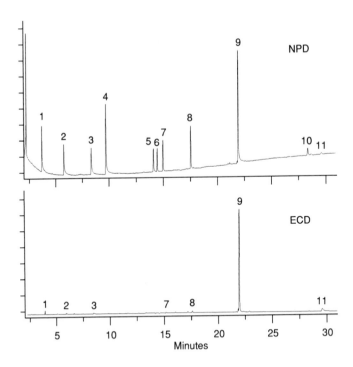

FIGURE 1. Mixed pesticide standard (1 ng each) by HRGC using column effluent splitting to ECD and NPD: 1, methamidaphos; 2, acephate; 3, omethoate; 4, dicrotophos; 5, carbaryl; 6, metalaxyl; 7, methiocarb; 8, triadimenol; 9, carbophenothion; 10, bitertanol; and 11, prochloraz. Split-splitless injection 220°, 25 m × 0.20 mm HP5 (0.33 μm) FSOT column, 85° 1 min, 40°/min to 150°, 2 min, 5°/min to 250°.

Disadvantages are:

1. Extracts must receive some cleanup to ensure reproducible operation of the injection, column, and detection systems.
2. The less specific ECD and NPD give a range of responses that require detailed interpretation and correlation to distinguish matrix and pesticide peaks so residues can be positively identified. However this procedure has been automated.[63]

Injection is the most critical process is obtaining and maintaining good HRGC performance for MR analysis of foods. Many studies have established the superior performance of cold on-column injection (OCI) compared to standard split-splitless injection (SSI) for the analysis of standard solutions of thermally unstable pesticides such as N-methylcarbamates[54,64] and phenylureas.[65] However, OCI deposits all coextractives on the head of the column.

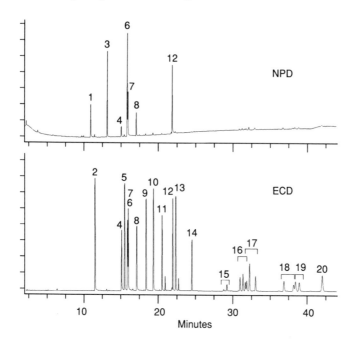

FIGURE 2. Mixed pesticide standard (1 ng each) by HRGC using column effluent splitting to ECD and NPD: 1, simazine; 2, lindane; 3, pirimicarb; 4, bromacil; 5, aldrin; 6, triadimefon; 7, parathion ethyl; 8, penconazole; 9, endosulfan-a; 10, dieldrin; 11, endosulfan-b; 12, carbophenothion; 13, *p,p*-DDT; 14, dicofol; 15, permethrin; 16, cyfluthrin; 17, cypermethrin; 18, fenvalerate; 19, fluvalinate; and 20, deltamethrin. See Figure 1 for conditions.

The build-up of the less volatile components leads to progressive, and sometimes rapid, deterioration of chromatographic performance (resolution and sensitivity) for pesticides in food extracts.[65] These effects can be reduced but not eliminated by a short, deactivated, uncoated precolumn that distributes coextractives in a less concentrated band as well as acting as a retention gap.[65,67] The programmed temperature vaporizer (PTV) can be configured as a cold on-column injector except that the temperature and heating rate may be controlled independently from the column oven. Stan and Muller reported[64,66,67] that use of a low-volume glass insert in the PTV gave similar, but slightly more variable, results compared to an OCI for OP and *N*-methylcarbamate pesticides. The results were much poorer if the insert was packed with a small amount of glass wool, particularly at low injection temperatures.

The experiences with various HRGC injectors lead to the hypothesis that a short residence time and a low surface area in the injection region are more critical than a low temperature for the stability of the majority of pesticides. High surface areas in hot SSI can lead to losses and degradation. On the other hand, coextractive build-up in OCI or the PTV can lead to longer residence

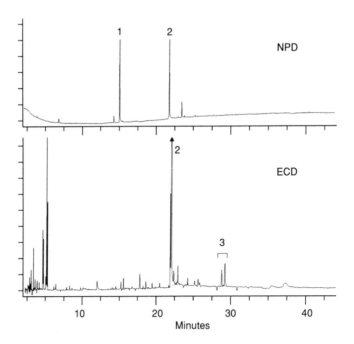

FIGURE 3. HRGC analysis of extract of tomatoes with incurred residues: 1, pirimiphosmethyl 0.17 mg/kg; 2, carbophenothion internal standard 0.4 mg/kg; and 3, permethrin 0.03 mg/kg. See Figure 1 for conditions.

of pesticides in a reactive region. The most practical compromise may be the use of SSI with low volume inserts and a wide-bore precolumn. Figures 1 and 2 illustrate the performance of SSI with 1-µl injections into an empty 0.2-ml insert at 220°. Excellent results were maintained for sensitive compounds such as carbaryl, acephate, and azinphos-methyl by replacement of the insert and precolumn after analysis of about 100 fruit or vegetable extracts (ex-GPC cleanup). However, methomyl and aldicarb were almost completely lost. HRGC of these very temperature-sensitive pesticides require OCI or PTV techniques.

c. Other Detector and Confirmatory Techniques

Gas chromatography/mass spectrometry (GC/MS) has become the preeminent technique for confirmation of pesticide residues.[68,69] Relatively inexpensive bench-top instruments have made the technique more routinely available and have led to the use of GC/MS as a primary screening tool. Stan[70] demonstrated that a small quadrapole mass selective detector could reliably detect 76 pesticides in food extracts at the 0.4-mg/kg level using full scan spectra, reducing to 0.01 mg/kg in the selected ion mode (SIM). The use of

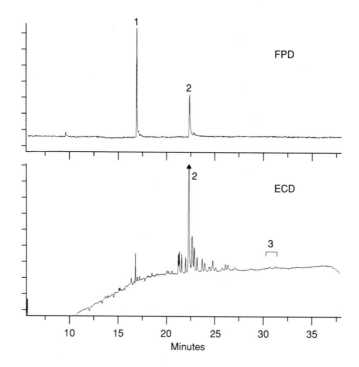

FIGURE 4. Confirmatory HRGC analysis of tomato extract. See Figure 3 for identification. Split-splitless injection 250°, 30 m × 0.25 mm DB17 (0.5 μm) FSOT column, 85°/min, 40°/ min to 150°, 2 min, 10°/min to 280°.

full scan spectra for screening is desirable because of their higher information content. It is also complex to set up and maintain the multiple time/mass "windows" required for sensitive detection of a large number of pesticides in a single run by SIM. The ion trap detector has a higher inherent sensitivity, and this allowed screening of fruit and vegetable extracts for a limited, but representative, range of pesticides to 0.05 mg/kg using full scan chemical ionization spectra.[71,72]

Recent technical advances in the atomic emission detector (AED)[73,74] and the incorporation of it into a commercial instrument offer a new approach to GC detection for MR analysis. The AED can carry out matrix- and compound-structure-independent analyses for a wide range of elements at atomic sensitivities in the range 1 to 50 pg/sec. All elements cannot be detected simultaneously due to the limited size of the photodiode array detector. Thus, two separate GC runs will detect the following sets of elements: N, P, S, C and Br, Cl, H, C. The AED provides similar sensitivity to the FPD for P and better sensitivity for low levels of S without the quadratic response. For N, Br, Cl, and F it is less sensitive than other selective GC detectors, but the selectivity is higher. Applications to residue analysis of food have been demonstrated, but full MR procedures incorporating the AED have not been reported.

2. High Performance Liquid Chromatography

The principal applications of HPLC to MR analysis have been for subclasses of pesticides not readily determined by GC. These include the examples listed below, which have been tested on a range of fruits and vegetables. Developments in LC/MS have provided a confirmatory technique[69,75] that has also been demonstrated as a primary MR screening tool.[72,76]

N-**Methylcarbamates** — The Krause method is based on reversed-phase separation followed by postcolumn hydrolysis and fluorometric determination of the methylamine using *o*-pthalaldehyde derivatization.[45] Modification include solid-phase hydrolysis[57] and the use of UV photolysis as an alternative to hydrolysis in forming amines from nitrogen-containing pesticides.[77]

Benzimidazoles and related polar fungicides — Benomyl and its hydrolysis product MBC were determined (as MBC) with thiophanate methyl and thiabendazole by ion pair reversed-phase HPLC using fluorescence detection.[78] HPLC with UV detection could determine the above fungicides plus *o*-phenylphenol, biphenyl, and diphenylamine,[79] but required more thorough cleanup.

Pyrethroids — A multiresidue procedure has been reported for pyrethroid insecticides in fruits and vegetables.[80] HPLC/UV determination gave lower detection limits than GC-ECD for the nonhalogenated pyrethroids bioresmethrin, cismethrin, phenothrin, and resmethrin.

Phenylurea herbicides — Reversed-phase HPLC was the basis for a comprehensive method for phenylureas using postcolumn UV photolysis to produce amines that were detected by fluorescence of *o*-pthalaldehyde derivatives.[46] HPLC with electrochemical detection gave sensitive and selective detection of the anilines produced from phenylureas in crops by steam distillation under basic conditions.[81] Detection limits were 0.02 mg/kg.

Sulfonylurea herbicides — A high degree of cleanup has been required in single compound methods that reach the low detection limits required for these very active herbicides in plant material. Adequate MR procedures have not been reported, but LC/MS (thermospray) has demonstrated promising results.[82] Detection limits of less than 0.1 mg/kg on wheat were achieved without cleanup of extracts.

3. Thin-Layer Chromatography

Developments in thin-layer chromatography (TLC) to the mid-1980s have been reviewed.[3,83,84] The technique has been thoroughly revised to modern standards by Gardyan and Thier[85] using high performance (HP) TLC systems including reversed phases. Detection limits were 0.01 to 0.05 mg/kg for over

160 pesticides in several fruit or vegetable crops using the standard German Federal extraction (acetone/CH_2Cl_2) and cleanup (GPC) procedure. Three visualization techniques were applied: (1) cholinesterase inhibition for OP and carbamate insecticides, (2) silver nitrate/UV for halogenated compounds, and (3) toluidine for compounds yielding chloramines.

Inhibition of the Hill reaction has been used to detect a range of herbicide residues by TLC.[86]

IV. FUTURE DEVELOPMENTS

Miniaturization and automation have long been recognized as the most promising approaches to improving the productivity and lowering the costs of MR analyses.[22,24,52,87] Some advances have been made in miniaturization, principally by carrying smaller volumes of primary extract through cleanup and concentration. Automation of GPC cleanup systems is now routine, as is the use of auto-injectors for GC and HPLC. SPE column and HPLC cleanups can also be automated. However, comprehensive MR methodology designed to incorporate all significant advances in these areas is still lacking. Some of the technical features of such future methods are likely to be reduced sub-sample sizes, super critical fluid or ultrasonic aided extraction (Chapter 2), solid-phase partitioning, fully automated GPC/concentrator or bonded-phase SPE column (Chapter 2) cleanup systems, HRGC with short ultra-high efficiency columns (0.1 mm I.D.), multiple internal standards, and full automation of data acquisition/preliminary interpretation.

Problems of inadequate selectivity in instrumental analysis will be aided by wider use of LC/MS, GC/MS and GC-AED. Multiple wavelength or diode array UV detection and other more selective detection systems such as photoconductivity will become more important for HPLC.

Increasing the scope of MR analyses can most readily be achieved by incorporated new pesticides into standard MR methods. This is likely to depend on efficient and selective microderivatizations for difficult GC pesticides or for improved HPLC detectability. Enzyme-linked immunosorbent assay (ELISA) will increasingly be used to rapidly screen for individual pesticides that do not readily fit into MR schemes. Immunoaffinity columns also have potential for cleanup of particular classes of pesticides.

Although MR analyses have reached a high degree of sophistication and are generating large volumes of data for food, there is a need to constantly evaluate the scope and direction of the monitoring programs in which they are employed. More assessment of priorities is needed based on toxicological concerns, past monitoring data, and the detailed residue trial data gathered in support of registration. Perhaps the greatest cost-benefit improvements in MR monitoring will be obtained not by improved technology, but rather by more careful targeting of the testing and from harmonization of residue regulations.

REFERENCES

1. **Luke, M. A. and Masumoto, H. T.,** in *Analytical Methods for Pesticides and Plant Growth Regulators,* Vol. 15, Zweig, G. and Sherma, J., Eds., Academic Press, San Diego, 1986, 161.
2. **Walters, S. M.,** in *Analytical Methods for Pesticides and Plant Growth Regulators,* Vol. 15, Zweig, G. and Sherma, J., Eds., Academic Press, San Diego, 1986, 3, 67.
3. **Ambrus, A. and Thier, H.-P.,** *Pure and Appl. Chem.,* 58, 1035, 1986.
4. **Wessel, J. R. and Yess, N. J.,** *Rev. Environ. Contam. Toxicol.,* 120, 83, 1991.
5. **Ekstrom, G. and Akerblom, M.,** *Rev. Environ. Contam. Toxicol.,* 114, 24, 1990.
6. Food and Drug Administration Pesticide Program, *J. Assoc. Off. Anal. Chem.,* 73, 127A, 1990.
7. **Andersson, A. and Bergh, T.,** *Fresenius Z. Anal. Chem.,* 339, 387, 1991.
8. **Pennington, J. and Gunderson, E.,** *J. Assoc. Off. Anal. Chem.,* 70, 772, 1987.
9. **Watts, B. B. and Holland, P. T.,** *Proc. 33rd New Zealand Weed and Pest Control Conf.,* New Zealand Weed and Pest Control Society, Palmerston North, NZ, 1980, 114.
10. **Okumura, D., Melnicoe, R., Jackson, T., Drefs, C., Maddy, K., and Wells, J.,** *Rev. Environ. Contam. Toxicol.,* 118, 87, 1991.
11. **Luke, M. A., Masumoto, H. T., Cairns, T., and Hundley, H. K.,** *J. Assoc. Off. Anal. Chem.,* 71, 415, 1988.
12. **Reed, D. V.,** *J. Assoc. Off. Anal. Chem.,* 68, 122, 1985.
13. Regulating Pesticides in Food: The Delaney Paradox, National Research Council, National Academy Press, Washington, D.C., 1987.
14. U.S. Congress Office of Technology Assessment, Pesticide Residues in Food: Technologies for Detection, OTA-F-398, U.S. Government Printing Office, Washington, D.C., 1988.
15. **Frank, R., Braun, H. E., and Ripley, B. D.,** *Food Additives Contaminants,* 7, 637, 1990.
16. Pesticide Analytical Manual (1968 and revisions), Vols. 1 and 2, Food and Drug Administration, Washington, D.C.
17. **McLeod, H. A. and Graham, R. A., Eds.,** *Analytical Methods for Pesticide Residues in Foods,* Ministry of National Health and Welfare, Ottawa, 1986.
18. **Thier, H.-P. and Zeumer, H.,** *Manual of Pesticide Residue Analysis,* Vol. 1, VCH Publishers, Weinheim, Germany, 1987.
19. Analytical Methods for Residues of Pesticides in Foodstuffs, 4th ed., Ministry of Welfare, Health and Cultural Affairs, Leidschendam, The Netherlands, 1985.
20. Association of Official Analytical Chemists, *Official Methods of Analysis,* Vol. 1, 15th ed., 1990.
21. Codex Alimentarius Commission CAC/PRG, *Guide to Codex Recommendations Concerning Pesticides Residues,* Part 6, Food and Agriculture Organization/World Health Organization, Rome, 1984.
22. **Hemingway, R. J.,** *Pure and Appl. Chem.,* 56, 1131, 1984.
23. **Egli, H. and Bohm, K. H.,** Poster 8A-17, Abstracts, Vol. 3, presented at 7th Int. Congr. of Pesticide Chemistry, Hamburg, 1990.
24. **Steinwandter, H.,** in *Analytical Methods for Pesticides and Plant Growth Regulators,* Vol. 17, Sherma, J., Ed., Academic Press, San Diego, 1989, 35.
25. **Luke, M. A., Froberg, J. E., Doose, G. M., and Masumoto, H. T.,** *J. Assoc. Off. Anal. Chem.,* 64, 1187, 1981.
26. **Luke, M. A., Langham, W. S., Kodama, D. M., Masumoto, H. T., Froberg, J. E., and Doose, G. M.,** in *Proc. 7th Int. Congr. Pesticide Chem. (IUPAC),* Frehse, H., Ed., VCH Publishers, Weinheim, Germany, 1991, 373.
27. **Ambrus, A., Lantos, J., Visi, E., Csatlos, I., and Sarvari, L.,** *J. Assoc. Off. Anal. Chem.,* 64, 733, 1981.

28. **Specht, W. and Tillkes, M.,** *Fresenius Z. Anal. Chem.,* 322, 443, 1985.
29. **Daft, J. L.,** *J. Assoc. Off. Anal. Chem.,* 71, 748, 1988.
30. **Hopper, M. L.,** *J. Assoc. Off. Anal. Chem.,* 71, 731, 1988.
31. **Di Muccio, A., Ausili, A., Camoni, I., Dommarco, R., Rizzica, M., and Vergori, F.,** *J. Chromatogr.,* 456, 149, 1988.
32. **Gretch, F. M. and Rosen, J. D.,** *J. Assoc. Off. Anal. Chem.,* 70, 109, 1987.
33. **Storherr, R. W., Ott, P., and Watts, R. R.,** *J. Assoc. Off. Anal. Chem.,* 54, 513, 1971.
34. **Lee, S. M., Papathakis, M. L., Feng, H.-M. C., Hunter, G. F., and Carr, J. E.,** *Fresenius Z. Anal. Chem.,* 339, 376, 1991.
35. **Watts, R. R., Storherr, R. W., Pardue, J. R., and Osgood, T.,** *J. Assoc. Off. Anal. Chem.,* 52, 522, 1969.
36. **Watts, R. R.,** *J. Assoc. Off. Anal. Chem.,* 54, 953, 1971.
37. Panel on determination of residues of certain OP pesticides in fruits and vegetables, *Analyst,* 102, 858, 1977.
38. **Hill, A. R. C., Wilkins, J. P. G., Findlay, N. R. I., and Lontay, K. E. M.,** *Analyst,* 109, 483, 1984.
39. **Brauckhoff, S. and Thier, H.-P.,** *Z. Lebensm. Unters. Forsch.,* 184, 91, 1987.
40. **Roos, A. H., Van Munsteren, A. J., Nab, F. M., and Tuinstra, L. G. M. Th.,** *Anal. Chim. Acta,* 196, 95, 1987.
41. **Andersson, A., Blomkvist, G., and Ohlin, B.,** *Var Foeda,* in press.
42. **Andersson, A. and Ohlin, B.,** *Var Foeda,* Suppl. 2, 79, 1986.
43. **Andersson, A. and Palsheden, H.,** *Fresenius Z. Anal. Chem.,* 339, 365, 1991.
44. **Wheeler, W. B., Thompson, N. P., Andrade, P., and Krause, R. T.,** *J. Agric. Food Chem.,* 26, 1333, 1978.
45. **Krause, R. T.,** *J. Assoc. Off. Anal. Chem.,* 63, 1114, 1980.
46. **Luchtefeld, R. G.,** *J. Assoc. Off. Anal. Chem.,* 70, 740, 1987.
47. **Holland, P. T. and McGhie, T. K.,** *J. Assoc. Off. Anal. Chem.,* 66, 1003, 1983.
48. **Ishii, Y., Sakamoto, T., Asakura, K., Adachi, N., and Taniuchi, J.,** *J. Pest. Sci.,* 15, 205, 1990.
49. **Walters, S. M.,** *Anal. Chim. Acta,* 236, 77, 1990.
50. **Froberg, J. E. and Doose, G. M.,** in *Analytical Methods for Pesticides and Plant Growth Regulators,* Vol. 14, Zweig, G. and Sherma, J., Eds., Academic Press, San Diego, 1986, 41.
51. **Chamberlain, S. J.,** *Analyst,* 115, 1161, 1990.
52. **Vogelgesang, J. and Thier, H.-P.,** *Z. Lebensm. Unters. Forsch.,* 182, 400, 1986.
53. **Gillespie, A. M. and Walters, S. M.,** *Anal. Chim. Acta,* 245, 259, 1991.
54. **Wuest, O. and Meier, W.,** *Z. Lebensm. Unters. Forsch.,* 177, 25, 1983.
55. **Goewie, C. E. and Hogendoorn, E. A.,** *J. Chromatogr.,* 404, 352, 1987.
56. **Newsome, W. H. and Collins, P.,** *J. Chromatogr.,* 472, 416, 1989.
57. **de Kok, A., Hiemstra, M., and Vreeker, C. P.,** *Chromatographia,* 24, 469, 1987.
58. **Holland, P. T. and Greenhalgh, R.,** in *Analysis of Pesticide Residues,* Moye, H. A., Ed., Wiley-Interscience, New York, 1981, 51.
59. **Ripley, B. D. and Braun, H. E.,** *J. Assoc. Off. Anal. Chem.,* 66, 1084, 1983.
60. **Zweig, G.,** in *Analytical Methods for Pesticides and Plant Growth Regulators,* Vol. 14, Zweig, G. and Sherma, J., Eds., Academic Press, San Diego, 1986, 75.
61. **Lawrence, J. F.,** *Int. J. Environ. Anal. Chem.,* 29, 289, 1987.
62. **Kiviranta, A.,** *Int. Lab.,* 17, 58, 1987.
63. **Lipinski, J. and Stan, H.-J.,** *J. Chromatogr.,* 441, 213, 1988.
64. **Muller, H.-M. and Stan, H.-J.,** *J. High Res. Chromatogr.,* 13, 760, 1990.
65. **Grob, K.,** *J. Chromatogr.,* 208, 217, 1981.
66. **Muller, H.-M. and Stan, H.-J.,** *J. High Res. Chromatogr.,* 13, 697, 1990.

67. **Stan, H.-J. and Muller, H.-M.,** *J. High Res. Chromatogr.,* 11, 140, 1988.

68. **Cairns, T., Siegmund, E. G., and Stamp, J. J.,** *Mass Spectrom. Rev.,* 8, 93, 1989.

69. **Holland, P. T.,** *Pure and Appl. Chem.,* 62, 317, 1990.

70. **Stan, H.-J.,** *J. Chromatogr.,* 467, 85, 1989.

71. **Mattern, G. C., Singer, G. M., Louis, J., Robson, M., and Rosen, J. D.,** *J. Agric. Food Chem.,* 38, 402, 1990.

72. **Mattern, G. C., Lin, C.-H., Louis, J. B., and Rosen, J. D.,** *J. Agric. Food Chem.,* 39, 700, 1991.

73. **Quimby, B. D. and Sullivan, J. J.,** *Anal. Chem.,* 62, 1027, 1990.

74. **Sullivan, J. J. and Quimby, B. D.,** *Anal. Chem.,* 62, 1034, 1990.

75. **Voyksner, R. D. and Cairns, T.,** in *Analytical Methods for Pesticides and Plant Growth Regulators,* Vol. 17, Sherma, J., Ed., Academic Press, San Diego, 1989, 119.

76. **Liu, C.-H., Mattern, G. C., Yu, X., Rosen, R. T., and Rosen, J. D.,** *J. Agric. Food Chem.,* 39, 718, 1991.

77. **Miles, C. J. and Moye, H. A.,** *Chromatographia,* 24, 628, 1987.

78. **Gilvydis, D. M. and Walters, S. M.,** *J. Assoc. Off. Anal. Chem.,* 73, 753, 1990.

79. **Ohlin, B.,** *Var Foeda,* Suppl. 2, 11, 1986.

80. **Baker, P. G. and Bottomley, P.,** *Analyst,* 107, 206, 1982.

81. **Kuhne, R. O., Egli, H., and Heinemann, G.,** *Fresenius Z. Anal. Chem.,* 339, 374, 1991.

82. **Shalaby, L. M. and George, S. W.,** in *LC/MS Applications in Agricultural, Pharmaceutical and Environmental Chemistry ACS Symp. Ser. No. 420,* Brown, M. A., Ed., American Chemical Society, Washington, D.C., 1990, 91.

83. **Sherma, J.,** in *Analytical Methods for Pesticides and Plant Growth Regulators,* Vol. 14, Zweig, G. and Sherma, J., Eds., Academic Press, San Diego, 1986, 1.

84. **Stahr, H. M.,** *CRC Crit. Rev. Anal. Chem.,* 17, 213, 1987.

85. **Gardyan, C. and Thier, H.-P.,** *Z. Lebensm. Unters. Forsch.,* 192, 40, 1991.

86. **Kovac, J., Tekel, J., and Kurucova, M.,** *Z. Lebensm. Unters. Forsch.,* 184, 96, 1987.

87. **Rado, J. and Gorbach, S.,** *Fresenius Z. Anal. Chem.,* 302, 15, 1980.

Chapter 5

MULTIRESIDUE METHODS FOR CARBAMATE PESTICIDES

Richard T. Krause

TABLE OF CONTENTS

I. INTRODUCTION

A large variety of carbamate pesticides may be used on agricultural crops; depending on their chemical structure, they may be insecticides, fungicides, or herbicides. Carbamate insecticides have an aryl or oxime *N*-methylcarbamate structure. Some carbamate fungicides have a benzimidazole carbamate structure, and other carbamate fungicides have a dithio- or bisdithiocarbamate basic structure. The herbicides have the *N*-alkylthiocarbamate or *N*-phenylcarbamate structure. Structures of representative carbamate pesticides are shown in Figure 1.

All pesticide residue methodology must meet certain basic criteria. The method must (1) quantitatively extract incurred pesticide residues, (2) remove interfering coextractives, (3) determine the residues by a selective and sensitive analytical technique, and (4) quantitatively recover the pesticide residues.

Incurred pesticide residues are generally extracted from food products by blending or homogenizing the analytical sample portion in the presence of an organic solvent for 1 to 2 min. Overnight Soxhlet extraction has been used but is not practical because of the long extraction time. Extraction efficiency studies[1,2] have shown that pesticide residues are most effectively extracted with water-miscible organic solvents such as acetonitrile, acetone, or methanol

INSECTICIDES

Aryl N-methylcarbamate

Carbaryl

Oxime N-methylcarbamate

$$CH_3-S-\underset{\underset{CH_3}{|}}{\overset{\overset{CH_3}{|}}{C}}-CH=NO-\overset{\overset{O}{||}}{C}-N\overset{CH_3}{\underset{H}{\diagup}}$$

Aldicarb

FUNGICIDES

Dithiocarbamate

$$\left[(CH_3)_2-N-\overset{\overset{S}{||}}{C}-S-\right]_3 Fe$$

Ferbam

Ethylenebisdithiocarbamate (EBDC)

$$\left[\begin{array}{l}CH_2-NH-\overset{\overset{S}{||}}{C}-S-\\ CH_2-NH-\underset{\underset{S}{||}}{C}-S-Mn\end{array}\right]_n$$

Maneb

Benzimidazole carbamate

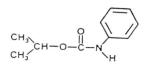

$$O=\overset{}{C}-N\overset{(CH_2)_3-CH_3}{\underset{H}{\diagup}}$$

Benomyl

HERBICIDES

N-Phenylcarbamate

$$\overset{CH_3}{\underset{CH_3}{\diagdown}}CH-O-\overset{\overset{O}{||}}{C}-N\overset{}{\underset{H}{\diagdown}}$$

Propham

N-Alkylthiocarbamate

$$\overset{Cl}{\underset{Cl}{\diagdown}}C=\overset{\overset{Cl}{|}}{C}-CH_2-S-\overset{\overset{O}{||}}{C}-N\overset{CH-(CH_3)_2}{\underset{CH-(CH_3)_2}{\diagup}}$$

Triallate

FIGURE 1. Examples of carbamate pesticides.

in the presence of water by using a blender or homogenizer. Burke and Porter[1] reported that the water-immiscible solvents ethyl acetate and methylene chloride did not satisfactorily remove the incurred residue. These two classical pesticide extraction studies[1,2] used the very nonpolar pesticides TDE and dieldrin, respectively, which severely tested the ability of water-miscible solvents to extract incurred pesticide residues.

Coextractives are removed from the filtered extract by using techniques such as liquid-liquid partitioning and/or adsorption chromatography. These techniques must not only remove the unwanted coextractives, but must (1) enable quantitative recovery of the residues through the procedure and (2) be compatible with the extraction and determinative procedures.

Pesticide residues are generally determined by using some form of chromatography to separate those residues that may be detected by a specific detector. The chromatographic techniques must provide a high degree of resolution among the compounds and not degrade the analytes of interest. The large number of pesticides in use, in addition to innumerable naturally occurring compounds, requires that the detector be selective in responding to the pesticides of interest. If the detector responds to an element or chemical functionality that is unique in comparison to the other materials in the analytical solution injected, then the detector has adequate selectivity. However, if the detector responds to a given element or functionality that is common not only to the analytes of interest but also to the innumerable other compounds in the analytical solution, then its selectivity is not adequate.

Multipesticide residue methods are generally preferred over single-residue methods because they effectively reduce analysis time and expense. Thus, pesticides of similar structure or common unique element can use the same extraction, coextractive removal, and determinative techniques, and be individually separated through chromatography.

The following discussion of carbamate pesticide residue methodology reflects these basic methodology criteria. However, differences in carbamate structure affect the specific analytical approaches used. Therefore, the carbamate pesticides in this chapter are subdivided into three groups: insecticides, fungicides, and herbicides.

II. CARBAMATE INSECTICIDES

Two series titled *Analytical Methods for Pesticides and Plant Growth Regulators*[3] and *Residue Reviews* (recently changed to *Reviews of Environmental Contamination and Toxicology*)[4] and the book *Carbamate Insecticides: Chemistry, Biochemistry and Toxicology*[5] contain reviews on analyses, metabolism, and/or residues of the carbamate insecticides.

The residue analytical chemist must understand the nature of the terminal carbamate residues that may be present in food products so that the appropriate analytical method is used. Therefore, information on carbamate metabolism, stability, solubility, and toxicity is important. The text by Kuhr and Dorough[5] contains a wealth of information on the chemistry, biochemistry, and toxicology of the carbamate insecticides. The series *Residue Reviews* contains similar information for aldicarb,[6-10] carbaryl,[6-9,11,12] carbofuran,[6-9,11] methomyl,[6,7,11] oxamyl,[7,9] propoxur,[6,7] and methiocarb.[6,11]

A. SUMMARY OF MULTIRESIDUE METHODOLOGY REVIEWS

Several investigators have reviewed the literature on multiresidue pesticide methods.[7,10,12–17] In 1971, Burke[13] reported a method for determining carbaryl, an *N*-methylcarbamate insecticide, that used an adaptation of an organohalogen and organophosphorus pesticide multiresidue method. Polarographic, thin-layer chromatographic, and spectrophotometric techniques were reported for determination of carbaryl.[12] The review by Williams[14] in the same year dealt solely with the methodology for carbamate insecticides and determination of residues directly or indirectly by gas-liquid chromatography (GLC). In 1975, Dorough and Thorstenson[15] reviewed GLC, thin-layer chromatography (TLC), fluorescence, high-performance liquid chromatography (HPLC), spectrophotometric methods, and enzymatic techniques for determination of the carbamates. Magallona[16] provided the first comprehensive review of carbamate methodology from the extraction through the GLC determinative procedures.

The importance of the extraction and cleanup steps of the analytical method were stressed by Magallona.[16] Although fortification studies provide information on the recovery of the carbamates through the steps of the method, such studies do not provide information on the ability of the extraction step to extract the incurred carbamate residues. Magallona stated that although numerous authors used chlorinated, water-immiscible solvents to extract the carbamate residues, publications on the extraction of the organochlorine and organophosphorus pesticides had shown these solvents to be generally inefficient in extracting incurred residues. Limited carbamate extraction studies available at the time seemed to confirm this. Magallona recommended that several extraction procedures be studied to determine their efficiency in extracting incurred carbamate residues. For such studies, radiolabeled compounds are generally applied to growing crops by using techniques that simulate field application methods.

Magallona stressed that the cleanup step is probably the most difficult and challenging part of the residue procedure. Cleanup techniques commonly used are solvent partitioning, chromatography, and/or coagulation. As mentioned earlier, these techniques must not only remove the crop coextractives that would interfere in the determination procedure, but also provide the interface between the extraction and determinative steps of the method.

Early carbamate methodology used GLC for determination of carbamate insecticide residues. The difficulties in determining the carbamate insecticides by GLC were discussed by several reviewers.[14–16,18] The carbamate insecticides are thermally labile and, thus, tend to degrade at the elevated temperature of a GLC column. Formation of thermally stable derivatives has been generally limited to the aryl *N*-methylcarbamates. The most difficult carbamate insecticides to chromatograph by GLC are the oxime *N*-methylcarbamate insecticides. The common entity of the carbamates that can be detected with a

GLC detector is nitrogen. GLC detectors that respond selectively to nitrogen are available; however, with the multitude of nitrogen-containing materials naturally present in crops and the large number of pesticides that contain nitrogen, this detector does not provide the needed selectivity.

Dorough and Thorstenson[15] were the first reviewers to report the potential of HPLC for the determination of *N*-methylcarbamate insecticides. In 1982, Lawrence[17] reviewed the HPLC multiresidue methods for the determination of carbamate insecticides in crops. In the method presented by Lawrence, the carbamates are extracted from the crops with acetone and then partitioned between hexane-dichloromethane and water. The fraction containing the insecticides is analyzed by HPLC. Adsorption chromatography is used for separation, followed by measurement of UV absorbance at 254 nm. Moye and Miles,[18] in their 1988 review, concluded that HPLC is the technique of choice for determination of the nonvolatile, thermally labile carbamate insecticides because this technique overcomes difficulties encountered with GLC and the limited sensitivity of TLC.

The UV detector will respond to the multitude of crop materials and pesticides that absorb UV light and, thus, does not provide the desired detection selectivity. The potential of the postcolumn hydrolysis-fluorometric derivatization detection technique developed by Moye et al.[19] was recognized by Lawrence,[17] and its specificity was pointed out by Moye and Miles.[18] This technique enables the selective detection of the carbamate moiety indirectly by postcolumn hydrolysis of the carbamate to methylamine, which is then reacted with *o*-phthalaldehyde and 2-mercaptoethanol. The resulting isoindole fluorophore is monitored with a fluorescence detector.

In summary, reviewers of the literature on carbamate residue methodology concluded that residues should be extracted with water-miscible solvents and determined by using a liquid chromatograph equipped with a selective, sensitive detector. Such a method was developed and has been collaboratively studied.[20] Initially, extraction efficiency studies were conducted to determine the most effective extraction solvent and technique for removal of field-incurred [14C]carbamate insecticides.[21] The data obtained from the study showed that methanol extracts up to 15% more of the [14C]carbamate insecticide residues than does acetone or acetonitrile. Therefore, the crops are extracted with methanol, and then to obtain adequate detection selectivity, the postcolumn fluorometric labeling technique developed by Moye et al.[19] is used. The analytical method has been accepted as the official method by the Association of Official Analytical Chemists (AOAC) and is presented as follows.

B. CARBAMATE INSECTICIDE MULTIRESIDUE METHOD FOR CROPS

1. Principle

The method determines both the oxime and aryl *N*-methylcarbamate insecticides and their toxic carbamate metabolites in crops. Methanol and a mechanical ultrasonic homogenizer are used to extract field-incurred residues from the crops. Water-soluble plant coextractives and nonpolar plant lipid materials are removed from the carbamate residues by liquid-liquid partitioning. Additional crop coextractives (e.g., carotenes and chlorophylls) are removed with a Nuchar S-N silanized Celite column. The carbamates are determined by the HPLC postcolumn fluorometric labeling technique originally developed by Moye et al.[19]

2. Reagents

Solvents — Acetonitrile, methanol, methylene chloride, petroleum ether, and toluene; distilled-in-glass grade (Burdick & Jackson Laboratories, Inc., Muskegon, MI).

HPLC acetonitrile — UV, distilled-in-glass grade (Burdick & Jackson Laboratories, Inc.). Before use, degas acetonitrile in glass bottles by applying vacuum and slowly stirring the solvent with a magnetic stirrer for 5 min.

Ultrapure water — Prepare, using Milli-Q water purification system (Millipore Corp., Bedford, MA). For use in HPLC, degas water as described for HPLC acetonitrile.

Sodium hydroxide solution (0.05 N) — Pipet 27 ml clear supernate sodium hydroxide in water (1 + 1) into 100-ml volumetric flask. Dilute to volume with water and mix (5 N sodium hydroxide). Pipet 10 ml of 5 N sodium hydroxide into a 1-l volumetric flask. Dilute to 1 l with degassed ultrapure water, and mix well but gently to minimize reincorporation of air into the solution.

Sodium sulfate (anhydrous granular) — Heat at 600°C overnight and then cool in a desiccator.

Sodium tetraborate solution (0.05 M) — Add 19.1 g of ACS grade sodium tetraborate decahydrate and ca. 500 ml of degassed ultrapure water to a 1-l volumetric flask. Heat the flask in a steam bath to dissolve the sodium borate, cool to room temperature, and dilute to volume with degassed ultrapure water. Mix well but gently to minimize reincorporation of air into the solution.

Reaction solution — Weigh 500 mg of *o*-phthalaldehyde (Fluoropa, Dionex Corp., Sunnyvale, CA), transfer to a 1-l volumetric flask, add 10 ml of methanol, and swirl the flask to dissolve *o*-phthalaldehyde. Add about 500 ml of 0.05 M sodium tetraborate solution and 1.0 ml of 2-mercaptoethanol (Aldrich Chemical Co., Inc., Milwaukee, WI) and dilute to volume with 0.05 M sodium tetraborate solution. Mix well but gently to minimize reincorporation of air into the solution.

Silanized Celite 545 — Slurry 150 g of Celite 545 (Johns-Manville Sales Corp., Lompoc, CA) with 1 l of hydrochloric acid-water (1 + 1) in a 2-l beaker, cover with a watch glass, and stir with a magnetic stirrer while boiling for 10 min. Cool the slurry, filter, and wash with distilled or ultrapure water until the filtrate in neutral. Wash the Celite with 500 ml of methanol followed by 500 ml of methylene chloride and then air-dry the Celite on a watch glass in a hood to remove the solvent. Transfer the Celite to a 1-l Erlenmeyer flask with a ground glass joint. Heat the unstoppered flask in a 120°C oven overnight and then cool the flask in a desiccator. Place the flask in a hood and carefully pipet 3 ml of dichlorodimethylsilane (Pierce Chemical Co., Rockford, IL) onto the Celite. Stopper the flask, mix well, and let the flask remain at room temperature for 4 h. Add 500 ml of methanol to the flask, mix, and let stand for 15 min. Filter the silanized Celite and wash with isopropanol until neutral. Air-dry the silanized Celite in a hood to remove isopropanol. Dry the silanized Celite in a 105°C oven for 2 h and cool in a desiccator. Store the silanized Celite in a glass-stoppered container. Test the Celite for total silanization by placing about 1 g in 50 ml of water and placing about 1 g in 20 ml of toluene saturated with methyl red. Silanized Celite should float on water and appear yellow with methyl red-toluene solution. Repeat the silanization of Celite with dichlorodimethylsilane if particles of Celite are dispersed in water and/ or appear pink with methyl red-toluene solution; these results indicate active sites.

Nuchar S-N — Slurry 100 g of Nuchar S-N (Kodak Laboratory and Research Products, Rochester, NY) with 700 ml of hydrochloric acid in a 2-l beaker, cover with a watch glass, and stir with a magnetic stirrer while boiling for 1 h. Add 700 ml of water, stir, and boil for an additional 30 min. Cool the slurry, filter, and wash with distilled or ultrapure water until neutral. Then wash the Nuchar S-N with 500 ml of methanol followed by 500 ml of methylene chloride, and air-dry the Nuchar S-N in a hood to remove the solvent. Dry Nuchar S-N in a 120°C oven for 4 h. Cool in a desiccator. Store Nuchar S-N in a glass-stoppered container.

Nuchar S-N/silanized Celite 545 chromatographic mixture — Mix Nuchar S-N with silanized Celite 545 (1 + 4 w/w). Test each batch of Nuchar S-N with a mixed carbamate standard solution (carbaryl, methiocarb, methiocarb sulfoxide, methomyl). (*Note:* Use freshly prepared mixed standard solution because methiocarb sulfoxide degrades in solution.) Prepare a mixed carbamate solution in methanol at a concentration of 5 μg each per ml. Pipet 5 ml of this solution into a 250-ml round-bottom flask, and 5 ml into a 25-ml actinic volumetric flask. Dilute the solution in the volumetric flask to 25 ml with methanol and use as the reference standard for HPLC. Evaporate the standard solution in the round-bottom flask just to dryness with a vacuum rotary evaporator as described in Section II.B.6. After the last trace of methanol has evaporated, remove the round-bottom flask from the evaporator and dissolve the carbamate residue in 10 ml of methylene chloride. Transfer the methylene chloride solution in the round-bottom flask to a prepared adsorbent column and elute as described in Section II.B.8. After evaporation of the eluate in the round-bottom flask, dissolve the residue in 25 ml of methanol. Filter 5 to 8 ml of this solution through a Swinny filter holder as described

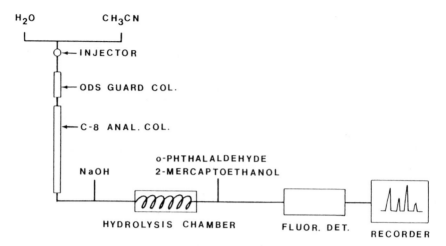

FIGURE 2. Carbamate insecticide HPLC system with in-line postcolumn fluorometric detector.

in Section II.B.8. Calculate the recovery of carbamates after determination by HPLC. Nuchar S-N is considered satisfactory if the average recovery of carbamates is ≥95%, with recovery of no one compound <90%.

3. General Apparatus

Homogenizer — Polytron Model PT 10-35 equipped with a PT 35K generator containing knives (Brinkmann Instruments, Inc., Westbury, NY).

Homogenizer jar — Four-sided glass quart jar (Tropicana Products, Inc., Bradenton, FL).

Vacuum rotary evaporator — Model RE rotavapor (Brinkmann Instruments, Inc.). Maintain the solution in the condensing coils and around the receiving flask at −15°C. (Refrigerated water-antifreeze solution works well.) Use a vacuum pump fitted with a manometer and needle valve to control the vacuum in the evaporator.

Chromatographic tubes — Chromaflex 30 cm × 22 mm I.D. column (size 233) with coarse porosity fritted glass disk and size 2 Varibor stopcock (No. K-420540-9042, Kontes, Vineland, NJ).

Swinny filter holder — 13-mm Filter size (No. XX3001200, Millipore Corp., Bedford, MA).

Mitex filters — 5-μm, 13-mm Diameter, white, plain filters (No. LSWP 01300, Millipore Corp.).

4. HPLC Apparatus

HPLC apparatus (Figure 2) must be capable of performing as described in Section II.B.5. Specific individual items of apparatus have been found to meet the operating parameters and are listed as a guide for the analyst, as follows.

Mobile-phase delivery system — Model 322 MP programmable gradient system (Beckman Instruments, Inc., Altex Div., Berkeley, CA).

Injector — Model 16AS-7000 automatic sampler with 100-µl injection loop (Valco Instruments Co., Houston, TX).

Guard column — 7-cm × 2.1-mm I.D. column containing 25- to 37-µm Co-Pell ODS packing (Whatman Inc., Clifton, NJ).

Analytical column — 25-cm × 4.6-mm I.D. column containing 6-µm Zorbax C-8 spherical particles (MAC MOD Analytical Inc., Wilmington, DE). An equivalent column should contain 5- or 6-µm spherical silica particles that have been bonded with a monofunctional octylsilane reagent to form a monomolecular bond.

Column oven — Custom-built forced draft oven (66 × 13 × 11 cm).

Carbamate hydrolysis chamber — Column bath (18 × 18 × 13 cm) from Model 5360 Barber Coleman gas chromatograph with Model 700-115 proportional temperature controller (RFL Industries, Inc., Boonton, NJ) containing 3-m × 0.48-mm I.D. No. 321 stainless steel tubing (Tubesales, Forest Park, GA).

Fluorescence detector — Model 650-10LC, with 20-µl cell (Perkin-Elmer Corp., Norwalk, NJ).

Recorder — Model 4000 microprocessor/printer plotter (Spectra-Physics, San Jose, CA).

Sodium hydroxide and reaction solution reservoirs — 60- × 25-cm I.D. glass columns with Teflon fittings (Glenco Scientific, Inc., Houston, TX). Pressurize reservoirs with nitrogen. Connect a 6-m × 0.5-mm I.D. Teflon restriction coil from the reservoir to 15-cm × 0.18-mm I.D. stainless steel tubing. Connect the stainless steel tubing to a 0.74-mm I.D. stainless steel reaction tee (No. ZVT-062, Valco Instruments Co.).

Connecting tubing — No. 304 stainless steel tubing (1.6 mm O.D. × 0.18 mm I.D.) to connect the injector, columns, and first tee.

5. HPLC Operating Parameters

Adjust the mobile-phase flow rate to 1.50 ± 0.02 ml/min at 50% acetonitrile in water. Equilibrate the system at 12% acetonitrile in water for 10 min, inject the analytical solution from the cleaned up extract, and begin a 30-min linear gradient to 70% acetonitrile in water. Adjust the flow rate of 0.05 N sodium hydroxide and reaction solution to 0.50 ± 0.02 ml/min each. Operate the column oven at 35°C and hydrolysis chamber at 100°C. Set the fluorescence excitation and emission wavelengths to 340 and 455 nm, respectively, and slit widths to 15 and 12 nm, respectively. Set the detector photomultiplier gain to low and time constant to 1 s. Adjust the sensitivity so that 10 ng of carbofuran produces a 50 ± 5% full-scale response on the recorder. The baseline noise should be <2% full-scale response. Carbamates should elute as shown in the chromatogram (Figure 3). (*Note:* If the system will not be used for several days, replace the water mobile phase with methanol

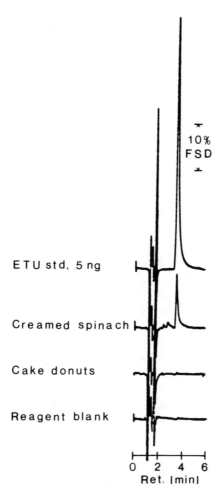

FIGURE 3. Example of ETU HPLC-EC detector chromatograms. FSD = full-scale recorder deflection.

and pump through the system, drain the sodium hydroxide and reaction solutions from the reservoirs, and wash the reservoirs and associated tubing first with water and then methanol.) When starting up the system, change the mobile phase to water, and wash the reaction reservoirs and associated tubing with water before adding the reaction solutions.

6. Extraction

For high moisture (>75% water) products, add 150 g of chopped composited sample and 300 ml of methanol to a homogenizer jar. Homogenize the analytical sample portion with a Polytron for 30 s at about half speed (setting of 7) and then 60 s at full speed. Vacuum-filter the homogenate through a 12-cm perforated Buchner funnel containing sharkskin or a 597 S & S filter, and collect the filtrate in a 500-ml filter flask. (*Note:* Reduce the vacuum during filtration if the filtrate begins to boil.) Transfer a portion of the filtrate equivalent to a 100-g analytical portion to a 2-l standard taper (**T**) 24/40 round-bottom flask. (*Note:* Volume of 100-g analytical portion = milliliters of water in 100-g analytical portion + 200 ml methanol − 10 ml contraction factor.) Add distilled or ultrapure water to the round-bottom flask to give a total of 100 ml of water. Also add a small star magnetic stirrer to the round-bottom flask.

Place a 250-ml **T** 24/40 trap on a 2-l round-bottom flask and attach to a vacuum rotary evaporator. Apply the vacuum slowly to minimize frothing. After the full vacuum is applied, slowly place the flask in a 35°C water bath. Concentrate the extract to 75 ml.

7. Coextractive Removal — Partitioning

Transfer the concentrated extract from the round-bottom flask to a 500-ml separatory funnel containing 15 g of sodium chloride. Shake the separatory funnel until the sodium chloride is dissolved. Wash the round-bottom flask with three 25-ml portions of acetonitrile, transfer each portion to the 500-ml separatory funnel, shake for 30 s, and let the layers separate for 5 min. Drain the aqueous phase into a 250-ml separatory funnel containing 50 ml of acetonitrile, shake for 20 s, let the layers separate, and discard the aqueous layer.

Add 25 ml of 20% aqueous sodium chloride solution to the acetonitrile in the 500-ml separatory funnel, shake for 20 s, let the layers separate, and transfer the aqueous solution to a 250-ml separatory funnel. Shake the 250-ml separatory funnel for 20 s, let the layers separate, and discard the aqueous layer.

Add 100 ml of petroleum ether to the 500-ml separatory funnel, shake for 20 s, let the layers separate, and drain the acetonitrile layer into a second 500-ml separatory funnel. Extract the mixture successively with 100-, 25-, and 25-ml portions of methylene chloride, shaking each for 20 s (shake the 25-ml portions gently). Drain the lower methylene chloride-acetonitrile layers through a 22-mm I.D. column containing about 5 cm of anhydrous sodium sulfate. Collect the eluate in a 1-l **T** 24/40 round-bottom flask. Evaporate the solution to dryness with a rotary evaporator as described. Remove the round-bottom flask from the evaporator immediately after the last traces of solution have evaporated, and then add 10 ml of methylene chloride to the round-bottom flask.

8. Coextractive Removal — Chromatographic

Fit a 1-hole No. 5 rubber stopper onto the tip of a chromatographic tube with Varibor stopcock, attach a **T** 14/40 side arm vacuum adaptor and connect to a 500-ml ⊤ 24/40 round-bottom flask, open the stopcock, and connect the apparatus to a vacuum line. Place 0.5 g of silanized Celite in the chromato-graphic tube, tamp, add 5 g of Nuchar S-N silanized Celite 545 (1 + 4) mixture, and tamp again. Add a 1- to 2-cm glass wool plug (Corning Glass Works, Corning, NY) on top of the adsorbent. Prewash the column with 50 ml of toluene-acetonitrile (1 + 3) eluting solution. Close the stopcock when the prewash solution is about 0.5 cm from the top of the glass wool. Disconnect the vacuum. Discard the eluate in the round-bottom flask, and reconnect the flask to the apparatus. Transfer the residue dissolved in 10 ml methylene chloride from the round-bottom flask to the column and elute at 5 ml/min. Wash the 1-l round-bottom flask with 10 ml of methylene chloride and then 25 ml of eluting solution. Transfer each separately to the column and elute each to the top of the glass wool before adding the next solution. Add 100 ml of eluting solution and elute at 5 ml/min. Close the stopcock when the top of the eluting solution reaches the top of the glass wool. Evaporate the solution in the 500-ml round-bottom flask just to dryness, using a vacuum. After all the solution has evaporated, immediately pipet 5 ml of methanol into the 500-ml round-bottom flask to dissolve the residue. Pour the meth-anolic solution into a 10-ml centrifuge tube or other suitable container. (*Note:* Approximately 4.5 ml of filtrate will be collected.) The volume of filtrate collected is not critical because the g analytical portion per ml of methanol is known. If the solution needs to be diluted, pipet an aliquot into another container and dilute to an appropriate volume.

9. Determination

Inject 10 μl of methanolic analytical solution onto the HPLC column, using the chromatographic apparatus and parameters described. Tentatively identify pesticide residue peaks on the basis of their retention times. Measure the peak area or peak height obtained from a known amount of appropriate reference material(s). To ensure valid measurement of analyte amount, the sizes of peaks from analyte and reference standard should match within ±25%. Chromatograph reference material(s) immediately after the analytical solution.

10. Carbamate Recovery and Method Sensitivity

Carbamate recovery values obtained in the collaborative study[20] averaged 95%, with a coefficient variation of 8.7% for seven carbamates and two carbamate metabolites. The low fortification level was 0.05 ppm and the high fortification level was the U.S. tolerance. Approximately 50% of aldicarb sulfoxide is recovered through the method. Krause[22] reported extension of the method to additional carbamates and crops in a subsequent publication. The estimated limit of quantitation is 0.01 ppm for high-moisture crops based on a chromatographic peak of 10% full-scale deflection.

11. Method Adaptations

Pardue[23] reported replacing the previous extraction and liquid-partitioning procedures with those used in the Luke et al. method[24] to reduce analysis time and improve recovery of aldicarb sulfoxide. Vannelli[25] modified the method for determination of total aldicarb carbamate residues by oxidizing the filtrate with peracetic acid to convert all aldicarb carbamate residues to the sulfone. McGarvey[26] simplified the postcolumn derivatization reaction to a single stage by combining the sodium hydroxide hydrolysis solution and *o*-phthalaldehyde/2-mercaptoethanol solution into one solution, thereby eliminating the need for two postcolumn pumps.

C. CARBAMATE INSECTICIDE RESIDUE CONFIRMATORY TECHNIQUES

Confirmatory techniques are often needed to validate the identity and the level of pesticide(s) found in the analytical sample. TLC is a rapid, simple, semiquantitative technique that can be used for the confirmation of carbamate insecticide residues, although adequate sensitivity for low levels may be a problem. Several authors[16,27–29] have reviewed the literature on the confirmation of carbamate insecticide residues by TLC. The developed plate is usually sprayed with a reagent that reacts with the carbamates to form a colored spot, or the carbamates are visualized by enzyme inhibition techniques. Appaiah et al.[30] have reported a new reagent — 4,4-diaminodiphenyl sulphone — for colorimetric visualization of several aryl *N*-methylcarbamates. TLC enables both aryl and oxime *N*-methylcarbamates to be detected by using the appropriate visualization technique.

Mass spectrometry (MS) is the technique of choice for the confirmation of identity of low molecular weight organic chemicals. For this technique to have practical use in the area of pesticide residues, the numerous components in the analytical sample must be separated by some form of chromatography and the chromatographic technique interfaced with the mass spectrometer. Separation by GLC prior to MS is not a satisfactory technique for confirmation of carbamate insecticide residues because of their thermally labile nature. HPLC and MS instrumentation can be interfaced by utilizing some device that can separate the mobile phase before introduction of the analyte into the mass spectrometer. Generally, in regular HPLC determinations, reversed-phase chromatography with a water/water-miscible organic solvent mobile phase is used. This type of mobile phase is difficult to evaporate in-line. Therefore, Wright[31] used a normal-phase HPLC mobile phase (isopropanol in hexane gradient) to facilitate rapid in-line evaporation of the solvents. Wright[31] reported the confirmation of aryl and oxime *N*-methylcarbamate insecticides as well as carbamate herbicides, using normal-phase HPLC/MS. Voyksner and Bursey[32] have reported an HPLC/MS interface that enables the use of reversed-phase mobile phases.

Supercritical fluid chromatography (SFC) has recently received attention as a technique that can be interfaced with MS for confirmation of carbamate insecticide residues.[33-36] Berry et al.[33,35] used a thermospray device to interface HPLC and MS. A silica column and 15% methanol in carbon dioxide were used to chromatograph the carbamate insecticides carbaryl and methiocarb. Kalinoski et al.[34] reported the use of direct supercritical fluid injection with capillary SFC. They chromatographed several aryl and oxime *N*-methylcarbamates and obtained ammonia and methane chemical ionization mass spectra. France and Voorhees[36] developed a solvent evaporation injection technique for obtaining mass spectra of bendiocarb and methiocarb that enabled increased injection volumes and eliminated the solvent.

An̈other approach for the confirmation of carbamate insecticide residue identity and level would be to detect a moiety of the molecule other than the carbamate moiety detected in the AOAC method. The aryl *N*-methylcarbamates also contain a phenolic moiety, which is the basis for their detection in several analytical methods. Johnson[37] reported a method for carbaryl in which this carbamate is hydrolyzed to 1-naphthol, the phenol reacted with *p*-nitrobenzenediazonium tetrafluoroborate, and the resulting product measured colorimetrically. Holden[38] reported a procedure in which the phenols of hydrolyzed carbamates are derivatized to dinitrophenol ethers, which are then determined by GLC with electron-capture detection. The metabolite of carbofuran, 3-hydroxycarbofuran, is not recovered through the method. Phenols can be selectively detected by oxidation with an electrochemical EC detector. Kissinger et al.[39] briefly described a precolumn hydrolysis technique for carbofuran in which the resulting phenol is chromatographed on a reversed-phase column with a slightly acidic mobile phase and detected with an EC detector. Olek et al.[40] applied this precolumn hydrolysis technique to several aryl *N*-methylcarbamates. The hydrolyzed phenols were subjected to liquid partitioning and solvent evaporation steps before separation by HPLC with an acidic mobile phase and detection in-line with an EC detector at 0.9 or 1.0 V. The acidic mobile phase suppresses ionization of the phenols, thereby increasing peak retention and eliminating or reducing peak tailing.[41]

Unfortunately, phenolic oxidations at low pH are difficult and require a high applied potential,[42] which can result in high background currents and baseline pump noise.[43] Kissinger et al.[42] proposed raising the mobile-phase pH by postcolumn addition of base to reduce the oxidation potential, thereby reducing background currents and pump noise. Krause[44] reported development of an HPLC–EC technique that selectively detects the phenolic moiety of the carbamates. With this technique, carbamates are chromatographed intact using the same C-8 column and acetonitrile/water mobile phase as in the previous method. The eluted carbamates are hydrolyzed in-line by postcolumn addition of base, which also serves as electrolyte in the EC detection of the resulting phenols. Thus, the technique, which eliminates the separate and manual predeterminative hydrolysis-derivatization steps of previous methods, reduces analysis time and requires minimal change in equipment and chemicals from those used in the AOAC HPLC postcolumn fluorometric determinative method.[20] The intact carbamates are separated on a reversed-phase HPLC column using a gradient acetonitrile-water mobile phase. The eluted carbamates are hydrolyzed in-line with dilute sodium hydroxide at 100°C, and the resulting phenols are measured with a coulometric EC detector.

III. DITHIO- AND ETHYLENEBISDITHIOCARBAMATE FUNGICIDES

The series titled *Analytical Methods for Pesticides and Plant Growth Regulators*[3] and *Residue Reviews*[4] contain reviews on the dithiocarbamate and ethylenebisdithiocarbamate (EBDC) fungicides and their metabolites in the areas of toxicity, metabolism, analysis, and/or residues of interest.

The toxicity of these fungicides has been reviewed by several authors.[45–47] Vettorazzi[45] reported that the dithiocarbamates thiram and ziram caused teratogenic effects, and ziram produced chromosomal aberrations. Ethylenethiourea (ETU), a degradation product of the EBDC fungicides, is reported to produce thyroid tumors in experimental animals. Toxicological studies were also reported that showed ETU to be teratogenic and mutagenic. Spynu[46] stated that from what is known about the chronic effects of fungicides, the carcinogenic effects are of greatest concern. He reported that experiments with models have revealed the tumorigenic danger of a number of dithiocarbamataes: thiram, zineb, ziram, and maneb. Lentza-Rizos,[47] in a review of EBDC fungicides, stated that although these fungicides are of low acute toxicity, some of these compounds (maneb, zineb, mancozeb, metiram, and nabam) can decompose to ETU. The reviewer further stated that because ETU has been shown to possess carcinogenic, mutagenic, goiterogenic, and teratogenic activities in animal tests, it has become a major human health concern. Thus, the need to have valid analytical data on the residue levels of these fungicides and metabolites is self-evident.

The review by Newsome[48] deserves special attention, as it contains a wealth of information on chemical and physical properties, toxicity, metabolism in plants, and residue methodology of EBDCs and their degradation products.

The following sections discuss multiresidue methods for the dithiocarbamate and EBDC fungicides, and methods for the determination of ETU.

A. SUMMARY OF DITHIOCARBAMATE MULTIRESIDUE METHODOLOGY REVIEWS AND METHODS

Several authors[47-49] have reviewed the methodology for dithiocarbamate and EBDC fungicide residues. The determination of residues using the method of Keppel[50,51] or modifications of his method are used most often. The fungicides are acid-hydrolyzed to carbon disulfide, which is then determined by colorimetry[50,51] or by headspace GLC.[52] In Great Britain, the Panel on Determination of Dithiocarbamate Residues[53] conducted several collaborative studies and recommended a GLC headspace procedure. Gustafsson and Thompson[49] stated that methods based on carbon disulfide evolution are neither specific for individual dithiocarbamates nor very accurate. Gustafsson and Fahlgren[54] reported an HPLC method for analyzing residues of the dithiocarbamate fungicides. Salts of the dithiocarbamates and EBDCs were transformed into water-soluble salts, which were subsequently converted to the methyl esters and determined by HPLC with UV detection. This reaction resulted in the conversion of ferbam and ziram to methyl *N,N*-dimethyldithiocarbamate and the conversion of nabam, zineb, maneb, and mancozeb to *N,N'*-ethylenebisdithiocarbamate. Recoveries of the dithiocarbamate fungicides from fortified crops ranged from 59 to 85% for the individual fungicides at the 0.5 ppm fortification level.

B. DITHIOCARBAMATE METHODOLOGY CONSIDERATIONS

The range of physical and chemical properties of the dithiocarbamates and EBDCs complicates the methodology for these compounds. Their solubilities, which range from soluble to insoluble in water and/or organic solvents, complicate the approach used for extraction of the residues from crops. In addition, because the EBDCs are generally polymers and do not have a finite molecular weight, their direct determination is very difficult at best. Also, as pointed out by Newsome,[48] EBDCs are inherently unstable. Therefore, development of analytical methods for these fungicides is much more difficult than for the *N*-methylcarbamate insecticides. Considering the physical and chemical properties of the dithiocarbamate and EBDC fungicides, it is unlikely that a method can be developed to determine the individual fungicides. Because no current analytical method for determination of these fungicides is totally satisfactory for all needs, it appears that selection of a method should be based on the purpose of the analytical data.

C. SUMMARY OF ETHYLENETHIOUREA METHODOLOGY REVIEWS AND METHODS

Bottomley et al.[58] presented a very comprehensive review of methods for the determination of ETU residues. Methods using paper chromatography and TLC as well as GLC and HPLC were described. TLC was reported to have been used successfully to determine ETU; however, considerable cleanup is required and ETU may decompose on the TLC plate.

Many workers experienced difficulties with the direct determination of ETU (without derivatization) by GLC, and some have shown that results must be treated with caution because of the possibility of decomposition of any EBDCs and intermediate breakdown products that are coextracted and/or formed during the analyses.[58] Most determinations of ETU residues have used packed-column GLC of ETU derivatives, although it was noted that derivatization techniques have the disadvantage of increased analysis times and the possibility of incomplete reactions, which give inconsistent yields. Capillary-column GLC of ETU derivatives was used in some situations. The authors pointed out that difficulties can occur when attempting to apply these procedures to substrates other than those for which they were originally developed.

Bottomley et al.[58] reported that several authors used HPLC instruments equipped with UV detectors to determine ETU residues in food products without the need for derivatization. These procedures were less sensitive than GLC in most situations.

After the review by Bottomley et al.,[58] Lentza-Rizos[47] reviewed the subsequently published ETU methods. The reviewer reported that the ETU method adopted by the AOAC[59] as official final action utilized GLC for determination of the derivatized ETU. The number of HPLC methods had increased dramatically since the review by Bottomley et al. Most methods of HPLC used a UV detector. Several authors used EC detectors.[60–63] Prince[60] and Dogan et al.[61] used carbon working electrodes. Krause and Wang[62] used an Hg/Au electrode and Wang et al.[63] used a detector with a copper electrode.

Lawrence,[17] in his review of HPLC methods, recommended the HPLC ETU method reported by Onley et al.,[64] which uses a UV detector. This method is similar to the official AOAC GLC method but uses an HPLC determinative technique.

D. ETU METHODOLOGY CONSIDERATIONS

Determination of ETU can be considered as a multiresidue determination because it is the common metabolite of all the EBDC fungicides. ETU is the residue of greatest concern because of its carcinogenic properties.[45–47]

The physical and chemical properties of ETU must be considered in the development and selection of a method for its determination. ETU is soluble in water; moderately soluble in methanol, alcohol, and ethylene glycol; and insoluble in acetone, ether, and chloroform. It is known to undergo oxidation with hydrogen peroxide and hypochlorite[55] and photolysis by light in the presence of oxygen or photosensitizers such as acetone or riboflavin[56] or on silica gel.[57] Thus, two very important considerations for ETU methodology are a need to ensure that the method (1) uses an extraction solvent in which

ETU is highly soluble, and (2) does not cause degradation of ETU as the analyte is taken through the method.

ETU cannot be determined by the dithiocarbamate method of Keppel,[50,51] because it does not produce carbon disulfide and was not reported to be determined by the method of Gustafsson and Fahlgren.[54] Because of its unique chemical and physical properties, this compound is generally determined by single-residue methods as noted in the previous methodology review section.

The determinative technique used should enable the selective and sensitive detection of ETU without causing its decomposition. Also, the chromatography or detector should not be adversely affected by crop coextractives. TLC generally does not provide the desired precision for the quantitation of ETU and, as noted by Bottomley et al.,[58] ETU can degrade on the TLC plate. GLC is a less than satisfactory technique because ETU generally chromatographs poorly on a GLC column unless a derivative is formed. Unfortunately, food coextractives can adversely affect the reaction yield of such derivatives, causing low and erratic ETU recoveries. ETU can be successfully chromatographed by HPLC without derivatization,[60–64] and thus it would seem to be the chromatographic technique of choice.

ETU has generally been directly chromatographed on C-8 or C-18 reversed-phase silica-based HPLC columns with little or no organic modifier in the mobile phase.[60,62,65–67] Krause[68] reported that the chromatographic separation of ETU on reversed-phase C-8 silica-based columns seemed to be due to residual silanols, which resulted in significant ETU retention-time differences between columns. A graphitized carbon column, which could be operated at low pH, produced a sharp, relatively symmetrical ETU peak and similar ETU retention times between columns.[68]

Several authors[58,64,67] have used UV detectors for monitoring ETU in the HPLC column effluent. Because of the multitude of UV-absorbing crop coextractives and pesticides, the UV detector does not provide adequate selectivity for detection of ETU residues. Hanekamp et al.[66] investigated the application of HPLC dropping mercury electrode (DME) and glassy carbon electrode amperometric detectors for detection of ETU. They concluded that for ETU, the DME provided superior selectivity but was significantly inferior in sensitivity to the glassy carbon electrode. For direct detection of thiols, Allison and Shoup[69] used an HPLC Hg/Au electrode amperometric detector, which provided the selectivity of the DME detector and the potential for improved sensitivity. Krause[68] investigated the Hg/Au electrode amperometric detector coupled with the carbon column and was able to demonstrate that satisfactory selectivity and sensitivity for ETU could be obtained with this system.

The official AOAC method for ETU was revised[70] to increase ETU recoveries and improve their consistency. The derivatization step was eliminated, and ETU was determined directly by an HPLC Hg/Au EC system.

Replacement of sodium chloride with sodium acetate controlled the pH of the concentrated aqueous extract for improved partitioning and recovery of ETU into the organic solvent. Also, silanization of the round-bottom flask used in evaporation of the organic solvent improved ETU recovery. This method is presented as follows.

E. ETU METHOD FOR FOOD PRODUCTS

1. Principle

The method determines ETU directly by HPLC with an EC detector in a variety of food products. ETU is extracted from food products with a methanol-aqueous sodium acetate solution. A portion of the concentrated filtrate is added to a column of diatomaceous earth, and ETU is eluted with 2% methanol in methylene chloride to separate it from food coextractives, which are retained on the column. The eluate is collected in a siliconized flask and evaporated, the residue is dissolved in water, and 20 μl of solution is injected onto an HPLC graphitized carbon column. ETU is eluted from the HPLC column with a mobile phase of acetonitrile-aqueous phosphoric acid $(0.1 M)$-water $(5 + 25 + 70)$, and the eluted ETU is detected by using an amperometric EC detector equipped with an Hg/Au working electrode.

2. Reagents

Aluminum oxide — Fisher No. A-540 alumina, adsorption, 80–200 mesh (Fisher Scientific, Pittsburgh, PA).

Antifoam B solution (1%) — Dilute Dow Corning Antifoam B emulsion (10%) (Dow Corning Corp., Midland, MI) with water $(1 + 9)$.

Solvents — Acetonitrile (UV grade), methanol, methylene chloride, and water; all distilled-in-glass grade (Burdick & Jackson Laboratories, Inc., Muskegon, MI). Use distilled-in-glass grade wherever a solvent, including water, is required.

ETU standard solution — Dissolve ETU standard (Environmental Protection Agency, Research Triangle Park, NC) in water to give a concentration of 0.25 μg/ml, or other concentrations as needed. Store the solution in actinic glassware. When not in use, store in a refrigerator.

Partitioning eluent (2% methanol in methylene chloride) — Pipet 20 ml of methanol into a graduated cylinder. Dilute to 1 l with methylene chloride and mix.

Filter aid — Celite 545 (Manville Sales Corp., Lompac, CA). Do not acid-wash; do not use acid-washed grade.

Siliconizing solution — Dilute SurfaSil siliconizing fluid (Pierce Chemical Co., Rockford, IL) with methylene chloride $(1 + 9)$.

Phosphoric acid solution (0.1 M) — Pipet 7 ml of HPLC-grade 85% phosphoric acid (Fisher Scientific) into 1-l graduated cylinder containing water. Dilute to volume with water, and mix.

Solid column support — Gas-Chrom S, 45–60 mesh (Alltech Associates, Inc./Applied Science Labs, Deerfield, IL). Use as is.

3. General Apparatus

Blender — Explosion-proof Waring Laboratory Blendor.

Chromatographic tubes — 30-cm × 22-mm I.D. Chromaflex column (size 223) with coarse porosity fritted glass disk and size 2 Varibor stopcock (No. K-420540-9042, Kontes, Vineland, NJ).

Siliconized round-bottom flasks — Siliconize flasks in a well-ventilated hood while wearing rubber gloves and eye protection. Add 50 ml of siliconizing solution to each new 500-ml ᵀ 24/40 round-bottom flask. Stopper and shake the flask; remove the stopper after shaking the flask to release pressure. Shake the flask two additional times to ensure complete siliconizing of the glass surface. Pour the siliconizing solution into a waste container. Let the flask air-dry and then cure the flask surface for 30 min in 105°C oven. Let the flask come to room temperature. Wash the flask with methylene chloride to remove unreacted SurfaSil and reaction by-products. (*Note:* After use, simply rinse the flask with methanol followed by methylene chloride and then air-dry before reuse. If the flask needs to be cleaned with soap and water and/or a scrub brush, the flask must be resiliconized. Low ETU recoveries may also indicate the need to resiliconize the flask.)

Filter paper — 11-cm Sharkskin (Arthur H. Thomas Co., Philadelphia, PA).

Pasteur pipets — Borosilicate glass, 9 in. (229 mm), disposable (Fisher Scientific). Do not use soda-lime type.

Swinny filter holder — 13-mm Filter size (No. XX3001200, Millipore Corp., Bedford, MA).

Membrane filters — Nylon-66, 0.45-μm pore size, 13-mm disk (Rainin Instrument Co., Inc., Woburn, MA).

Vacuum rotary evaporator — Model RE Rotavapor (Brinkmann Instruments, Inc., Westbury, NY). Maintain the evaporation water bath temperature at 35°C. Maintain the solution in the condensing coils and around the receiving flask at −15°C. (Refrigerated water-antifreeze solution works well.) Use a vacuum pump fitted with a vacuum gauge and needle valve to control the vacuum in the evaporator.

4. HPLC Apparatus

HPLC mobile-phase reservoirs — Ultraware HPLC solvent reservoirs (No. K953935-1000 [1 l] and K953935-2000 [2 l], Kontes). Replace Teflon lines with 1/16-in. (1.6 mm) O.D. × 0.040-in. (1.0 mm) I.D. stainless steel tubing to prevent reincorporation of air (oxygen) into the mobile phase. Degas solvents with helium (99.95%) purified with in-line Hydro-Purge II and Oxy-Purge traps (Alltech Associates, Inc./Applied Science Labs).

Mobile-phase delivery system — SP8700XR HPLC pump (Spectra-Physics, San Jose, CA).

Injector — SP8780XR autosampler fitted with a 20-μl loop (Spectra-Physics). Use methanol for wash solution.

HPLC column — Shandon Hypercarb graphitized 7-μm spherical carbon particles, 10 cm × 4.6 mm I.D. (Keystone Scientific, Inc., Bellefonte, PA). Condition a new column for 1 to 2 d by passing acetonitrile through the column at 0.5 to 1.0 ml/min. Conditioning removes impurities and improves ETU peak shape.

Column oven — 2080 HPLC forced-air column oven (Varian Associates, Inc., Palo Alto, CA).

EC detector — Use LC-17 thin-layer EC cell (with Hg/Au working electrode, Ag/AgCl reference electrode, stainless steel block auxiliary electrode, and 5-μm Teflon gasket) and LC-4B amperometric detector controller (No. MF-9094, Bioanalytical Systems, West Lafayette, IN). Prepare Hg/Au working electrode surface as follows: to remove old amalgam, place a few drops of 6 N nitric acid on the electrode surface. When a rusty yellow color appears, the old amalgam has been destroyed. Polish the gold surface first with diamond-polishing compound and then with alumina as directed by the manufacturer. In a tray, coat the new, highly polished gold surface with a few drops of high purity mercury; the electrode should be totally covered. (Do not touch the gold surface with a pipet or other sharp object because the surface may become scratched.) After 5 min, brush the excess mercury from the electrode surface with a soft paper tissue. Continue brushing and polishing the surface with the tissue until the surface is flat and without any signs of mercury puddling. The surface should have a silver sheen (a mirror-like surface indicates excessive mercury). Let the amalgam equilibrate in an air atmosphere for ≥2 d before the electrode is connected to the cell. (Response to ETU seems to improve if the Hg/Au surface is lightly polished with a dampened, soft paper tissue just before the electrode is connected to the cell.) Connect the cell to the HPLC unit after the mobile phase (see Section III.E.5) has been degassed for several hours and then passed through the column to eliminate oxygen from the system. Maintain the flow of mobile phase at 0.5 ml/min through the system and cell overnight with the detector turned off. Overnight equilibration is necessary to provide stable response and low baseline noise.

Computing integrator — SP4200 computing integrator (Spectra-Physics). Operate at a chart speed of 0.5 cm/min and attenuation of 8, and use proper peak-width and peak-threshold values.

5. HPLC Operating Parameters

Set the pump so that the solenoid switching valves produce a mobile-phase composition of acetonitrile-aqueous phosphoric acid (0.1 *M*)-water

TABLE 1
Size of Analytical Portion Used for
Extraction

Food product	Weight (g)
Applesauce, canned, sweetened	50
Cabbage, fresh, boiled	100
Celery, fresh	100
Collards, frozen, boiled	100
Donuts, cake, powdered sugar-coated	50
Grape jelly	50
Green beans, canned	100
Lettuce, fresh	100
Mushrooms, canned	100
Potatoes, baked	100
Spinach, canned	100
Spinach, canned, creamed, baby food	100
Tomatoes, canned	100

(5 + 25 + 70). Adjust the mobile-phase flow rate to 1.00 ± 0.02 ml/min. Operate the column oven at 35°C. After the EC cell has equilibrated overnight (see previous section), set the EC detector potential to 350 mV, turn the cell mode from "stby" to "on", and let the baseline stabilize. Adjust the microamp sensitivity setting so that 5 ng of ETU produces approximately 50% full-scale response on the integrator paper. If the HPLC system will not be used overnight, set the mobile-phase flow rate to 0.5 ml/min, turn the cell to "stby", and turn the detector off. If the system will not be used for several days, replace the mobile phase in the column and tubing with 100% acetonitrile or methanol, and store the reference electrode in 3 M sodium chloride.

6. Extraction

To a blender jar add a 100- or 50-g (Table 1) analytical portion, 15 g of sodium acetate, 20 g of Celite 545, 150 ml of water, and 200 ml of methanol. Blend the mixture for 2 min at high speed. Vacuum-filter the homogenate through a 10- to 15-g bed of Celite 545 spread evenly on two premoistened sharkskin filter papers in an 11-cm perforated Buchner funnel. Collect the filtrate in a 500-ml filter flask. Transfer a volume of filtrate equivalent to 20% of the analytical portion [(ml water in analytical portion + 150 ml water + 200 ml methanol)/5] to a preweighed 500-ml ẟ 24/40 round-bottom flask. Using a Pasteur pipet, add 5 drops of 1% Antifoam B solution to the round-bottom flask. Place a 250-ml ẟ 24/40 trap on the round-bottom flask and attach it to a vacuum rotary evaporator. Apply the vacuum slowly. After the full vacuum is applied, slowly place the flask in a 35°C water bath. Concentrate the extract to about 12 g and then add water to bring the weight to 13 g. Swirl the flask to dislodge (and dissolve, if possible) any residue adhering to the glass surface of flask. Proceed immediately to cleanup.

7. Cleanup

Add 10 g of Gas-Chrom S to a round-bottom flask containing the concentrated extract. Stopper and shake the flask. Let the flask stand for 2 to 3 min. Tap the flask on a cork ring to break up lumps. (The mixture should now be free-flowing and particles should not ahere to the wall of the flask.) Pour Gas-Chrom S into a chromatographic column containing 4.5 g alumina and settle the particles by gently tapping the side of the column. Place a 0.5-cm plug of glass wool on top of the Gas-Chrom S. Add 50 ml of partitioning eluent to the round-bottom flask. Transfer the solution to the column. Let the liquid rapidly flow through the column to remove entrained air and then adjust the flow rate to 5 ml/min. Rinse the flask with a second 50-ml portion of eluent, and add the rinse to the column after the last of the first portion has just touched the glass wool. Rinse the flask with two additional 50-ml portions of eluent, and add each rinse to the column. Collect the eluate in a 500-ml ℱ 24/40 round-bottom flask that was previously siliconized. Evaporate the eluate just to dryness, using a vacuum rotary evaporator. Immediately pipet 4.0 ml of water into the flask. Swirl the flask to dissolve the ETU residue. Pour the solution into a 10-ml glass syringe containing a Swinny filter holder with a 0.45-μm Nylon-66 filter. Push the solution through the filter with the syringe plunger and collect the filtrate in a 10-ml centrifuge tube or other suitable container. If diluted solution is needed, pipet an aliquot into another container and dilute to the desired volume.

8. Determination

Transfer the solution to be injected to autosampler vials with a borosilicate glass Pasteur pipet. Inject 20 μl of solution onto the HPLC carbon column, using the chromatographic apparatus and parameters described in Sections III.E.4 and 5. Measure the peak area and quantify the ETU residue by comparison to the peak area obtained from a known amount of ETU standard injected immediately after the purified extract. To ensure valid measurement of the ETU residue, peak areas of the purified extract and the ETU standard solution should match within ±2.5%.

9. ETU Recovery and Method Sensitivity

Average recoveries of ETU from 12 food products were 92%, with a standard deviation of 12% for low (0.05 and 0.1 ppm) fortification levels and 90% with a standard deviation of 6% for the higher (0.5 and 1 ppm) fortification levels. Raw celery was found to cause low ETU recoveries during the extraction step. The limits of quantification were 0.01 and 0.02 ppm for food products with low and high sugar content, respectively. Examples of chromatograms obtained by HPLC with an EC detector are shown in Figure 3.

F. CONFIRMATORY TECHNIQUES

The review by Bottomley et al.[58] on the determination of ETU also included a review of the literature for the confirmation of ETU residues. They reported that several authors[71–73] have developed GLC/MS techniques for the confirmation of ETU residues. These authors formed derivatives of ETU that would successfully chromatograph by GLC. Because of the very high percentage of water in the mobile phase used to chromatograph ETU by reversed-phase HPLC, confirmation of ETU by HPLC/MS would be difficult. SFC/MS may have potential as a tool in the confirmation of ETU residues.

IV. BENZIMIDAZOLECARBAMATE FUNGICIDES

A review of the benzimidazolecarbamate fungicide multiresidue methodology is presented in Chapter 7, "Multiresidue Methods for Organonitrogen Pesticides" by Ronald G. Luchtefeld.

V. CARBAMATE HERBICIDES

A literature search did not reveal a multiresidue method specifically for the carbamate herbicides, and the established multiresidue methods do not determine all of the carbamate herbicides. Chlorpropham, cycloate, diallate, propham, and triallate are recovered through the multiresidue method of Luke et al.[24,74] For cycloate, a nitrogen- or sulfur-selective GLC detector is used. For propham, a GLC nitrogen-selective detector is used. Chlorpropham, diallate, and triallate are recovered through the nonfatty food multiresidue method in the *Pesticide Analytical Manual*, Vol. 1.[75] Vernolate is partially recovered through this method. The *Pesticide Analytical Manual*, Vol. 2 contains single-residue methods for asulam[76] and butylate.[77] Asulam can be determined by colorimetry or TLC. Butylate can be determined by colorimetry or GLC.

REFERENCES

1. **Burke, J. A. and Porter, M. L.,** *J. Assoc. Off. Anal. Chem.,.* 49, 1157, 1966.
2. **Caro, J. H.,** *J. Assoc. Off. Anal. Chem.,* 54, 1113, 1971.
3. **Zweig, G. and Sherma, J., Eds.,** *Analytical Methods for Pesticides and Plant Growth Regulators,* Academic Press, San Diego.
4. **Gunther, F. A. and Davies Gunther, J., Eds.,** *Residue Reviews,* Springer-Verlag, New York.
5. **Kuhr, R. J. and Dorough, H. W.,** *Carbamate Insecticides: Chemistry, Biochemistry and Toxicology,* CRC Press, Cleveland, 1976.
6. **Schlagbauer, B. G. L. and Schlagbauer, A. W. J.,** in *The Metabolism of Carbamates,* Gunther, F. A. and Davies Gunther, J., Eds., *Residue Reviews,* Vol. 42, Springer-Verlag, New York, 1972, 1.
7. **Laskowski, D. A., Swann, R. L., McCall, P. J., and Bidlack, H. D.,** in *Soil Degradation Studies,* Gunther, F. A. and Davies Gunther, J., Eds., *Residue Reviews,* Vol. 85, Springer-Verlag, New York, 1983, 139.

8. **Quistad, G. B. and Menn, J. J.**, in *The Disposition of Pesticides in Higher Plants,* Gunther, F. A. and Davies Gunther, J., Eds., *Residue Reviews,* Vol. 85, Springer-Verlag, New York, 1983, 173.

9. **Rajagopal, B. S., Brahmaprakash, G. P., Reddy, B. R., Singh, U. D., and Sethunathan, N.**, in *Effect and Persistence of Selected Carbamate Pesticides in Soil,* Gunther, F. A. and Davies Gunther, J., Eds., *Residue Reviews,* Vol. 93, Springer-Verlag, New York, 1984, 1.

10. **Baron, R. L. and Merriam, R. L.**, in *Toxicology of Aldicarb,* Gunther, F. A. and Davies Gunther, J., Eds., *Reviews of Environmental Contamination and Toxicology,* Vol. 105, Springer-Verlag, New York, 1988, 1.

11. **Paris, D. F. and Lewis, D. L.**, in *Chemical and Microbial Degradation of Ten Selected Pesticides in Aquatic Systems,* Gunther, F. A. and Davies Gunther, J., Eds., *Residue Reviews,* Vol. 45, Springer-Verlag, New York, 1973, 95.

12. **Mount, M. E. and Oehme, F. W.**, in *Carbaryl: A Literature Review,* Gunther, F. A. and Davies Gunther, J., Eds., *Residue Reviews,* Vol. 80, Springer-Verlag, New York, 1981, 1.

13. **Burke, J. A.**, in *Development of the Food and Drug Administration's Method of Analysis for Multiple Residues of Organochlorine Pesticides in Foods and Feeds,* Gunther, F. A. and Davies Gunther, J., Eds., *Residue Reviews,* Vol. 34, Springer-Verlag, New York, 1971, 1.

14. **Williams, I. H.**, in *Carbamate Insecticide Residues in Plant Material: Determination by Gas Chromatography,* Gunther, F. A. and Davies Gunther, J., Eds., *Residue Reviews,* Vol. 38, Springer-Verlag, New York, 1971, 1.

15. **Dorough, H. W. and Thorstenson, J. H.**, *J. Chromatogr. Sci.,* 13, 212, 1975.

16. **Magallona, E. D.**, in *Gas Chromatographic Determination of Residues of Insecticidal Carbamates,* Gunther, F. A. and Davies Gunther, J., Eds., *Residue Reviews,* Vol. 56, Springer-Verlag, New York, 1975, 1.

17. **Lawrence, J. F.**, in *High-Performance Liquid Chromatography of Pesticides,* Zweig, G. and Sherma, J., Eds., *Analytical Methods for Peticides and Plant Growth Regulators,* Vol. 12, Academic Press, New York, 1982, 31.

18. **Moye, H. A. and Miles, C. J.**, in *Aldicarb Contamination of Groundwater,* Gunther, F. A. and Davies Gunther, J., Eds., *Reviews of Environmental Contamination and Toxicology,* Vol. 105, Springer-Verlag, New York, 1988, 99.

19. **Moye, H. A., Scherer, S. J., and St. John, P. A.**, *Anal. Lett.,* 10, 1049, 1977.

20. **Krause, R. T.**, *J. Assoc. Off. Anal. Chem.,* 68, 726, 1985.

21. **Krause, R. T.**, *J. Assoc. Off. Anal. Chem.,* 63, 1113, 1980.

22. **Krause, R. T.**, *J. Assoc. Off. Anal. Chem.,* 68, 734, 1985.

23. **Pardue, J. R.**, FDA Laboratory Information Bulletin, No. 3138, Food and Drug Administration, Rockville, MD, 1987.

24. **Luke, M. A., Froberg, J. E., Doose, G. M., and Masumoto, H. T.**, *J. Assoc. Off. Anal. Chem.,* 64, 1187, 1981.

25. **Vannelli, J. J.**, FDA Laboratory Information Bulletin, No. 2819, Food and Drug Administration, Rockville, MD, 1984.

26. **McGarvey, B. D.**, *J. Chromatogr.,* 481, 445, 1989.

27. **Mendoza, C. E.**, in *Analysis of Pesticides by the Thin-Layer Chromatographic-Enzyme Inhibition Technique,* Gunther, F. A. and Davies Gunther, J., Eds., *Residue Reviews,* Vol. 43, Springer-Verlag, New York, 1972, 105.

28. **Mendoza, C. E.**, in *Analysis of Pesticides by the Thin-Layer Chromatographic-Enzyme Inhibition Technique, Part II,* Gunther, F. A. and Davies Gunther, J., Eds., *Residue Reviews,* Vol. 50, Springer-Verlag, New York, 1974, 43.

29. **Sherma, J.**, in *Quantitative Thin-Layer Chromatography (TLC),* Zweig, G. and Sherma, J., Eds., *Analytical Methods for Pesticides and Plant Growth Regulators,* Vol. 11, Academic Press, New York, 1980, 79.

30. **Appaiah, K. M., Nag, U. C., Puranaik, J., Nagaraja, K. V., and Kapur, O. P.,** *Indian Food Packer,* 38, 28, 1984.
31. **Wright, L. H.,** *J. Chromatogr. Sci.,* 20, 1, 1982.
32. **Voyksner, R. D. and Bursey, J. T.,** *Anal. Chem.,* 56, 1582, 1984.
33. **Berry, A. J., Games, D. E., and Perkins, J. R.,** *J. Chromatogr.,* 363, 147, 1986.
34. **Kalinoski, H. T., Wright, B. W., and Smith, R. D.,** *Biomed. Mass Spectrom.,* 13, 33, 1986.
35. **Berry, A. J., Games, D. E., Mylchreest, I. C., Perkins, J. R., and Pleasance, S.,** *Biomed. Mass Spectrom.,* 15, 105, 1988.
36. **France, J. E. and Voorhees, K. J.,** *HRC & CC, J. High Resolut. Chromatogr., Chromatogr. Commun.,* 12, 753, 1989.
37. **Johnson, D. P.,** *J. Assoc. Off. Anal. Chem.,* 47, 283, 1964.
38. **Holden, E. R.,** *J. Assoc. Off. Anal. Chem.,* 58, 562, 1975.
39. **Kissinger, P. T., Bratin, S., King, W. P., and Rice, J. R.,** in *Electrochemical Detection of Picomole Amounts of Oxidizable and Reducible Residues Separated by Liquid Chromatography,* Harvey, J., Jr. and Zweig, G., Eds., *Pesticide Analytical Methodology, ACS Symposium Series No. 136,* American Chemical Society, Washington, D.C., 1980, 5.
40. **Olek, M., Blanchard, F., and Sudraud, G.,** *J. Chromatogr.,* 325, 239, 1985.
41. **Majors, R. E.,** in *Practical Operation of Bonded-Phase Columns in High-Performance Liquid Chromatography,* Horvath, C., Ed., *High-Performance Liquid Chromatography,* Academic Press, New York, 1980, 76.
42. **Kissinger, P. T., Bratin, K., Davis, G. C., and Lawrence, A. P.,** *J. Chromatogr. Sci.,* 17, 137, 1979.
43. **Anderson, J. I. and Chesney, D. J.,** *Anal. Chem.,* 52, 2156, 1980.
44. **Krause, R. T.,** *J. Chromatogr.,* 442, 333, 1988.
45. **Vettorazzi, G.,** in *State of the Art of the Toxicological Evaluation Carried Out by the Joint FAO/WHO Expert Committee on Pesticide Residues. III. Miscellaneous Pesticides Used in Agriculture and Public Health,* Gunther, F. A. and Davies Gunther, J., Eds., *Residue Reviews,* Vol. 66, Springer-Verlag, New York, 1977, 137.
46. **Spynu, E. I.,** in *Predicting Pesticide Residues to Reduce Crop Contamination,* Ware, G. W., Ed., *Reviews of Environmental Contamination and Toxicology,* Vol. 109, Springer-Verlag, New York, 1989, 89.
47. **Lentza-Rizos, C.,** in *Ethylenethiourea (ETU) in Relation to Use of Ethylenebisdithiocarbamate (EBDC) Fungicides,* Ware, G. W., Ed., *Reviews of Environmental Contamination and Toxicology,* Vol. 115, Springer-Verlag, New York, 1990, 1.
48. **Newsome, W. H.,** in *Ethylenebisdithiocarbamates and Their Degradation,* Zweig, G. and Sherma, A., Eds., *Analytical Methods for Pesticides and Plant Growth Regulators,* Vol. 11, Academic Press, New York, 1980, 197.
49. **Gustafsson, K. H. and Thompson, R. A.,** *J. Agric. Food Chem.,* 29, 729, 1981.
50. **Keppel, G. E.,** *J. Assoc. Off. Anal. Chem.,* 52, 162, 1969.
51. **Keppel, G. E.,** *J. Assoc. Off. Anal. Chem.,* 54, 528, 1971.
52. **McLeod, H. A. and McCully, K. A.,** *J. Assoc. Off. Anal. Chem.,* 52, 1226, 1969.
53. Report by the Panel on Determination of Dithiocarbamate Residues, *Analyst (London),* 106, 782, 1981.
54. **Gustafsson, K. H. and Fahlgren, C. H.,** *J. Agric. Food Chem.,* 31, 461, 1983.
55. **Marshall, W. D.,** *J. Agric. Food Chem.,* 27, 295, 1979.
56. **Ross, R. D. and Crosby, D. G.,** *J. Agric. Food Chem.,* 21, 335, 1973.
57. **Cruickshank, P. A. and Jarrow, H. C.,** *J. Agric. Food Chem.,* 21, 333, 1973.
58. **Bottomley, P., Hoodless, R. A., and Smart, N. A.,** in *Review of Methods for the Determination of Ethylenethiourea (Imidazolidine-2-thione) Residues,* Gunther, F. A. and Davies Gunther, J., Eds., *Residue Reviews,* Vol. 95, Springer-Verlag, New York, 1985, 45.

59. **Williams, E., Ed.**, *Official Methods of Analysis, Sections 29.119–29.125,* 14th ed., Association of Official Analytical Chemists, Arlington, VA, 1984, 554.

60. **Prince, J. L.**, *J. Agric. Food Chem.,* 33, 93, 1985.

61. **Dogan, S., Corvi, C., and Vogel, J.**, *Chimia,* 39, 110, 1985.

62. **Krause, R. T. and Wang, Y.**, *J. Liq. Chromatogr.,* 11, 339, 1988.

63. **Wang, H., Pacakova, V., and Stulik, K.**, *J. Chromatogr.,* 457, 398, 1988.

64. **Onley, J. H., Guiffrida, L., Ives, N. F., Watts, R. R., and Storherr, R. W.**, *J. Assoc. Off. Anal. Chem.,* 60, 1105, 1977.

65. **Farrington, D. S.**, *Meded. Fac. Landbouwwet. Rijksuniv. Gent.,* 44, 901, 1979.

66. **Hanekamp, H. B., Bos, P., and Frei, R. W.**, *J. Chromatogr.,* 186, 489, 1979.

67. **Kobayashi, H., Matano, O., and Goto, J.**, *J. Pest. Sci. (Nippon Noyaku Gakkaishi),* 11, 81, 1986.

68. **Krause, R. T.**, *J. Liq. Chromatogr.,* 12, 1635, 1989.

69. **Allison, L. A. and Shoup, R. E.**, *Anal. Chem.,* 55, 8, 1983.

70. **Krause, R. T.**, *J. Assoc. Off. Anal. Chem.,* 72, 975, 1989.

71. **Autio, K.**, *Finn. Chem. Lett.,* 1–2, 10, 1983.

72. **Nitz, S., Moza, P. N., and Korte, F.**, *J. Agric. Food Chem.,* 30, 593, 1982.

73. **Uno, M., Ueda, E., Okada, R., and Onji, Y.**, *Shokuhin Eiseigaku Zasshi.,* 189, 53, 1977.

74. **McMahon, B. M. and Hardin, N. F., Eds.**, *Pesticide Analytical Manual,* Vol. 1, Food and Drug Administration, Washington, D.C., sec. 232.41.

75. **McMahon, B. M. and Hardin, N. F., Eds.**, *Pesticide Analytical Manual,* Vol. 1, Food and Drug Administration, Washington, D.C., sec. 212.

76. **Marcotte, A. L. and Bradley, M., Eds.**, *Pesticide Analytical Manual,* Vol. 2, Food and Drug Administration, Washington, D.C., sec. 180.360.

77. **Marcotte, A. L. and Bradley, M., Eds.**, *Pesticide Analytical Manual,* Vol. 2, Food and Drug Administration, Washington, D.C., sec. 180.232.

Chapter 6

RESIDUE ANALYSIS OF ORGANOPHOSPHORUS PESTICIDES

Damià Barceló and James F. Lawrence

TABLE OF CONTENTS

I. INTRODUCTION

Organophosphorus compounds probably represent the largest single class of substances used as pesticides at present. There are perhaps over 100 such compounds available throughout the world. They have several advantages over other types of pesticides, including high acute toxicity to target organisms, but they are not persistent in the environment as are organochlorines, and they decompose to nontoxic products. However, their acute toxicity is of concern, and stringent regulations have been implemented in most countries limiting the maximum levels of residues on food for human consumption. In addition to human toxicity, organophosphates are toxic to a wide variety of animal life. Thus, there is a general concern about their levels in the environment.

In order to ensure that these substances are not present at hazardous levels in food, water, and environmental samples, analytical methods have been developed and implemented for many years. To date, these primarily involve gas or liquid chromatography with various means of detection. The determination of low concentrations of organophosphorus pesticides in different matrices requires, in addition to highly selective and sensitive detection techniques, the application of efficient extraction and thorough cleanup procedures to enable a quantitative as well as qualitative determination to be implemented. Several review articles have been published discussing the importance of sample preparation methods in the analysis of pesticides in different matrices. Analytical methods most commonly employed are generally off-line, with either liquid-liquid extraction (LLE) or liquid-solid extraction (LSE) procedures for the isolation, cleanup, and preconcentration of organophosphorus pesticides from water[1] and other environmental samples[2] including food matrices.[3] The use of on-line precolumn techniques in liquid chromatography (LC) and LC/mass spectrometry (MS) in environmental pesticide analysis has been reported in a recent review article.[4] In this chapter, the different approaches for sample handling and analysis of organophosphorus pesticides in environmental and food matrices will be described, with special emphasis on the newest analytical techniques available.

II. EXTRACTION METHODS

A number of studies have evaluated many different off-line extraction techniques for organic environmental contaminants, including organophosphorus pesticides, and these have been reviewed in detail.[1-6] The type of extraction chosen depends upon the matrix analyzed. The following discusses those methods employed for liquid (water) and solid (soils, sediments, biological) samples.

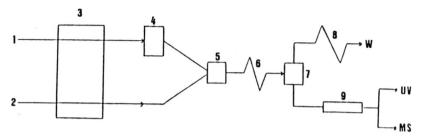

FIGURE 1. Schematic diagram of the manifold of a completely continuous flow analysis system combined with LC-UV and LC/MS: (1) water stream used as displacing solvent, (2) aqueous sample stream, (3) peristaltic pump, (4) displacement flask containing *n*-heptane, (5) segmenter, (6) extraction coil, (7) phase separator, (8) restrictor, and (9) LC system, (W) waste. (Reprinted from Farran, A., Cortina, J. L., De Pablo, J., and Barceló, D., *Anal. Chem. Acta*, 234, 119, 1990. With permission.)

A. WATER SAMPLES

The identification and quantification of pesticides in matrices such as drinking and surface water are required not only for health reasons but for measuring environmental waste levels as well. Various preconcentration methods based on different physicochemical principles have been used for these purposes. Among them, LLE, dynamic and static headspace analysis, solid-phase extraction, and membrane processes are quite commonly used and have been recently reviewed.[1]

A variety of extracting solvents have been used for LLE. Thus, concentration of organophosphorus pesticides from water samples can be accomplished by using organic solvents such as *n*-hexane,[7] dichloromethane with[8] or without[9] prior acidification to avoid hydrolysis of the pesticides, and chloroform.[10] The use of these and other extracting solvents, such as ethyl acetate and acetonitrile, has been discussed in detail in a monograph on the analysis of pesticides in water.[11]

In SPE, the water sample is passed through a short bed of packing material, which may contain functional groups of different polarity such as C-8- or C-18-bonded phase, graphitized carbon black, or Amberlite XAD resins. C-8- and C-18-bonded phase cartridges[12–14] have been used for the analysis of various pesticides, including organophosphorus compounds in sea and surface waters. Amberlite XAD resins have been widely used in off-line analyses for the extraction of organophosphorus pesticides from water samples at the parts per billion level.[15]

Other methods for the removal of organophosphorus pesticides from water samples include on-line systems, either with preconcentration using precolumns, as has been discussed in Chapter 13 of this book or with on-line continuous flow-extraction.[16] Figure 1 shows a continuous flow-extraction manifold for the on-line extraction of organophosphates. Standard Tygon

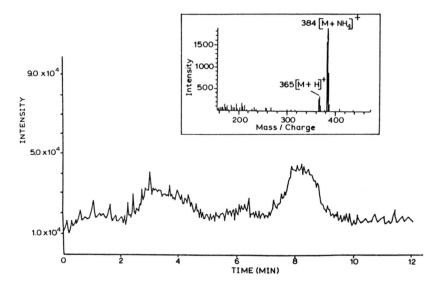

FIGURE 2. Reconstructed ion chromatogram using the system described in Figure 1 with LC/MS of a river water spiked with 1 ppm of tetrachlorvinphos. Extraction coil, 4 m; total flow rate, 5.6 ml/min; aqueous/organic volume ratio, 6; liquid chromatographic eluent, methanol-water (70:30) + 0.1 *m* ammonium formate at 1 ml/min. (Reprinted from Farran, A., Cortina, J. L., De Pablo, J., and Barceló, D., *Anal. Chem. Acta*, 234, 119, 1990. With permission.)

pump tubing is used for the sample (aqueous) phase, and the organic solvent (*n*-heptane) is delivered by means of a displacement flask with water used as the displacement liquid. Polytetrafluoro-ethylene (PTFE) tubing (0.5 mm I.D.) was used throughout the system. The sample is introduced by means of a peristaltic pump and merges with *n*-heptane in a Y-piece segmenter. The extraction takes place in the coil. The two phases are separated with the aid of a 0.5 μm PTFE membrane placed in a phase separator. Phase separation is better than 90%. The aqueous phase is directed to waste through a restrictor (5 mm × 0.5 mm I.D. PTFE tube), and the organic phase passes through the sample loop of the injection value of the LC system. An extraction recovery between 60 and 70% was achieved using this system for several organophosphorus pesticides. Figure 2 shows a reconstructed ion chromatogram obtained using the setup in Figure 1 with positive ion mode-LC/MS detection of a river water sample spiked with 1 ppm of tetrachlorvinphos. Considering the preconcentration step, the total amount injected was 70 ng. This injection was made under full-scan MS conditions. If selected ion monitoring were employed, the detection limit could be reduced 100 to 1000 times to the picogram level, which makes the method suitable for on-line organic trace analysis of water samples.

B. SOIL, SEDIMENT, AND BIOLOGICAL SAMPLES

Freeze-drying is quite a common method for conditioning soil, sediment, and biological samples for extraction of pesticides.[17] Freeze-drying of biological material destroys cell membranes and, hence, increases contact between the sample matrix and the extraction solvent. Despite its tediousness, Soxhlet extraction allows efficient removal of pesticides from complex environmental samples. Efficiency depends, among other things, on suitable choice of solvent for each group of pesticides. The matrix is also of great importance in this respect. Some pesticides (e.g., triazines) in sediment or soil samples can be retained by complex physical and chemical adsorption mechanisms that hinder extraction of the residues.[17,18] Other serious problems encountered in the analysis of organic compounds in sediments are caused by elemental sulfur, often present in such materials, and by the varying amounts of lipids occurring in many types of biological samples.

A wide variety of solvents are currently used for extraction purposes. For example, acetone, acetone-water, dichloromethane, ethyl acetate,[11] 4:1 and 6:4 acetone-hexane[7,19,20] (also suitable for pyrethroids), 1:1 acetone-dichloromethane[21] (also employed for carbamates), and 9:1 methanol-water have often been used for extraction of organophosphorus pesticides from soil matrices.[17] Extraction of the pesticide residues from biological samples containing substantial amounts of lipids is facilitated by strong acids and alkalis. However, some pesticides, including organophosphorus compounds, may decompose as a result of such harsh treatment. Thus, organic solvents have been employed most frequently to remove the pesticides from solid matrix components. Ethyl acetate is one of the most commonly used solvents for the extraction of organophosphorus pesticides from biological materials.[9,11,22,23] Acetonitrile,[24] *n*-hexane,[5] acetone, and methanol[11] have also been used for this purpose. For food crops, combinations such as acetonitrile-chloroform,[10] acetone-methanol,[25] acetone-water,[26] and acetonitrile-water,[27] as well as acetone alone,[28,29] have been evaluated.

III. SAMPLE CLEANUP METHODS

Sample cleanup is almost always required to some degree before the actual quantitative determination of residues of organophosphorus pesticides, even with the use of selective detection systems such as gas chromatography with flame photometric or nitrogen-phosphorus thermionic detectors. Classical cleanup techniques have involved liquid-liquid partitions between immiscible solvents followed by adsorption chromatography on silica gel or Florisil-type packing materials.[1-3,19,30,31] The U.S. Environmental Protection Agency (EPA) cleanup methods[31] employ Florisil column chromatography with elution of the pesticides using mixtures of ethyl ether in *n*-hexane. A similar procedure has been used for organophosphates and triazines in soil samples, in which the compounds of interest were eluted with a 1:1 mixture of ethyl ether and *n*-hexane. Disposable solid-phase extraction cartridges containing a wide variety of sorbents are now commercially available. Florisil and C-18 bonded-phase have been evaluated for organophosphate pesticide analysis in rice[32] and human urine and plasma.[33] Gel permeation chromatography (GPC) (or size exclusion chromatography) has been found to be effective in removing high molecular weight coextractives, particularly lipid material from sample extracts. Basically, large biogenic compounds, such as lipids, are excluded from the pores of the polymeric packing material and eluted before smaller analytes, which are retained in the pores. Separation mechanisms other than size exclusion, that is, adsorption and partition, may also be involved. The prevalence of one type of mechanism over the others is largely determined by the mobile phase and packing pore size chosen. With large pore GPC packings (1000 to 2000 molecular weight exclusion) such as Bio-Beads SX-3, SX-4, and SX-8, both size exclusion and adsorption occur in the presence of poorly solvating mobile phases. This is the case with Bio-Beads SX-3 (2000 molecular weight exclusion limit), the GPC packing most frequently used to clean up extracts of pesticides in conjunction with a variety of eluting solvents, such as ethyl acetate-cyclohexane,[9,17,34,35] acetone-cyclohexane, acetone-petroleum ether,[36] cyclohexane-dichloromethane,[37,38] and dichloromethane-*n*-hexane.[39] A wide range of pesticides, including organophosphorus compounds, phenoxy acids, and nitrogen-containing pesticides, have been successfully isolated from a variety of environmental matrices. The approach has been particularly successful for the cleanup of food extracts.[37,39] Table 1 shows the recovery values for 14 organophosphorus pesticides spiked in a vegetable oil and lettuce at levels of 0.1 to 0.25 ppm using either dichloromethane-cyclohexane (1:1) or dichloromethane-acetone (7:3) for elution. With the exception of ethion and fenthion in the vegetable oil, all recoveries were very good. It is possible that the elution patterns for ethion and fenthion are different and that they could be recovered in greater yield with a change in solvent composition or elution volume.

Although GPC is one of the most common cleanup techniques for isolation of organophosphorus pesticides from sediment and biological samples, in some instances further cleanup may need to be done. This has been the case when analyzing biological samples with high lipid content.[22] An example is shown in Figure 3. The GC/nitrogen-phosphorus thermionic detector (NPD) chromatogram corresponding to the GPC fractionation of an extract of the bivalve *Tapes semidecussatus* (A) shows a group of interfering peaks, probably phospholipid derivatives associated with the matrix. These were eliminated using the combination of GPC and Florisil column chromatography (B). Thus, a two-step cleanup procedure may be required with certain types of samples, particularly those with high lipid content.

TABLE 1
Organophosphates Evaluated for GPC Cleanup

Compound	Vegetable oil[a]		Lettuce[b]	
	Level added (ppm)	Recovery (%)	Level added (ppm)	Recovery (%)
Diazinon	0.25	82	0.10	96
Parathion	0.25	95	0.10	88
Ethion	0.25	68	0.10	91
Fenthion	0.10	46	0.10	109
Ronnel	0.10	107	—	—
Malathion	0.15	100	—	—
Chlorpyrifos	—	—	0.10	110
Carbophenothion	—	—	0.10	99
Dimethoate	—	—	0.10	79
Dimethoxon	—	—	0.10	94
Fonofos	—	—	0.10	103
Methamidophos	—	—	0.10	81
Fensulfothion	—	—	0.10	104
Phosphamidon	—	—	0.10	106

[a]   Vegetable oil: $MeCl_2$/cyclohexane (1 + 1).

[b]   Lettuce: $MeCl_2$/acetone (7 + 3).

Reproduced from Lawrence, J. F., *Int. J. Environ. Anal. Chem.*, 29, 289, 1987. With permission.

A less common approach to performing cleanup on samples from environmental sources to be analyzed for pesticides involves the use of normal phase LC[40,41] with scaled-up silica columns and dichloromethane-*n*-hexane or dichloromethane-*n*-pentane as eluents. Matrix interferences, usually lipids in the case of biological material, appear at the end of the chromatogram, and the pesticides are eluted earlier. This approach has been successfully applied to the isolation of organochlorine and organophosphorus pesticides from a variety of biological samples containing lipids. However, a major drawback is that the more polar organophosphorus pesticides coelute with the fat peak and are thus poorly resolved from the lipid matrix.[40] An alternative is the use of an octadecylsilyl (ODS)-bonded reversed-phase LC system where the long-chain lipids are highly retained by the LC column.[42] Several organophosphorus pesticides were tested, e.g., acephate, metamidophos, parathion-ethyl, chlorpyrifos, and ronnel among others; and the order of elution of analytes was generally the opposite of that found for the silica column, as would be expected, with the highly polar compounds eluting first and the nonpolar compounds last. In this case, the separation of the analytes from the lipid material was much better than by normal phase. After the cleanup

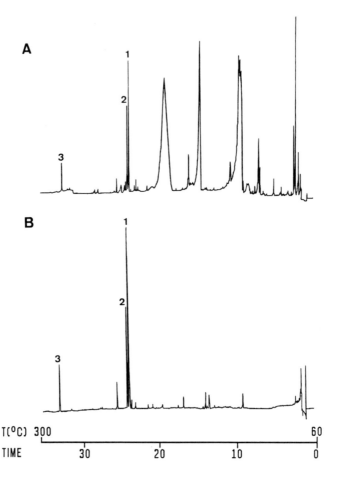

FIGURE 3. GC-NPD chromatograms of a bivalve sample of *Tapes semidecussatus* after gel permeation chromatographic cleanup (A) and after gel permeation and Florisil column chromatography (B). Compounds identified correspond to: (1) fenitrothion, (2) malathion, and (3) phosmet. (Reprinted from Barceló, D., Solé, M., Durand, G., and Albaigés, J., *Fresenius J. Anal. Chem.*, 339, 676, 1991. With permission.)

is completed, the retained lipids are stripped from both the ODS-bonded column and the silica column by reverse flushing with dichloromethane, in preparation for the cleanup of subsequent samples.

Solid-phase partition chromatography is becoming a popular means of carrying out liquid-liquid partitions without the problems associated with the classical approach of mixing immiscible liquids in a separatory funnel. The principle normally involves adding a certain volume of sample extract to a column containing an adsorbent such as macroporous kieselguhr. The sample extract adsorbs completely onto the packing material. The pesticide can then be eluted from the column with a variety of solvents. The technique was successfully applied to organophosphate residues in vegetable extracts,[43] lipid material,[44] and total diet samples.[45]

Sweep codistillation is a method that has not found widespread use for sample cleanup. However, it does offer a simple means of removing lipid material from biological samples and has been applied to the cleanup of organophosphorus pesticide residues in animal fat.[46-48] The procedure is actually a crude form of GC and involves the injection of fat samples into a heated glass column filled with glass beads. The fat melts and coats the beads, while the pesticides are flushed from the tube by continuous passage of nitrogen gas. A Florisil trap at the outlet of the tube collects the eluted pesticides, which are subsequently removed by solvent elution for chromatographic analysis.

IV. ANALYSIS

A. GAS CHROMATOGRAPHIC METHODS

Capillary gas chromatography (GC) in conjunction with selective detectors — mainly nitrogen-phosphorus thermionic (NPD), electron capture (ECD), flame photometric (FPD), and mass spectrometry (MS) — is the most common technique for the determination of environmental pesticide residues. The low detection limits, high selectivity, and affordability of GC instrumentation is rather appealing to most laboratories involved in pesticide residue analysis. Several articles on the use of GC/NPD and GC/ECD[49,50] and GC/MS in various operational modes such as electron impact (EI), positive chemical ionization (PCI), and negative chemical ionization (NCI) have been published.[51-53]

GC-NPD analyses can be performed in the phosphorus or the nitrogen mode. The phosphorus mode, which is roughly one order of magnitude more sensitive than the nitrogen mode, allows organophosphorus pesticides to be routinely determined at concentrations of a few nanograms per gram in different environmental matrices.[3,9,17,19,21,22,31,35,37,49] Other detectors used for organophosphorus pesticide GC analysis include the FPD[23,37] and alkali flame ionization detectors (AFID).[23] In a comparison of the FPD and NPD detectors on the same extracts, it was found that the FPD was more selective than the NPD for residues of methamidophos, acephate, ethion, and ethion oxon in green peppers or oranges.[37]

GC/MS is widely used at present by environmental laboratories involved in pesticide residue analysis for confirmation of pesticide identity.[14,51-53] The most common practice in this context is to perform GC/MS in the electron impact mode with a library search for the unequivocal identification of the pesticide or with a second injection to check for coelution with an authentic standard of the pesticide of interest. The use of GC/MS with NCI is a selective approach, particularly suitable for pesticides containing electron withdrawing groups (e.g., chlorine, nitro group), which can stabilize negative charges. The main advantages of NCI are its high selectivity and sensitivity to organochlorine and organophosphorus pesticides. GC/NCI/MS has been used to confirm the presence of organophosphorus compounds.[7,51,53-55] GC/MS employing PCI and NCI with selected ion monitoring (SIM) of two to three characteristic fragment ions of each analyte permitted the unequivocal identification of organophosphorus pesticide residues in a variety of environmental matrices.[55] Selectivity and sensitivity can be further enhanced by using various ionization modes. However, only a few libraries of standard mass spectra are available for PCI and NCI GC/MS, so each laboratory must create their own. This is one of the chief drawbacks of this technique and arises from the fact that instrumental parameters such as the source temperature and reagent gas pressure have a critical influence on the relative ion intensities of mass spectra obtained under chemical ionization conditions.

Figure 4 illustrates spectra obtained by GC/MS with EI, PCI, and NCI for fenitrothion.[53] Typical fragments indicating the functional group structure of fenitrothion obtained under GC/MS with EI corresponds to m/z 109 $[(CH_3O)_2PO]^+$ and to m/z 125 $[(CH_3O)_2PS]^+$. Other main ions for organophosphorus pesticides correspond to losses of (OH) at m/z 260. The molecular weight fragment exhibited a relative abundance of 100%. With PCI, [M + H]$^+$ at m/z 278 represents only the base peak with no other fragments, whereas with NCI, the formation of $[M]^-$ at m/z 277 ion as the base peak is a consequence of its aromatic structure, which is easily stabilized under negative ion conditions by the nitro group.[54] Another anion corresponded to the thiophenolate ion $[SC_7H_6NO_2]^-$ at m/z 168, which was due to the transfer of the aromatic moiety from the oxygen to the sulfur atom. This is a common occurrence for phosphorothionate pesticides containing an aromatic moiety with electronegative groups. All these spectra correspond to 10 ng injected under full-scan conditions. The best sensitivity was obtained under NCI conditions, with a detection limit of 10 pg with selected ion monitoring, whereas in EI the limit of detection was 500 pg under the same conditions. The use of PCI is not recommended, because (as can be observed in Figure 4) the sensitivity was 3 orders of magnitude less than NCI and 20 times less than EI. Consequently, for the characterization of organophosphorus pesticides, the NCI mode of operation is recommended over EI owing to its higher selectivity and sensitivity. In addition, NCI offers enough structural information, with two or three diagnostic ions that can be easily used for monitoring pesticides in environmental matrices.

B. LIQUID CHROMATOGRAPHIC METHODS

Systems of LC used for environmental pesticide analyses have been extensively reviewed.[4,56] The increasing use of LC for pesticides is chiefly the

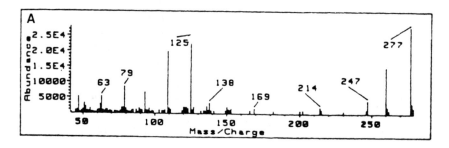

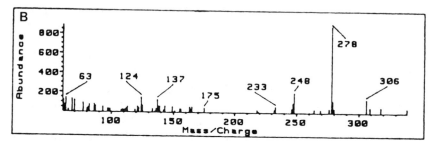

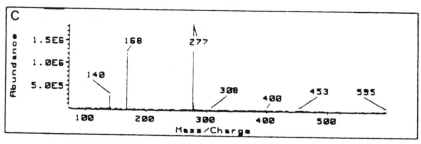

FIGURE 4. GC/MS of fenitrothion with (A) electron impact, (B) positive chemical ionization, and (C) negative chemical ionization. Amount injected: 5 to 10 ng. (Reprinted from Durand, G. and Barceló, D., *Anal. Chim. Acta,* 243, 259, 1991. With permission.)

result of its suitability for thermally labile and polar pesticides that require derivatization prior to GC analysis. LC methods of analysis also have a major advantage over those of GC in that on-line precolumn and postcolumn reaction systems are compatible with LC. In addition, MS has also been successfully coupled to LC instruments.

UV detectors are the most common choice for detection of organophosphates by LC. The wavelength is set according to the pesticide to be analyzed. The use of a UV detector in LC in conjunction with off-line sample preparation is still the most common choice in environmental pesticide analysis. Information on specific wavelengths, LC eluents, and columns for over 200 pesticides is available in the literature. UV detection has been used to analyze organophosphorus compounds at wavelengths such as 205, 210, 254, and 280 nm.[17,57,58] By using LLE, 15 organophosphorus pesticides were extracted from water samples and analyzed by LC-UV with a limit of quantification of 0.5 ppb.[58]

The on-line combination of liquid chromatography and mass spectrometry (LC/MS) plays an important role in environmental pesticide analysis, since most of the compounds can be successfully detected. This technique, when compared with GC/MS, offers major advantages for analyzing thermally labile, low volatility, and polar organophosphorus pesticides.[4,16,59–61] The advantage of LC/MS is enhanced by the fact that not all organophosphorus pesticides can be satisfactorily detected by UV because of a lack of a chromophore, or inadequate sensitivity or selectivity.[4]

Three main approaches have been used for interfacing LC with MS for the characterization of organophosphorus pesticides: direct liquid introduction (DLI),[59] thermospray (TSP),[60,61] and moving belt.[4,62] At present, thermospray is the most frequently used interfacing system in LC/MS for pesticide analysis. For organophosphorus pesticides, the analysis of different oxo metabolites, which are major degradation products, presents great difficulties by GC methods. Generally, there are problems in achieving good gas chromatographic resolution without peak tailing and the formation of artifacts or decomposition products. Thus, in these cases, the use of LC is particularly advantageous.[61]

For LC/MS, difficulties often arise when comparing the fragmentation obtained for a given pesticide using the different interfacing systems and when comparing such results with data obtained by GC/MS. In most cases, the combination of LC/MS provides a mechanism of ion formation similar to chemical ionization. Table 2 compares two different interfacing systems of LC/MS with GC/MS in terms of the relative abundances of the different fragment ions of several organophosphates obtained in the negative ion mode (TSP LC/MS, DLI LC/MS, and GC/NCI/MS).[61] The base peak for the different pesticides corresponds to $[M]^{\tau}$ or to the phenolate or thiophenolate anion, and in the case of bromophos-ethyl, to the adduct of acetic acid with Br^-. The abundances of $[M]^{\tau}$ varied from 80 to 100% for fenitrothion and parathion-ethyl and their oxygen analogs as well. These results, in close agreement with GC/NCI/MS, can be explained by the fact that $[M]^{\tau}$ is only found with higher intensities if the moiety contains an aromatic ring that can easily stabilize the negative charge by electron delocalization.[53,54] This has been particularly noticeable for compounds of the parathion type, such as fenitrothion, containing strong electron-withdrawing groups. For fenchlorphos and bromophos-ethyl, which also have an aromatic ring and contain chlorine or chlorine and bromine substituents, the $[M]^{\tau}$ ion is absent.

The relatively high abundance of the phenolate or thiophenolate anions of the compounds in Table 2 is also of interest. In the case of the parent pesticides, the thiophenolate anion is most abundant, although the phenolate

TABLE 2
Comparison of Relative Abundances in Negative Ion Mode Mass Spectrometry Using LC/MS with Thermospray (TSP) or Direct Liquid Introduction (DLI) Interfaces and Conventional Negative Chemical Ionization (NCI) MS

Mol wt	Compound m/z Tentative identification	TSP LC/MS	DLI LC/MS	NCI GC/MS
261	Fenitrooxon			
	125 [FG]$^-$	n.r.	n.r.	17
	152 [OC$_7$H$_6$NO$_2$]$^-$	100		
	246 [M − R]$^-$	7		
	261 [M]$^-$	80		100
	321 [M + CH$_3$COOH]$^-$	3		
277	Fenitrothion			
	141 [FG]$^-$	15	n.r.	29
	152 [OC$_7$H$_6$NO$_2$]$^-$	10		
	168 [SC$_7$H$_6$NO$_2$]$^-$	15		100
	262 [M − CH$_3$]$^-$	15		
	277 [M]$^-$	100		84
275	Paraoxon-ethyl			
	153 [FG]$^-$		100^a	100
	198 [OC$_6$H$_4$NO$_2$ + CH$_3$OOH]$^-$	50		
	246 [M − R]$^-$	10	2^a	
	275 [M]$^-$	100		100
291	Parathion-ethyl			
	154 [SC$_6$H$_4$NO$_2$]$^-$	7	100	100
	262 [M − R]$^-$	10	0.1	
	291 [M]$^-$	100		28
304	Fenchlorphos oxon			
	125 [FG]$^-$	10	n.r.	100
	197 [OC$_6$H$_2$Cl$_3$]$^-$	100		
	218 [M − CH$_3$ − HCl$_2$]$^-$			50
	254 [M − CH$_3$ − Cl]$^-$	6		27
320	Fenchlorphos			
	141 [FG]$^-$	40	100	37
	211 [SC$_6$H$_2$Cl$_3$]$^-$	100	90	100
	270 [M − CH$_3$ − Cl]$^-$	14	0.1–5	13
	305 [M − CH$_3$]$^-$		2	
376	Bromophos-ethyl oxon			
	79, 81 [Br]$^-$	n.r.	n.r.	63
	139, 141 [CH$_3$COOH + Br]$^-$	30		
	153 [FG]$^-$			50
	241 [OC$_6$H$_2$Cl$_2$Br]$^-$	100		60
	298, 301 [CH$_3$COOH + OC$_6$H$_2$Cl$_2$Br]$^-$	5		
	340 [M − HCl]$^-$			100

TABLE 2 (continued)
Comparison of Relative Abundances in Negative Ion Mode Mass
Spectrometry Using LC/MS with Thermospray (TSP) or Direct Liquid
Introduction (DLI) Interfaces and Conventional Negative Chemical
Ionization (NCI) MS

	Compound m/z	TSP	DLI	NCI
Mol wt	Tentative identification	LC/MS	LC/MS	GC/MS
392	Bromophos-ethyl			
	79 [Br]$^-$		100[b]	22
	139, 141 [CH$_3$COOH + Br]$^-$	100		
	255, 257 [SC$_6$H$_2$Cl$_2$Br]$^-$	20	25[b]	100
	284 [M − C$_2$H$_5$Br]$^-$	2	2[b]	11
	357, 359 [M − Cl]$^-$	7		59

Note: n.r. = not reported.

[a] Corresponds to paraoxon-methyl, thus $m/z = 125$ for [FG]$^-$ and $m/z = 232$ for [M − R]$^-$
[b] Corresponds to the same fragment as for bromophos-methyl. [FG] = specific group fragment of organophosphorus pesticides corresponding to (RO)$_2$P = S,O for the thionates and to (RO)$_2$P = O,O for the oxygen analog. R = methyl or ethyl.

Reproduced from Durand, G., Sanchez-Baseza, F., Messeguer, A., and Barceló, D., *Biol. Mass Spectrom.*, 20, 3, 1991. Reprinted by permission of John Wiley & Sons, Ltd.

anion can also be formed, e.g., for fenitrothion. The thiophenolate anion results from a rearrangement caused by electron capture, which is followed by fragmentation with transfer of the aromatic moiety from the oxygen to the sulfur atom. It is known that the thiophenolate anion exhibits greater electron affinity than does the phenolate anion, and, therefore, forms the more stable anion. This has been noticed in conventional GC/NCI/MS.[53,54] This anion is especially useful for the identification of the phosphorothionate pesticides vs. their oxygen analogs. It is obvious that when working with the oxygen analog from synthetic origin, where no sulfur is present in the molecule, only the phenolate anion will be obtained. This behavior can be observed in Table 2, where the results provide useful information for the unequivocal identification of the oxygen analogs of the phosphorothionates. The abundances of these anions vary from 20 up to 100%, except in the case of parathion-ethyl, which has an abundance of 7% and where the formation of [M]$^-$ is much more favored. Other fragments at lower intensity corresponded to dissociative electron-capture mechanisms such as [M-R]$^-$, [M-CH$_3$Cl]$^-$, or [M-Cl]$^-$, which are expected under NCI/MS conditions.[53,54] In the case of bromophos-ethyl and its oxygen analog, the [CH$_3$COOH + Br]$^-$ ion is formed with the dominating bromide anions (Br$^-$, m/z = 79 and 81). In this context, the high sensitivity and selectivity of the NI mode in monitoring brominated compounds provides a means of detection with minimal interference from other halogenated compounds.

Residue analysis of malathion in red mullet by TSP LC/MS is shown in Figure 5.[60] The positive and negative ion TSP LC/MS chromatograms of a sample containing 0.1 ppm of malathion are presented. It is clear that TSP LC/MS in the positive ion mode is more sensitive and selective than in the negative ion mode. The sensitivity difference can be observed from the noise level, which is at least 30 times higher in the negative ion mode than in the positive ion mode. For selectivity, the negative ion mode will likely encounter problems with analytes with short retention times. In both cases, the most important ions monitored correspond to [M + NH$_4$]$^+$ in the positive ion mode and to [M]$^-$ in the negative ion mode. Considering the amount injected (5 ng of malathion) together with the signal, the positive ion mode exhibits minimal susceptibility to matrix effects and would be very suitable for environmental analysis.

C. OTHER METHODS

Novel methods of analysis for organophosphorus pesticides continue to be developed. Multidimensional chromatographic techniques, for example, using on-line coupling of LC to GC, have become important during the past few years. With these systems, the pesticides are isolated from their matrices by LC and then transferred to a gas chromatograph for separation and determination. A recent review article[63] has described the potential of this technique. Application of the approach to the determination of chlorpyrifos in rat feed was accomplished by using a normal-phase system of LC with an eluent containing heptane-methyl-*t*-butyl ether (95:5) followed by on-line transfer of the analyte to a capillary GC system equipped with an ECD. Detection limits of 20 ppb were obtained.[63]

Radio-column chromatography, by definition, involves the separation of radioelements or radiolabeled compounds by ion exchange, size exclusion, partition, or adsorption LC. Using this technique, ^{14}C-labeled parathion was separated by reversed-phase LC. The detection took place after postcolumn extraction and segmentation of the eluate with a water-miscible liquid scintillator. During the separation, the segmented stream is stored in a capillary storage loop. After the separation, the content of the loop is introduced into the β-detector at varying flow rates for counting.[64]

Supercritical fluid chromatography (SFC) is becoming a technique useful for analyzing thermally labile and/or polar pesticides.[4] For the analysis of organophosphorus pesticides, it has the additional advantage in that the SFC system can be equipped with selective detectors, e.g., the NPD. The most powerful coupling involving SFC is SFC/MS. This novel approach has been

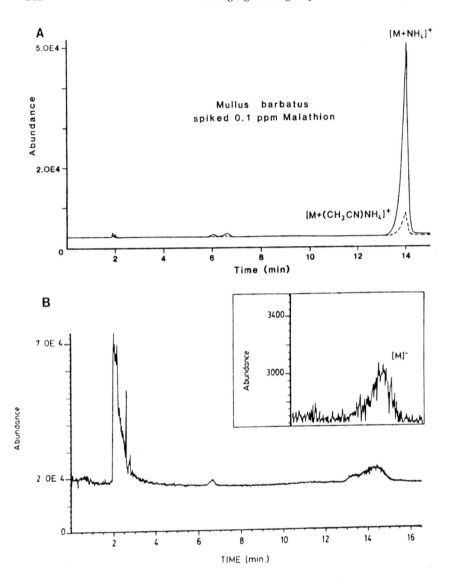

FIGURE 5. Thermospray LC/MS chromatogram using SIM in positive (A) and negative (B) ion modes of a fish sample spiked with 0.1 ppm of malathion. Injected amount: 5 ng. Eluent composition: acetonitrile-0.1 *M* ammonium acetate programmed from 50 to 80% acetonitrile in 15 min at 1 ml/min. Thermospray temperatures: stem, 105 to 114°C; tip, 178°C; and vapor and ion source, 194°C. (Reprinted from Barceló, D., *Biomed. Environ. Mass Spectrom.*, 17, 363, 1988. Reprinted by permission of John Wiley & Sons.)

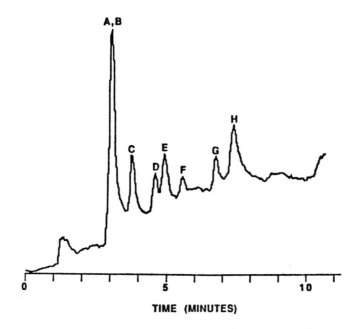

FIGURE 6.   Total ion chromatogram of the SFC/MS separation of an extract of cherries spiked with 20 to 30 ppm each of eight organophosphorus pesticides: chlorpyrifos-methyl (A), chlorpyrifos (B), iodofenphos (C), leptophos (D), metidathion (E), tetrachlorvinphos (F), phosmet (G), and famphur (H). An amino-phase microbore column and a CO_2/2-propanol supercritical mobile phase was utilized. Chemical ionization with 2-propanol was used for mass spectrometric detection. (Reprinted with permission from Kalinoski, H. T. and Smith, R. D., *Anal. Chem.*, 60, 529, 1988. Copyright 1988 American Chemical Society.)

applied to the determination of organophosphorus pesticides.[4,65] Figure 6 shows an SFC/MS analysis of an extract of cherries spiked with 20 to 30 ppm of 8 organophosphorus insecticides. The separation was achieved by using a mixture of CO_2 and 2-propanol, and the MS characterization was achieved using 2-propanol as the chemical ionization gas.

SFC has also been evaluated for desorption of pesticides, including parathion, from the porous polymer Tenax-GC for subsequent determination by GC or other chromatographic procedures.[66] SFC was found to be much superior to the commonly used thermal desorption for parathion in terms of recovery and elution of volume.

Thin-layer chromatography (TLC) continues to be used for organophosphorus pesticide analysis. It is mainly used either as a rapid screening technique or for confirmation of pesticide identity.[67-70] TLC in combination with enzyme inhibition spray reagents offers selectivity along with sensitivity (low nanogram quantities are detectable).[67] Because of their simplicity and low cost, such methods will continue to find a place in residue laboratories in countries where more sophisticated and expensive equipment is not available.

Inhibition of enzyme activity, particularly cholinesterase, has also been exploited as a means of organophosphorus pesticide determination. A field test kit has been developed for foliar pesticide residues.[71] Although the technique is only semiquantitative, it offers a rapid means of screening. The method involves oxidation of the pesticides to their respective oxons, then mixing with a color reagent and acetylcholinesterase. The degree of inhibition by the pesticide was related to the lack of color development and used to estimate the amount of pesticide present.

A potentiometric method for determination of organophosphorus pesticides in water samples was developed employing an immobilized cholinesterase electrode.[72] The electrode is very selective for cholinesterase inhibiting pesticides and thus may prove useful for monitoring levels in agricultural or industrial run-off waters, although no practical applications were presented.

D. CONFIRMATION TECHNIQUES

Confirmation of analytical results is often an important part of determination of pesticide residues in environmental samples. It is necessary to minimize misidentification of, for example, peaks eluting from a GC or an LC system. This is particularly important with organophosphorus pesticides since there are many such pesticides in use, and along with the many possible metabolites, errors in identification are a real possibility even with selective detectors such as the FPD or NPD. Mass spectrometry associated with either GC or LC as described earlier in this chapter offers the best means of confirmation of pesticide identity. However, because of their expense, not all laboratories have easy access to them.

There are other means of confirmation, and one of the most used in chromatography is to compare results obtained on different columns or with different detectors. However, such comparisons do not constitute a foolproof means of confirmation, although they do provide strong evidence for unknown identification. One example of data comparison for confirmation involves the use of linked response data from parallel flame photometric and electron-capture detectors with retention data from linear temperature programming.[73] This approach has been successfully used to identify a wide range of organophosphorus, sulfur-containing, and organochlorine pesticides.

Another approach is the use of chemical derivatization, a technique that has found substantial use in pesticide residue analysis when other means of confirmation are not available. A variety of reagents and chemical reactions have been evaluated for many different pesticides including organophosphorus compounds.[74,75] A typical example is shown in Figure 7 for the chemical conversion of azodrin (monocrotophos) to its trifluoroacetate under acidic conditions.[76] The derivative, having different physical and chemical characteristics from azodrin itself, elutes with a different retention time and has a somewhat different detector response depending upon the detector used. For example, Figure 8 shows gas chromatograms of azodrin found in

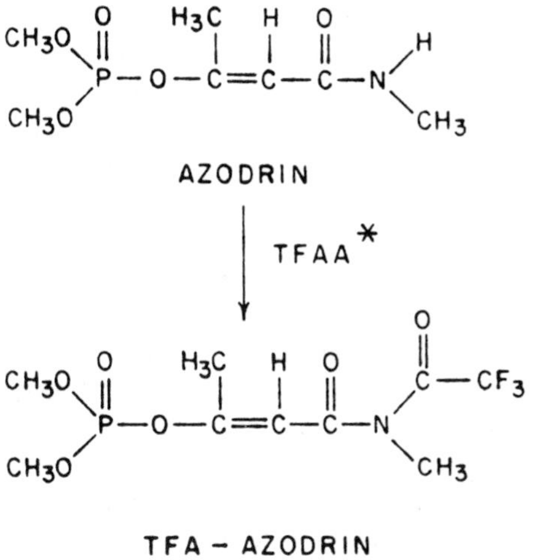

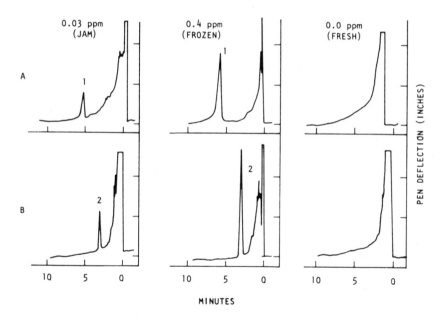

FIGURE 7. Reaction of azodrin with trifluoracetic anhydride (TFAA). (Reprinted from Lawrence, J. F., *Int. J. Environ. Anal. Chem.*, 29, 289, 1987. With permission.)

FIGURE 8. Determination of azodrin in fresh and frozen strawberries and strawberry jam. (A) Upper chromatograms, direct analyses; (B) lower chromatograms, after TFAA reaction; 1, azodrin; 2, TFA-azodrin. (Reprinted from Lawrence, J. F., *Int. J. Environ. Anal. Chem.*, 29, 289, 1987. With permission.)

strawberry products. As can be seen after trifluoroacetylation, the original azodrin peaks disappear, and new peaks appear corresponding to its trifluoroacetate derivative.

Little work has been done on chemical confirmation techniques for organophosphates by LC, although the LC characteristics of fluorescent derivatives of several organophosphates, including fenthion, crufomate, and fenchlorphos, have been reported.[77]

An enzymatic confirmation technique has been evaluated for certain organophosphorus pesticides.[78] The method is based on the selective degradation of certain organophosphates by the bacterial enzyme phosphotriesterase, which is ineffective in degrading other types of pesticides such as carbamates or organochlorines. Thus, for application, an extract that was found to be positive by GC for an organophosphate pesticide would be treated with the enzyme. If the initial result were truly due to an organophosphorus pesticide, reanalysis after enzyme treatment should indicate a loss of the peak. Not all organophosphates are degraded by the enzyme, but the method does have potential.

V. CONCLUSIONS

Extraction of organophosphorus pesticides from water samples can be achieved by either LLE or LSE. In the last few years, LSE has become popular for the following reasons: greater sample throughput, simplicity, less use of hazardous solvents, and potential for automation, (e.g., in the case of coupling SPE techniques on-line with LC).

For the isolation of organophosphorus pesticides from sediment and biological matrices containing interferring compounds (e.g., lipids), cleanup procedures are usually required. The two most common methods involve Florisil and/or GPC. The cleanup step in such matrices is of great importance, since phospholipid compounds can easily interfere with chromatographic determinations. For routine analysis, the determinative step usually is carried out by GC-NPD or GC-FPD. Of the available techniques, GC/MS in the negative ion mode gives lower detection limits and, thus, it is preferred. However, negative ion operation is not available on most of the commercial bench-top GC/MS instruments, and no MS libraries are available.

New approaches such as LC/MS are becoming especially important in this field, since most of the organophosphorus pesticides are easily detected. The limit of detection needs to be improved, but the use of novel interfacing approaches, such as plasmaspray, electrospray, or particle beam, should be of great help in this respect. In addition, couplings of SFC (e.g., SFC-NPD and SFC/MS) are now commercially available and appear to have potential as useful tools for the determination of thermally labile organophosphorus pesticides.

REFERENCES

1. **Barceló, D.,** *Analyst (London),* 116, 681, 1991.
2. **Poole, S. K., Dean, T. A., Ondegema, J. W., and Poole, C. F.,** *Anal. Chim. Acta,* 236, 3, 1990.
3. **Sharp, G. S., Brayan, J. G., Dilli, S., Haddad, P. R., and Desmarchelier, J. M.,** *Analyst (London),* 133, 1493, 1988.
4. **Barceló, D.,** *Chromatographia,* 25, 928, 1988.
5. **Das, K. G., Ed.,** *Pesticide Analysis,* Marcel Dekker, New York, 1981.
6. **Zweig, G. and Sherma, J., Eds.,** *Analytical Methods for Pesticides and Plant Growth Regulators,* Vol. 6, Academic Press, New York, 1977.
7. Organophosphorus pesticides in river and drinking water 1980, Tentative method, Her Majesty's Stationery Office, London, 1983, 1.
8. **Wang, T. C., Lenahan, R. A., and Tucker, J. W., Jr.,** *Bull. Environ. Contam. Toxicol.,* 38, 226, 1987.
9. **Barceló, D., Porte, C., Cid, J., and Albaigés, J.,** *Int. J. Environ. Anal. Chem.,* 38, 199, 1990.
10. **Neicheva, A., Kovacheva, E., and Marudov, G.,** *J. Chromatogr.,* 437, 249, 1988.
11. **Chau, A. S. Y. and Afghan, B. K.,** *Analysis of Pesticides in Water,* Vol. 1, 2, 3, CRC Press, Boca Raton, FL, 1982.
12. **Hinckley, D. A. and Bidleman, T. F.,** *Environ. Sci. Technol.,* 23, 995, 1989.
13. **Mães Vinuesa, J., Moltó Cortés, J. C., Igualada Cañas, C., and Font Pérez, G.,** *J. Chromatogr.,* 472, 365, 1989.
14. **Bangnati, R., Benfenati, E., Davoli, E., and Fanelli, R.,** *Chemosphere,* 17, 59, 1988.
15. **Verweij, A., Van Liempt, M. A., and Boter, H. L.,** *Int. J. Environ. Anal. Chem.,* 21, 63, 1985.
16. **Farran, A., Cortina, J. L., De Pablo, J., and Barceló, D.,** *Anal. Chim. Acta,* 234, 119, 1990.
17. **Durand, G., Forteza, R., and Barceló, D.,** *Chromatographia,* 28, 597, 1989.
18. **Battista, M., Di Corcia, A., and Marchetti, A.,** *J. Chromatogr.,* 454, 233, 1988.
19. **Kjolholt, J.,** *Chemosphere,* 14, 1763, 1985.
20. **Elhag, F. A., Yule, W. N., and Marshall, W. D.,** *Bull. Environ. Contam. Toxicol.,* 42, 172, 1989.
21. **Belisle, A. A. and Swineford, D. M.,** *Environ. Toxicol. Chem.,* 7, 749, 1988.
22. **Barceló, D., Solé, M., Durand, G., and Albaigés, J.,** *Fresenius J. Anal. Chem.,* 339, 676, 1991.
23. **McLeese, D. W., Zitko, V., and Sergeant, D. B.,** *Bull. Environ. Contam. Toxicol.,* 22, 800, 1979.
24. **Aquatic Biology Group,** *Chemosphere,* 13, 19, 1984.
25. **Bottomley, P. and Baker, P. G.,** *Analyst (London),* 109, 85, 1984.
26. Committee for Analytical Methods for Residues of Pesticides and Veterinary Products in Food Stuffs, *Analyst (London),* 110, 765, 1985.
27. **Adachi, K., Ohokuni, N., and Mitsuhashi, T.,** *J. Assoc. Off. Anal. Chem.,* 67, 798, 1984.
28. **Blaha, J. J. and Jackson, P. J.,** *J. Assoc. Off. Anal. Chem.,* 68, 1095, 1985.
29. **Sawyer, L. D.,** *J. Assoc. Off. Anal. Chem.,* 68, 64, 1985.
30. **Lores, E. M., Moore, J. C., and Moody, P.,** *Chemosphere,* 16, 1065, 1987.
31. **Pressley, T. A. and Longbottom, J. E.,** The Determination of Organophosphorus Pesticides in Industrial and Wastewaters, Method 614, U.S. Environmental Protection Agency, Cincinatti, OH, 1982, 1.
32. **Brayan, J. G., Haddad, P. R., Sharp, G. J., Dilli, S., and Desmarchelier, J. M.,** *J. Chromatogr.,* 447, 249, 1988.

33. **Liu, J., Suzuki, O., Kumarzawa, T., and Seno, H.,** *Forensic Sci. Int.,* 41, 67, 1989.
34. **Thier, H. P. and Zeumer, H.,** *Manual of Pesticide Residue Analysis,* Vol. 1, VCH Publishers, Weinheim, Germany, 1987, 75.
35. **Roos, A. H., Van Munsteren, A. J., Nab, F. M., and Tuinstra, L. G. M.,** *Anal. Chim. Acta,* 196, 95, 1987.
36. **Steinwandter, H.,** *Fresenius Z. Anal. Chem.,* 331, 499, 1988.
37. **Lawrence, J. F.,** *Int. J. Environ. Anal. Chem.,* 29, 289, 1987.
38. **Venant, A., Boorel, S., Mallet, J., and Van Neste, E.,** *Analusis,* 17, 64, 1989.
39. **Chamberlain, S. J.,** *Analyst (London),* 115, 1161, 1990.
40. **Petrick, G., Schulz, D. E., and Duinker, J. C.,** *J. Chromatogr.,* 435, 241, 1988.
41. **Gillespie, A. M. and Walters, S. M.,** *J. Liq. Chromatogr.,* 9, 2111, 1986.
42. **Waters, S. M.,** *Anal. Chim. Acta,* 236, 77, 1990.
43. **DiMuccio, A., Cicero, A. M., Camoni, I., Pontecorvo, D., and Dommarco, R.,** *J. Assoc. Off. Anal. Chem.,* 70, 106, 1987.
44. **DiMuccio, A., Ausili, A., Camoni, I., Dommarco, R., Rizzica, M., and Vergori, F.,** *J. Chromatogr.,* 456, 149, 1988.
45. **Hopper, M. L.,** *J. Assoc. Off. Anal. Chem.,* 71, 731, 1988.
46. **Luke, B. G., Richards, J. C., and Dawes, E. F.,** *J. Assoc. Off. Anal. Chem.,* 67, 295, 1984.
47. **Luke, B. G. and Richards, J. C.,** *J. Assoc. Off. Anal. Chem.,* 67, 902, 1984.
48. **Brown, R. L., Farmer, C. N., and Millar, R. G.,** *J. Assoc. Off. Anal. Chem.,* 70, 442, 1987.
49. **Onuska, F. I.,** *J. High Resolut. Chromatogr. Chromatogr. Commun.,* 7, 660, 1984.
50. **Mansour, M., Hustert, K., and Müller, R.,** *Intern. J. Environ. Anal. Chem.,* 37, 83, 1989.
51. **Levsen, K.,** *Org. Mass Spectrom.,* 23, 406, 1988.
52. **Stan, H. J.,** *J. Chromatogr.,* 467, 85, 1989.
53. **Durand, G. and Barceló, D.,** *Anal. Chim. Acta,* 243, 259, 1991.
54. **Stan, H. J. and Kellner, G.,** *Biomed. Mass Spectrom.,* 9, 483, 1982.
55. **Stan, H. J. and Kellner, G.,** *Biomed. Environ. Mass Spectrom.,* 18, 645, 1989.
56. **Lawrence, J. F.,** in *Analytical Methods for Pesticides and Plant Growth Regulators,* Vol. 12, Zweig, G. and Sherma, J., Eds., Academic Press, New York, 1982, 39.
57. **Greve, P. A. and Goewie, C. E.,** *Int. J. Environ. Anal. Chem.,* 20, 29, 1985.
58. **Mallet, V. N., Duguay, M., Bernier, M., and Trottier, N.,** *Int. J. Environ. Anal. Chem.,* 39, 271, 1990.
59. **Barceló, D., Maris, F. A., Geerdink, R. B., Frei, R. W., De Jong, G. J., and Brinkman, U. A. Th.,** *J. Chromatogr.,* 394, 65, 1987.
60. **Barceló, D.,** *Biomed. Environ. Mass Spectrom.,* 17, 363, 1988.
61. **Durand, G., Sanchez-Baseza, F., Messeguer, A., and Barceló, D.,** *Biol. Mass Spectrom.,* 20, 3, 1991.
62. **White, K. D., Min, Z., Brumley, W. C., Krause, R. T., and Sphon, J. A.,** *J. Assoc. Off. Anal. Chem.,* 66, 1358, 1983.
63. **Cortes, H. J.,** in *Multidimensional Chromatography. Techniques and Applications,* Cortes, H. J., Ed., Marcel Dekker, New York, 1990, 251.
64. **Veltkamp, A. C.,** in *Selective Sample Handling and Detection in High-Performance Liquid Chromatography,* Part B, Zech, K. and Frei, R. W., Eds., Elsevier, Amsterdam, 1989, 133.
65. **Kalinoski, H. T. and Smith, R. D.,** *Anal. Chem.,* 60, 529, 1988.
66. **Raymer, J. H. and Pellizzari, E. D.,** *Anal. Chem.,* 59, 1043, 1987.
67. **Mendoza, C. E.,** in *Pesticide Analysis,* Das, K. G., Ed., Marcel Dekker, New York, 1981, 1.
68. **Fodor-Csorba, K. and Dutka, F.,** *J. Chromatogr.,* 365, 309, 1986.

69. **Kovacs, G. H.,** *J. Chromatogr.,* 303, 309, 1984.
70. **Patil, V. B., Padalikar, S. V., and Kawale, G. B.,** *Analyst (London),* 112, 1756, 1987.
71. **Blewett, T. C. and Krieger, R. I.,** *Bull. Environ. Contam. Toxicol.,* 45, 120, 1990.
72. **Durand, P., Nicaud, J. M., and Mallevialle, J.,** *J. Anal. Toxicol.,* 8, 112, 1984.
73. **Stan, H. J. and Mrowetz, D.,** *J. Chromatogr.,* 279, 173, 1983.
74. **Lawrence, J. F.,** in *Pesticide Analysis,* Das, K. G., Ed., Marcel Dekker, New York, 1981, 425.
75. **Cochrane, W. P.,** in *Chemical Derivatization in Analytical Chemistry,* Vol. 1, Frei, R. W. and Lawrence, J. F., Eds., Plenum Press, New York, 1981, 1.
76. **Lawrence, J. F. and McLeod, H. A.,** *J. Assoc. Off. Anal. Chem.,* 59, 639, 1976.
77. **Lawrence, J. F., Renault, C., and Frei, R. W.,** *J. Chromatogr.,* 121, 343, 1976.
78. **Chiang, T., Dean, M. C., and McDaniel, C. S.,** *Bull. Environ. Contam. Toxicol.,* 34, 809, 1985.

Chapter 7

MULTIRESIDUE METHODS FOR ORGANONITROGEN PESTICIDES

Ronald G. Luchtefeld

TABLE OF CONTENTS

I. INTRODUCTION

A multiresidue method (MRM) for determining pesticide residues can be defined as a method that is able to detect more than one pesticide during the analysis of a sample.[1] In addition, an MRM should be sensitive, precise, and accurate enough for regulatory purposes. Examples include the procedure of Mills for determining pesticide residues in fatty foods[2] and the methods of Mills et al.,[3] Luke et al.,[4,5] and Storherr et al.[6] for assaying residues in nonfatty foods.

Generally, these procedures include extraction, cleanup, and determinative steps designed for the analysis of a particular group of compounds in a specific type of product or products (e.g., chlorinated hydrocarbons in nonfatty foods). Each step in the process of analysis can make the procedure more selective for the compound group to be analyzed while eliminating various chemicals from analysis. For example, the Luke procedure[4,5] may be capable of extracting a wide polarity range of chemicals, but not all of these compounds are necessarily detectable by gas chromatography (GC) as some may be susceptible to decomposition. As a consequence, more specific MRMs are developed for particular classes of chemicals. These MRMs, because they are more specific, do not usually encompass the large numbers of chemicals that are detected by methods such as the Luke procedure, but they do permit the accurate analysis of a particular class of chemicals not previously assayed. For example, the method of Luchtefeld[7] was developed to fill a need for an assay procedure for phenylurea residues. Likewise, other MRMs have been developed for specific classes of organonitrogen compounds, as many of these residues were not assayable by one of the existing procedures. The following brief literature review indicates examples of specific MRMs for organonitrogen residues.

A. BENZIMIDAZOLES

The majority of methods for determining benzimidazoles use high performance liquid chromatography (HPLC) with reversed-phase or reversed-phase ion pairing, although normal-phase chromatography is also used. The following are examples of some of the applications reported in the literature: the determination of benomyl, carbendazim, thiabendazole, thiophanate, and thiophanate methyl in seven different crops, including fruits and rice, by HPLC with ion pairing;[8] analysis of peach trees for benomyl, captan, and thimet by HPLC;[9] analysis of apples, pears, and their pulp for benomyl, carbendazim, and thiophanate by HPLC using a diol column with a hexane-isopropyl alcohol mobile phase;[10] the determination of benomyl, carbendazim (MBC), and nine other pesticides in water using HPLC with on-line preconcentration;[11] and the analysis of post-harvest-treated apples for benomyl, carbendazim, thiabendazole, and thiophanate using HPLC.[12]

B. BIPYRIDINIUM

A second example of a specific class of compounds for which specific MRMs were developed is the bipyridinium herbicides, which, because of their quaternary amine structure, are not usually extracted by existing procedures or, if extracted, are not easily detected. These compounds seem well suited for HPLC analysis. GC has also been utilized, but requires prior hydrogenation of the bipyridiniums or pyrolysis GC. Recent methods published for these residues include: the determination of paraquat and diquat residues in potatoes using HPLC with ion pairing and UV or fluorescence detection;[13] the use of reverse-phase HPLC with an amino (NH$_2$)-type column for determining paraquat and diquat in agricultural products;[14] the analysis of food crops for bipyridinium herbicides by GC using hydrogenation with sodium borohydride and nickel chloride;[15] and the GC determination of paraquat and diquat in crops, also with hydrogenation prior to GC.[16]

C. DICARBOXIMIDES

The dicarboximides captan, captofol, and folpet have been analyzed using specific MRMs, as well as with general MRMs, with GC as the primary method for determining the residues. A multiresidue method for determining fungicides in fruits and vegetables included the dicarboximides, which were assayed using solid-phase extraction and GC analysis.[17] Two multiresidue methods[3,4] were evaluated for their ability to extract and cleanup captan, folpet, and captofol for analysis by GC.[18] This study indicated that the Luke procedure gave better overall recoveries for captan, folpet, and captofol. Thin-layer chromatography was used to determine captan, folpet, and captofol in water, apples, and lettuce,[19] and HPLC was used to determine captan, folpet, and captofol in plant material.[20]

D. SULFONYLUREAS

The sulfonylureas are a relatively new class of organonitrogen herbicides, and as a result there are few multiresidue procedures available for these compounds. Three methods were found in the literature, and these included the HPLC determination of chlorsulfuron, sulfometuron-methyl, and imazapyr in water using solid-phase extraction followed by HPLC determination;[21] the analysis of chlorsulfuron and sulfometuron-methyl by liquid chromatography/mass spectrometry;[22] and a bioassay for determining six sulfonylureas in soil.[23]

E. DITHIOCARBAMATES

The dithiocarbamate fungicides, which include the metal chelates ferbam, maneb, zineb, and ziram, are a group of residues that are often assayed by determining one of the degradation products, such as carbon disulfide. Several methods have been reported that use HPLC as a determinative step to assay the individual components. A multiresidue procedure was developed for the analysis of the metal chelate fungicides in fruits and vegetables using HPLC.[24] Bardarov et al.[25] reported the analysis of zineb, maneb, and their degradation products in air using normal-phase HPLC. Thiram and ferbam were determined in strawberries, maize, and tobacco using normal-phase liquid chromatography (LC).[26] Traces of ferbam and its three degradation products were determined in soil with reverse-phase and normal-phase HPLC.[27]

F. TRIAZINES

Methods for the analysis of triazine herbicides have included enzyme-linked immunosorbent assay,[28,29] HPLC,[30–33] GC,[34–36] and isotachophoresis.[37] In addition, Hajslova et al.[38] compared GC and HPLC for determining triazines in water and Xu et al.[39] compared GC and LC for the analysis of triazines in soil.

MRMs for specific residues, such as the organonitrogen compounds, are developed with consideration of the chemistry of the compounds in order to utilize specific detection systems based on the available functional groups. This is illustrated in the carbamate procedure and the Luchtefeld method[7] for analysis of phenylureas.

The procedures for the analysis of both carbamates[40,41] and phenylureas[7] utilize HPLC, as GC has a tendency to cause decomposition of these compounds. To meet the levels of detection required for trace analysis, postcolumn derivatization and fluorescence detection was used for analyzing the respective residues. These detection systems add sensitivity and selectivity to both procedures. The carbamate procedure utilizes alkaline hydrolysis followed by reaction with *o*-phthalaldehyde-mercaptoethanol (OPA-MERC) to produce a fluorescent derivative that can be both sensitively and selectively determined. The phenylurea procedure employs HPLC with a postcolumn detection technique that involves the use of photodegradation followed by derivatization with OPA-MERC.

The carbamate procedure relies on alkaline hydrolysis to produce a primary amine from the carbamates (Equation 1), however, alkaline hydrolysis does not produce primary amines with the phenylureas (Equation 2).

$$\text{R–O–}\overset{\overset{\displaystyle O}{\|}}{\text{C}}\text{–}\overset{\overset{\displaystyle H}{|}}{\text{N}}\text{–CH}_3 \quad \xrightarrow[\text{OH}^-]{\text{Hydrolysis}} \quad \text{H}_2\text{N(CH}_3) \tag{1}$$

$$\text{R–}\overset{\overset{\displaystyle H}{|}}{\text{N}}\text{–}\overset{\overset{\displaystyle O}{\|}}{\text{C}}\text{–N(CH}_3)_2 \quad \xrightarrow{\text{OH}^-} \quad \text{No Primary Amine} \tag{2}$$

$$\text{R–N–}\overset{\overset{\displaystyle O}{\|}}{\text{C}}\text{–N(CH}_3)_2 \quad \xrightarrow[\text{h}\nu]{\text{Photolysis}} \quad \text{H}_2\text{N(CH}_3) \tag{3}$$

$$\underset{R-O-C-N-CH_3}{\overset{O\quad H}{\underset{\|\quad\,|}{}}} \xrightarrow{\quad h\nu \quad} H_2N(CH_3) \tag{4}$$

Photodegradation (hν) as used in the phenylurea procedure generates primary amines with both phenylureas (Equation 3)[42,43] and carbamates[44] (Equation 4), as well as other organonitrogen compounds.

This reaction, when coupled with OPA-MERC derivatization, provides an HPLC detection scheme for a variety of organonitrogen compounds. Thirty-five compounds were evaluated using gradient elution and postcolumn photolysis with derivatization to determine elution characteristics and response factors. In addition, the extraction and cleanup procedure used for the phenylureas was evaluated to determine its application to the analysis of the same compounds. This has resulted in a multiresidue procedure capable of determining a wide variety of organonitrogen compounds.

II. EXPERIMENTAL

A. APPARATUS AND REAGENTS

The configuration of the HPLC and postcolumn derivatization apparatus used for the analysis of organonitro compounds is shown in Figure 1. The HPLC instrument consisted for a Hewlett Packard Model 1090 capable of gradient elution, an Alltech Adsorbosphere C-8, 5-μm cartridge column, 4.6 × 150 mm, and an Alltech Adsorbosphere C-8 guard cartridge. The column and guard column were maintained at 40°C.

Two mobile phases were used: (1) methanol and water with a 20 to 80% methanol linear gradient in 25 min and (2) acetonitrile and water with a 15 to 70% acetonitrile linear gradient in 25 min. The flow rate for both systems was 1.0 mL/min. The eluate from the column was passed directly through a photolysis coil prepared from standard 1/16 in. O.D. Teflon tubing with an I.D. of 0.5 mm and a length of 4 m and woven into a sleeve as described in Reference 45. A suitable substitute photolysis coil, 0.5 mm I.D. × 10 ft × 1/16 in. may be obtained already woven from Supelco Inc., Bellefonte, PA. The coiled Teflon sleeve is placed over a UV lamp, 17 cm × 9 mm I.D., obtained from BHK Inc., Model 80-11781-01, with the appropriate power supply, Model 90-0001-01.

The eluate, after leaving the coil, is mixed with OPA-MERC in a Valco Mixing Tee, Part No. ZT1C. The OPA-MERC is added via a Milton-Roy Mini Pump, Model 396-31, at a flow rate of 0.2 mL/min. The OPA-MERC solution is prepared by dissolving 75 to 80 mg of *o*-phthalaldehyde (Sigma Chemical Co.) in 2 mL of methanol. This is added to 500 mL of 0.010 *M* pH 10.5 borate buffer. The solution is mixed, and 0.5 mL of mercaptoethanol (Kodak) is added and mixed. The reaction mixture leaves the mixing tee and

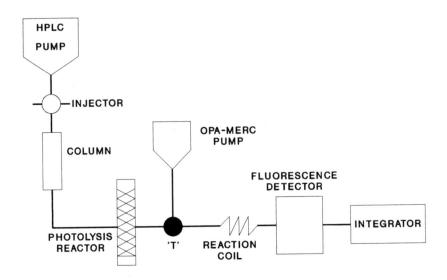

FIGURE 1. Configuration of HPLC and postcolumn derivatization apparatus for analysis of organonitro compounds.

is passed through a reaction coil consisting of 10 ft of Teflon tubing, 1/16 inch O.D. × 0.3 mm I.D., made into a three-dimensional coil.[45]

The fluorescent products are detected with a Shimadzu RF 535 fluorescence detector set at an excitation wavelength of 340 nm and an emission wavelength of 455 nm. A computing integrator is used to record and calculate the data from the detector.

All standard materials were obtained from the U.S. Environmental Protection Agency Standards Repository, Research Triangle Park, NC; 1-mg/mL stock solutions were prepared in methanol; all further dilutions were prepared in acetonitrile and water (1 + 1).

B. METHOD FOR SUBSTITUTED UREA HERBICIDES[46]
1. Principles
The sample is extracted with methanol, and coextractives are removed by shaking with hexane in the presence of salt. Residues are transferred by partitioning to methylene chloride, which is cleaned up on a Florisil column eluted with acetone/methylene chloride. The concentrated extract is dissolved in (1 + 1) acetonitrile + water for determination by HPLC on a C-18 reversed-phase column using a water-methanol gradient mobile phase. Eluted residues are measured in-line by postcolumn photodegradation, chemical derivatization with *o*-phthalaldehyde, and spectrofluorometric detection. Residues are confirmed by rechromatographing with a different mobile phase.

Extract analyte from
product with methanol
↓
Partition with hexane;
discard hexane
↓
Partition residues into
methylene chloride
↓
Florisil cleanup
↓
HPLC determination: reversed-
phase separation with
photodegradation and chemical
derivatization to permit
fluorometric determination
↓
Confirmation: HPLC reversed-
phase separation with
acetonitrile/water mobile phase

Limits of detection and quantitation for selected compounds are included in Table 1. The method has been shown to be applicable to determination of at least 14 phenylurea herbicides at concentrations of 0.05, 0.5, and 1.0 ppm in 14 nonfatty food products.

2. Standard Reference Materials

Prepare stock solutions (1 mg/mL) of each standard in LC grade isopropyl alcohol. Combine appropriate amounts of stock standard solutions and further dilute with acetonitrile-water (1 + 1) for working standards. Use standards at a concentration of 1.25 μg/mL for HPLC determination. Keep stock solutions refrigerated prior to use and prepare working standards weekly.

3. Apparatus

The apparatus for extraction and cleanup includes:

- **Homogenizer.** Polytron Model 10-35 with PT-20ST generator (Brinkmann Instruments, Inc., Westbury, NY) or equivalent
- **Centrifuge.** Explosion proof; to accommodate 500-mL centrifuge bottles
- **Centrifuge bottles.** 500 mL
- **Chromatographic column.** 10 mm I.D. × 300 mm, with Teflon stopcock and coarse porosity fritted disc

TABLE 1
Pesticides Recovered Through the Method for Substituted Urea Herbicides

Chemical	Methanol/water mobile phase[a]			Acetonitrile/water mobile phase[b]		
	Retention time (relative to diuron)	L_D[c] (ppm)	L_S[d] (ppm)	Retention time (relative to diuron)	L_D[c] (ppm)	L_S[d] (ppm)
Fenuron	0.42			0.49		
Metoxuron	0.62			0.67		
Monuron	0.72			0.75		
Chlorotoluron	0.87			0.91		
Fluometuron	0.87	0.002	0.006	0.93	0.001	0.003
Monolinuron	0.91			0.99		
Metobromuron	0.91	0.004	0.015	1.04	0.004	0.014
Isoproturon	0.96			1.00		
Diuron	1.00	0.002	0.007	1.00	0.001	0.003
Siduron	1.08			1.16		
Linuron	1.12	0.004	0.014	1.23	0.005	0.017
Chlorbromuron	1.16	0.006	0.022	1.28	0.003	0.011
Chloroxuron	1.25	0.002	0.008	1.28	0.001	0.003
Neburon	1.34			1.43		

Note: All of these compounds are completely recovered through the method described in Section II.B. Percent recoveries and coefficients of variation developed during two intralaboratory and one interlaboratory validation studies of this method. Commodities fortified for these recovery tests include asparagus, baby food (mixture of oatmeal, applesauce, and bananas), carrots, creamed corn, grape juice, onions, oranges, canned peas, potatoes, baked potatoes, and strawberries. Only oranges contained compounds that interfered with the analysis.

a   Econosphere ODS column, 25 cm × 4.6 mm I.D.; 1 mL/min elution with linear gradient of 40% methanol/water to 80% methanol/water in 30 min.

b   Econosphere ODS column, 25 cm × 4.6 mm I.D.; 1 mL/min elution with linear gradient of 30% acetonitrile/water to 80% acetonitrile/water in 30 min.

c L_D, limit of detection, is calculated as the amount of phenylurea required to cause a response three times the baseline noise.

d L_8, limit of quantitation, is calculated as the amount of phenylurea required to cause a response ten times the baseline noise.

- **Cylinders, graduated (graduates).** 250 mL
- **Funnel.** Glass
- **Kuderna-Danish concentrators.** 250 mL and 500 mL, with Snyder column and receiving flask (Kontes Glass Co., Cat. Nos. K-570000 or K-570050, sizes 425 or 1025, and K-621400, size 525 or equivalent)
- **Separatory funnels (separators).** 500 mL, 1 L

The apparatus for HPLC (see Figure 1) includes:

- **Dual pump gradient system.** With 40 μL valve loop injector (Model 322, Beckman Altex Instrument Co., Berkeley, CA)
- **Guard column.** Direct-connect cartridge system containing a prepacked C-18 cartridge (Cat. No. 28013, Alltech Associates, Deerfield, IL)
- **Analytical column.** Octadecylsilane (ODS), 25 cm × 4.6 mm I.D. Econosphere C-18, containing spherical particles with a monomeric bonded layer (Alltech Associates)
- **Photodegradation apparatus.** (1) UV lamp for photodegradation, 17 cm × 9 mm O.D. (Model 80-1178-01) with power supply (Model 90-0001-01) (BHK Inc., Pomona, CA); (2) Teflon sleeve for photodegradation lamp. Prepare by weaving 3.7 m × 0.5 mm I.D. Teflon tubing into a coil according to Reference 45 or purchase 10 ft × 0.5 mm I.D. delay coil, Cat. No. 5-9206, from Supelco, Inc. (Supelco Park, Bellefonte, PA). Place the coiled Teflon tubing over the lamp and connect one end to the mixing tee and the other end to the column
- **Mixing tee.** Stainless steel, 0.25-mm bore. 1/16 in. standard fittings (No. ZT1C, Valco Instruments, Inc., Houston, TX)
- **Low flow-rate pump.** For OPA-MERC solution (Model 396-31, LCD/ Milton Roy, Riviera Beach, FL). Connect a pulse dampener, consisting of a cone-shaped coil of 13 ft × 1/8 in. O.D. stainless steel tubing, between the pump and 0.5 mm I.D. Teflon tubing, which is connected to the tee
- **Reaction coil.** 3.0 m × 0.3 mm I.D. Teflon tubing, made into a three-dimensional coil according to Reference 45 or purchase a 10-ft × 0.5-mm I.D. delay coil from Supelco, Inc. (Cat. No. 5-9206). (A 0.3-mm I.D. reaction coil is preferable to minimize band spreading, but the 0.5-mm delay coil, which has the smallest I.D. commercially available, gives acceptable results.)
- **Fluorescence detector.** (Model 650 10S, Perkin-Elmer Corp., Norwalk, CT)
- **Computing integrator.** (Model 4290, Spectra-Physics, Arlington Heights, IL)

4. Reagents

Reagents for extraction and cleanup are:

- **Boiling chips.** Carborundum, 20 mesh, or other suitable boiling chips
- **Florisil.** PR grade, 60/100 mesh (Floridin Co., Berkeley Springs, WV), activated by heating at 130°C overnight in an open container
- **Glass wool.** Pyrex fiberglass, Corning #3950 or VWR #32848-003; use as is without further treatment
- **Sodium sulfate.** Anhydrous, granular, reagent grade
- **Sodium chloride.** Reagent grade, prepared as a saturated solution in water
- **Solvents.** Acetone, hexane, methanol, methylene chloride: distilled from all glass apparatus; UV-grade acetonitrile + HPLC-grade water (1 + 1) to dissolve final extract for HPLC determination

Reagents for HPLC are:

- **Methanol.** Distilled-in-glass grade (Burdick & Jackson Laboratories, Inc., Muskegon, MI)
- **HPLC acetonitrile, acetone.** UV grade (Fisher Scientific, St. Louis, MO)
- **HPLC methanol.** Omnisolv (EM Science, Cherry Hill, NJ)
- **HPLC water.** Obtain from a Milli-Q water purification system (Millipore Corp., Bedford, MA)
- **Acetic acid, glacial.** Baker Analyzed Reagent grade (J. T. Baker Chemical Co., No. 9507-03)
- **Sodium borate buffer.** Dissolve 19.1 g ACS-grade sodium tetraborate in approximately 500 mL HPLC water. Dilute to 1 L with HPLC water and adjust the pH to 10.5 with sodium hydroxide, using a pH meter. Filter through a 0.2 μm membrane filter
- **o-Phthalaldehyde (OPA).** Reagent grade (Eastman Kodak, Rochester, NY). Prepare a stock solution by dissolving 300 mg of OPA in 100 mL methanol
- **Mercaptoethanol (MERC).** Reagent grade (Eastman Kodak). Prepare a stock solution by diluting 10 mL MERC to 100 mL with methanol
- **Potassium phosphate monobasic.** Certified ACS grade (Fisher No. P-285). Prepare a 0.002 M solution by dissolving 0.27 g in 1 L HPLC-grade water
- **OPA-MERC solution.** Transfer approximately 250 mL degassed sodium borate buffer to a 500-mL volumetric flask. Add 25 mL OPA stock solution and 5 mL MERC stock solution. Dilute to volume with borate buffer with mixing

5. HPLC Operating Parameters

- To prepare the system for the analysis of samples, elute the column overnight at 0.3 mL/min with 0.002 *M* monobasic potassium phosphate and methanol (90 + 10), using a dual pump system. (Or prepare a solution consisting of 10% methanol and 90% 0.002 *M* monobasic potassium phosphate, and elute this solution with a single pump.) Change the mobile phase to 10% methanol in water and elute for 30 min at 1.0 mL/min. If after several days of use chromatogram baselines begin to drift, repeat the above procedure.
- Equilibrate the system using 40% methanol in water at a flow rate of 1.0 mL/min, with OPA-MERC solution introduced at 0.2 mL/min. Allow the detector to stabilize after starting the OPA-MERC solution flow and turning on the UV photodegradation lamp.
- After injecting a sample or standard, begin a 30-min linear gradient from 40% methanol in water to 80% methanol in water. Operate the detector at an excitation wavelength of 340 nm and an emission wavelength of 455 nm. Adjust the detector sensitivity to obtain 50% recorder or integrator deflection when 40 μL 1.0 μg/mL diuron solution is injected. Diuron elutes in approximately 24 min under these conditions.
- Flush the pump and tubing used for addition of the OPA-MERC solution daily after use by pumping 3% glacial acetic acid in water through the system at 1.0 mL/min for about 20 min.
- After use each day, elute the HPLC column with 80% methanol/water for 20 min. Column may be stored in this mobile phase overnight and for longer periods.

6. System Suitability Test

- Prepare a solution containing 1 μg/mL each of metobromuron and diuron.
- Using the HPLC conditions of Section II.B.5, chromatograph the solution three times. Retention times are approximately 22 and 24 min for metobromuron and diuron, respectively.
- Determine the relative standard deviation (RSD) (standard deviation/ mean) of the peak heights measured in the three chromatograms; also determine the resolution between the two peaks, according to the formulas in Figure 1.
- RSD will be <3% and resolution will be ≥1.5 on a system performing adequately.

7. Extraction

- Weight a 50-g sample into a 500-mL centrifuge bottle. Add 100 mL methanol and homogenize the mixture using a Polytron for 1.5 min with the speed set at 8.
- Centrifuge the homogenate at 1500 rpm for 5 min and decant 80 to 100 mL through a funnel fitted with a glass wool plug into a 250-mL graduated cylinder.
- Add 100 mL methanol to the material in the centrifuge bottle and homogenize for 1.0 min with the speed set at 8.
- Centrifuge the homogenate and decant as much liquid as possible through the funnel into the 250-mL cylinder.
- Record the volume of the combined methanol extract.

8. Partition

- Transfer the entire methanol extract to a 500-mL separatory funnel.
- Add 30 mL saturated sodium chloride solution and 50 mL hexane and shake the mixture for 30 s. Let the layers separate.
- Transfer the lower aqueous layer to a 1-L separatory funnel containing 500 mL water and 30 mL saturated sodium chloride solution. Discard the hexane.
- Add 200 mL methylene chloride to a 1-L separatory funnel and shake for 1 min.
- Drain the methylene chloride through a glass tube (approximately 30 mm I.D. × 15 cm) containing 5 cm sodium sulfate into a 500-mL Kuderna-Danish concentrator fitted with a 5-mL collection tube.
- Reextract the aqueous phase two more times with 100 mL each of methylene chloride. Drain each extract through sodium sulfate into a Kuderna-Danish concentrator.
- Rinse the sodium sulfate with 50 mL methylene chloride, adding rinse to Kuderna-Danish concentrator.
- Place a Snyder column on the Kuderna-Danish concentrator and evaporate the methylene chloride on a steam bath to about 3 mL.

9. Florisil Cleanup

- Prepare the chromatographic column by placing 4 g activated Florisil in a 10-mm I.D. × 30-cm glass column; top with a 1-cm layer of anhydrous sodium sulfate. Wash the column with 30 mL methylene chloride and discard the wash.

- When the methylene chloride reaches the top of the sodium sulfate, place a 250-mL Kuderna-Danish concentrator under the column. Transfer the sample extract to the column. Rinse the container with about 3 mL methylene chloride and add the rinse to the column.
- When the sample extract reaches the top of the sodium sufate, rinse the column with two 3-mL portions of methylene chloride, allowing each rinse to reach the top of the sodium sulfate.
- When the final rinse reaches the top of the sodium sulfate, elute the column with 50 mL of 20% acetone/methylene chloride.
- Evaporate the solvent on a steam bath to about 3 mL. Remove the receiving flask and evaporate the solvent to dryness, using a stream of nitrogen and a water bath at 40°C. Pipet 2 mL acetonitrile + water (1 + 1) into the receiving flask.
- Calculate the sample equivalent in the final cleaned up extract according to the following formula:

$$\frac{\text{g sample}}{\text{mL}} = \frac{\text{mL methanol filtrate collected}}{200 \text{ mL} + (50\text{g} \times \% \text{ water in sample}) - 5}$$
$$\times \frac{50 \text{ g}}{2 \text{ mL final extract}}$$

10. Determination

The determinative step is very selective for phenylureas in the presence of food coextractives that remain after partition and Florisil column cleanup steps. Figure 2 compares chromatography from this system to that using UV detection at 245 nm.

- Filter both the sample solution and the reference standard solution(s) through 0.45 μm membrane filters; inject 40 μL filtrate into the HPLC system operating as previously described.
- Compare the detector responses (peak height is preferred) to the sample and to a reference standard using the same chromatographic conditions.
- Use acetonitrile-water (1:1) to make further dilutions of the sample, as necessary to make the peak heights of the sample and standard match closely.

11. Confirmation

Confirm the residue identity by chromatographing the sample using the same HPLC column, but with an acetonitrile-water gradient mobile phase instead of the methanol-water gradient. Table 1 lists retention times relative to diuron for 14 phenylureas and Figure 3 shows chromatograms for a mixture of 6 phenylureas in each of these systems. Since the two systems provide a

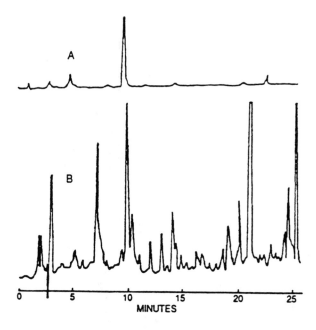

FIGURE 2. HPLC chromatograms of carrot extract. (A) HPLC determinative step; (B) detection by UV detector at 245 nm. The large peak in chromatogram A is caused by an interference from the carrot sample. (From Luchtefeld, R. G., *J. Assoc. Off. Anal. Chem.*, 70, 740, 1987. With permission.)

different elution order for the herbicides, they are useful for confirmation of residue identity based on retention times.

III. RESULTS AND DISCUSSION

The OPA-MERC reagent reacts with primary amines to produce a fluorescent isoindole derivative that is detected fluorometrically.[47] If a compound can be hydrolyzed or photodegraded to produce a primary amine, it can be detected using the OPA-MERC reagent. Moye et al.[40] applied this principle to the postcolumn detection of carbamates by converting them into primary amines using alkaline hydrolysis and reacting the primary amines with OPA-MERC to produce the fluorescent derivative. As Krause demonstrated,[41] the alkaline hydrolysis reaction does not produce primary amines with amides, dithiocarbamates, thiocarbamates, or phenylureas. These compounds, if hydrolyzed, produce dialkyl amines and do not react with OPA-MERC. The pathways of phenylurea photolysis were elucidated by several researchers[42,43] who indicated the possibility of primary amines being produced from the phenylureas during the process of photodegradation. Using this information, Luchtefeld[48] developed a postcolumn detection system based on photolysis

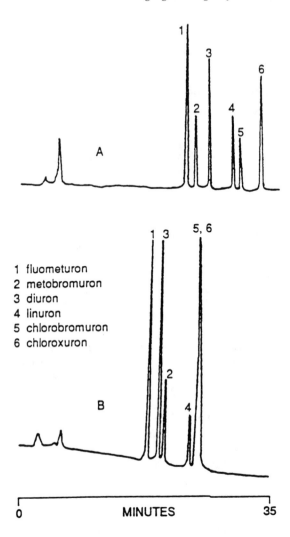

1 fluometuron
2 metobromuron
3 diuron
4 linuron
5 chlorobromuron
6 chloroxuron

FIGURE 3. Chromatography of six substituted urea herbicides on Econosphere ODS column, 25 cm × 4.6 mm I.D. with two different mobile phases. (A) Linear gradient from 40 to 80% methanol in water in 30 min; 1 mL/min flow rate; (B) linear gradient from 30 to 80% acetonitrile in water in 30 min; 1 mL/min flow rate. (From Luchtefeld, R. G., *J. Assoc. Off. Anal. Chem.*, 70, 740, 1987. With permission.)

and derivatization. Postcolumn photodegradation was used to produce primary amines from phenylureas, and the primary amines were reacted with OPA-MERC to produce isoindole derivatives suitable for spectrofluorometric detection. This same photodegradation and derivatization procedure was investigated for its application to other nitrogen-containing chemical entities. These compounds and their general structures are listed in Table 2. As indicated, there are representative compounds from several organonitrogen groups, including amides, dithiocarbamates, thiocarbamates, carbamates, and ureas.

An analytical scheme to assay this variety of chemicals was developed using HPLC with the two gradient mobile phases and the postcolumn photolysis-derivatization system shown in Figure 1. The two mobile phases described in Section II.A were designed to assay chemicals that ranged from polar (aldicarb sulfone) to nonpolar (butylate).

Table 3 lists the 35 compounds that were assayed and their respective relative retention times for each of the gradients used. As can be seen by the similarity of some of the relative retention times, all of the compounds were not separated using a single mobile-phase gradient. However, with a combination of the two gradients it is possible to assay each of the residues and obtain a confirmation. Figure 4 shows the chromatograms of a mixture of several of the organonitrogen compounds representing each of the chemical groups assayed. The standards were chromatographed using the two gradient systems: chromatogram A was obtained using the acetonitrile and water mobile phase and chromatogram B was obtained using the methanol and water mobile phase. For both chromatograms, the amount of each compound assayed is the same. As is indicated, the acetonitrile and water system enhances the response of the phenylureas (peak 3), the amide (peak 4), the thiocarbamate (peak 5), and the dithiocarbamate (peak 6). The two gradients provide different retention times for the compounds, thus allowing for confirmation. The use of photolysis makes it easy to add an additional confirmatory step, i.e., the lamp can be turned off and the assays rerun to determine if the response is obtained with the lamp on and off. Since the organonitro compounds studied only respond when the lamp is on, a response obtained with the lamp off indicates the peak detected does not represent any of the compounds listed.

All of the compounds were assayed to determine their limits of detection. These limits were based on three times the baseline noise level.[49] Table 4 lists this data for all of the compounds studied. These values were intended to be used as a guide to determine approximate levels of determination for each compound. The values will vary from system to system based on the column and detector used.

For some of the residues, there are considerable detection limit differences for the same compound in the two solvent systems. This appears to be a function of the effect of the solvent on the amount of photolysis of the compound assayed. All of the phenylureas show lower detection limits (better response) in acetonitrile-water mobile phases than in methanol-water. The amides diphenamide and CDAA are considerably more sensitive in acetonitrile-water than in methanol-water. The response of three of the carbamates, carbaryl, aldicarb sulfoxide, and aldicarb sulfone, is much less than the other compounds, but is still enough to allow analysis of these compounds at 0.05 ppm in samples.

TABLE 2
Compounds Assayed

Amides

$$\begin{array}{c} O \\ \parallel \\ R-C-N(R)_2 \end{array}$$

CDAA
Diphenamide

Dithiocarbamates

$$\begin{array}{c} S \\ \parallel \\ R-S-C-N(R)_2 \end{array}$$

CDEC
Thiram

Thiocarbamates

$$\begin{array}{c} O \\ \parallel \\ R-S-C-N(R)_2 \end{array}$$

Butylate
Cycloate
Thiobencarb

Carbamates

$$\begin{array}{c} O \quad H \\ \parallel \quad | \\ R-O-C-N-CH_3 \end{array}$$

Aldicarb
Aldicarb sulfone
Aldicarb sulfoxide
Bufencarb
Carbaryl
Carbofuran
Dioxacarb
3-Hydroxycarbofuran
Methiocarb
Methomyl
Oxamyl
Propoxur

Ureas

$$\begin{array}{c} O \\ \parallel \\ R-NH-C-N(R)_2 \end{array}$$

Chlorbromuron
Chlortoluron
Chloroxuron
Dichlorophenylmethylurea
Diuron
Fenuron
Fluometuron
Isoproturon
Linuron
Metobromuron
Metoxuron
Monolinuron
Monuron
Neburon
Siduron
Tebuthiuron

Note: Compounds are listed by chemical class and general structure.

TABLE 3
Relative Retention Times for Chemical Residues on Adsorbosphere C-8 4.6 mm × 150 mm Using Two Mobile Phases

	Mobile phase 1	Mobile phase 2
Aldicarb sulfone	0.32	0.35
Oxamyl	0.39	0.38
Methomyl	0.40	0.40
Aldicarb sulfoxide	0.51	0.51
Fenuron	0.51	0.52
Dioxacarb	0.52	0.53
3-Hydroxycarbofuron	0.53	0.50
Thiram	0.55	0.60
Aldicarb	0.67	0.69
CDAA	0.70	0.72
Metoxuron	0.72	0.73
Monuron	0.76	0.77
Propoxur	0.80	0.84
Carbofuran	0.82	0.86
Carbaryl	0.85, 0.87	0.91
Tebuthiuron	0.87	0.67
Monolinuron	0.89	0.94
Metobromuron	0.94	0.99
Chlortoluron	0.94	0.92
Fluometuron	0.94	0.97
Dichlorophenylmethylurea	0.95	0.91
Diuron	1.00[a]	1.00[b]
Isoproturon	1.01	0.99
Diphenamide	1.07	1.10
Linuron	1.10	1.15
Siduron	1.11, 1.13	1.13, 1.16
Mesurol	1.12	1.17
Chlorbromuron	1.13	1.18
Chloroxuron	1.21	1.24
CDEC	1.29	1.44
Bufencarb	1.29, 1.30	1.34
Neburon	1.30	1.37
Thiobencarb	1.39	1.46
Cycloate	1.40	1.46
Butylate	1.45	1.56

Note: Mobile phases: (1) Methanol and water gradient: 20 to 80% MeOH in 25 min. (2) Acetonitrile and water gradient: 15 to 70% MeCN in 25 min.

[a] Diuron retention time = 17.44 min.
[b] Diuron retention time = 15.18 min.

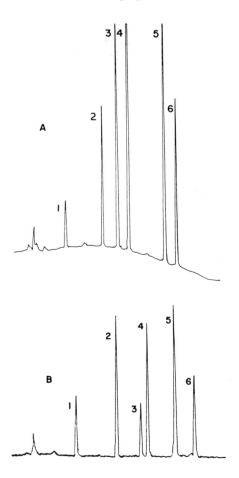

FIGURE 4. Chromatograms of six organonitrogen compounds representing each of the chemical groups assayed: (1) aldicarb sulfoxide, 150 ng; (2) carbofuran, 15 ng; (3) diuron, 15 ng; (4) diphenamide, 3.75 ng; (5) CDEC, 6 ng; and (6) butylate, 15 ng. Chromatogram A was obtained using acetonitrile and water gradient 15 to 70% acetonitrile in 25 min. Chromatogram B was obtained using methanol and water gradient 20 to 80% methanol in 25 min.

To demonstrate the application of this technique to real samples, onions and potatoes were used as matrices for recovery studies. Recoveries were determined at 0.05 ppm and at 1.0 ppm for all of the compounds in both products. Duplicate determinations were used to obtain the values listed in Table 5. The recovery table indicates low recoveries for several compounds. This problem is caused by the use of Florisil in the cleanup step, and research is continuing to remedy this difficulty. This method appears to be readily adaptable to other organonitrogen compounds and, with some modification of the cleanup steps, will give excellent recoveries for all the residues.

TABLE 4
Limits of Detection for Chemical Residues on Adsorbosphere C-8 4.6 mm × 150 mm Using Two Mobile Phases

	Mobile phase 1 (ng)	Mobile phase 2 (ng)
Aldicarb sulfone	0.53	2.7
Oxamyl	0.07	0.08
Methomyl	0.008	0.17
Aldicarb sulfoxide	2.5	7.3
Fenuron	0.81	0.28
Dioxacarb	0.64	0.78
3-Hydroxycarbofuran	0.17	0.17
Thiram	0.45	0.17
Aldicarb	0.04	0.03
CDAA	1.5	0.50
Metoxuron	0.65	0.18
Monuron	0.44	0.09
Propoxur	0.19	0.21
Carbofuran	0.2	0.19
Carbaryl	3.75	5.8
Tebuthiuron	0.45	1.2
Monolinuron	0.58	0.19
Metobromuron	0.79	0.3
Chlortoluron	0.52	0.23
Fluometuron	0.37	0.21
Dichlorophenylmethylurea	0.22	0.34
Diuron	0.45	0.15
Isoproturon	0.29	0.15
Diphenamide	0.1	0.044
Linuron	0.625	0.36
Siduron	1.5	1.3
Mesurol	0.16	0.18
Chlorbromuron	1.98	0.68
Chloroxuron	0.4	0.17
CDEC	0.15	0.23
Bufencarb	0.25	0.25
Neburon	0.39	0.125
Thiobencarb	0.16	0.03
Cycloate	1.1	0.15
Butylate	0.68	0.23

Note: Mobile phases: (1) Methanol and water gradient: 20 to 80% MeOH in 25 min. (2) Acetonitrile and water gradient: 15 to 70% MeCN in 25 min. Detection limit is based on three times the baseline noise level.

TABLE 5

Recovery (in Percent) of Organonitrogen Residues from Potatoes and Onions Spiked at 0.05 and 1.0 ppm

	Potatoes		Onions	
	0.05	**1.0**	**0.05**	**1.0**
Aldicarb sulfone	86	74	76	78
Oxamyl	102	77	76	89
Methomyl	92	90	94	90
Aldicarb sulfoxide	58	38	47	52
Fenuron	95	91	110	100
Dioxacarb	110	95	89	90
3-Hydroxycarbofuran	78	82	76	78
Thiram	96	75	85	84
Aldicarb	82	83	95	93
CDAA	70	79	80	88
Metoxuron	98	99	107	96
Monuron	101	97	89	95
Propoxur	98	97	97	97
Carbofuran	110	95	99	102
Carbaryl	88	94	90	103
Tebuthiuron	98	96	99	102
Monolinuron	97	97	94	99
Metobromuron	97	100	100	102
Chlortoluron	95	95	109	112
Fluometuron	94	95	90	89
Dichlorophenylmethylurea	84	87	90	93
Diuron	98	94	107	97
Isoproturon	93	94	96	88
Diphenamid	97	92	111	101
Linuron	96	98	71	103
Siduron	92	94	87	82
Mesurol	82	87	89	91
Chlorbromuron	94	102	79	101
Chloroxuron	99	99	109	98
CDEC	8	18	20	21
Bufencarb	76	90	87	85
Neburon	97	100	115	96
Thiobencarb	18	37	22	27
Cycloate	16	27	15	21
Butylate	23	25	20	23

Note: Each value is an average of two determinations.

IV. CONCLUSION

Multiresidue methods should be capable of determining more than one residue as well as be sensitive, precise, and accurate. These methods include those that are able to assay a wide variety of chemical compounds and methods for specific classes of chemicals, such as the organonitrogen residues.

A specific MRM is presented for the analysis of 35 organonitrogen compounds including phenylureas, carbamates, thiocarbamates, dithiocarbamates, and amides. The procedure utilizes HPLC with postcolumn photolysis and derivatization. The availability of two gradient mobile phases lends versatility as well as the ability to confirm any detected residues. The method is intended for the analysis of residues in foods at 0.05 ppm and above.

REFERENCES

1. Pesticide Residues in Food: Technologies for Detection, OTA-F-398, U.S. Government Printing Office, Washington, D.C., 1988.
2. **Mills, P. A.,** *J. Assoc. Off. Anal. Chem.,* 42, 734, 1959.
3. **Mills, P. A., Onley, J. H., and Gaither, R. A.,** *J. Assoc. Off. Anal. Chem.,* 46, 186, 1963.
4. **Luke, M. A., Froberg, J. E., and Masumoto, H. T.,** *J. Assoc. Off. Anal. Chem.,* 58, 1020, 1975.
5. **Luke, M. A., Froberg, J. E., Doose, G. M., and Masumoto, A. T.,** *J. Assoc. Off. Anal. Chem.,* 64, 1187, 1981.
6. **Storherr, R. W., Ott, P., and Watts, R. R.,** *J. Assoc. Off. Anal. Chem.,* 54, 513, 1971.
7. **Luchtefeld, R. G.,** *J. Assoc. Off. Anal. Chem.,* 70, 740, 1987.
8. **Gilvydis, D. M. and Walters, S. M.,** *J. Assoc. Off. Anal. Chem.,* 73, 753, 1990.
9. **Sanchez-Brunet, C., DeCal, A., Melgarejo, P., and Tadeo, J. L.,** *Int. J. Environ. Anal. Chem.,* 37, 35, 1989.
10. **Bicchi, C., Belliardo, F., and Cantamessa, L.,** *Pestic. Sci.,* 25, 355, 1989.
11. **Marvin, C. H., Brindle, I. D., Singh, R. P., Hall, C. D., and Chiba, M.,** *J. Chromatogr.,* 518, 242, 1990.
12. **Cano, P., DeLa Plaza, J. L., and Munoz-Delgado, L.,** *J. Agric. Food Chem.,* 35, 144, 1987.
13. **Worobey, B. L.,** *Pestic. Sci.,* 18, 245, 1987.
14. **Nagayama, T., Maki, T., Kan, K., Iida, M., and Nashima, T.,** *J. Assoc. Off. Anal. Chem.,* 70, 1008, 1987.
15. **Davidek, J., Hajslova, J., Davidek, T., and Cuhra, P.,** *Agric. Food Chem. Consum. Proc. Eur. Conf. Food Chem.,* 5, 268, 1989.
16. **Hajslova, J., Cuhra, P., Davidek, T., and Davidek, J.,** *J. Chromatogr.,* 479, 243, 1989.
17. **Newsome, W. H. and Collins, P.,** *J. Chromatogr.,* 472, 4118, 1989.
18. **Gilvydis, D. M. and Walters, S. M.,** *J. Assoc. Off. Anal. Chem.,* 67, 909, 1984.
19. **Sherma, J. and Stellmacher, S.,** *J. Liq. Chromatogr.,* 8, 2949, 1985.
20. **Buettler, B. and Hoermann, W. D.,** *J. Agric. Food Chem.,* 29, 257, 1981.

21. **Wells, M. J. M. and Michael, J. L.,** *J. Chromatogr. Sci.,* 25, 345, 1987.
22. **Shalaby, L. M.,** *Biomed. Mass Spectrom.,* 12, 261, 1985.
23. **Rahman, A.,** *Hydrobiologia,* 367, 188, 1989.
24. **Gustafsson, K. H. and Fahlgren, C. H.,** *J. Agric. Food Chem.,* 31, 461, 1983.
25. **Bardarov, V., Zaikov, C., and Mitewa, M.,** *J. Chromatogr.,* 479, 97, 1989.
26. **Brandsteterova, E., Lehotay, J., Liska, O., and Garaj, J.,** *J. Chromatogr.,* 354, 375, 1986.
27. **Brandsteterova, E., Lehotay, J., Liska, O., Garaj, J., and Zacsik, I.,** *J. Chromatogr.,* 286, 339, 1984.
28. **Harrison, R. O., Goodrow, M. H., and Hammock, B. D.,** *J. Agric. Food Chem.,* 39, 122, 1991.
29. **Thurman, E. M., Meyer, M., Pomes, M., Perry, C. A., and Schwab, A. P.,** *Anal. Chem.,* 62, 2043, 1990.
30. **Ohno, H., Hattori, M., and Aoyama, T.,** *Eisei Kagaku,* 35, 454, 1989.
31. **Stahl, M., Luehrmann, M., Kicinski, H. G., and Kettrup, A.,** *Z. Wasser Abwasser Forsch.,* 22, 124, 1989.
32. **Froehlich, D. and Meier, W.,** *J. High Resolution Chromatogr.,* 12, 340, 1989.
33. **Battista, M., DiCorcia, A., and Marchetti, M.,** *J. Chromatogr.,* 454, 233, 1988.
34. **Stankova, O., Cap, L., Smicka, J., and Lanikova, J.,** *Acta Univ. Palacki. Olomuc. Fac. Rerum Nat.,* 88, 167, 1987.
35. **Tonogai, Y., Hasegawa, Y., Nakamura, Y., and Itoh, Y.,** *Eisei Kagaku,* 34, 421, 1988.
36. **Grandet, M., Weil, L., and Quentin, K. E.,** *Z. Wasser Abwasser Forsch.,* 21, 21, 1988.
37. **Stransky, Z.,** *J. Chromatogr.,* 320, 219, 1985.
38. **Hajslova, J., Ryparova, L., Viden, I., Davidek, J., Kocourek, V., and Zemanova, I.,** *Int. J. Environ. Anal. Chem.,* 38, 105, 1990.
39. **Xu, Y., Lorenz, W., Pfister, G., Bahadir, M., and Korte, F.,** *Fresenius Z. Anal. Chem.,* 325, 377, 1986.
40. **Moye, H. A., Scherer, S. J., and St. John, P. A.,** *Anal. Lett.,* 10, 1049, 1977.
41. **Krause, R. T.,** *J. Assoc. Off. Anal. Chem.,* 63, 1114, 1980.
42. **Kotzias, D. and Korte, F.,** *Ecotoxicol. Environ. Saf.,* 5, 503, 1981.
43. **Mazzocchi, P. H. and Rao, M. P.,** *J. Agric. Food Chem.,* 20, 957, 1972.
44. **Miles, C. J. and Moye, H. A.,** *Anal. Chem.,* 60, 220, 1988.
45. **Engelhardt, H. and Neue, U. D.,** *Chromatographia,* 15, 403, 1982.
46. National Technical Information Service, *Pesticide Analytical Manual,* Vol. 1, U.S. Department of Commerce, May 1990, sec. 242.4.
47. **Roth, M. A.,** *Anal. Chem.,* 43, 880, 1971.
48. **Luchtefeld, R. G.,** *J. Chromatogr. Sci.,* 23, 516, 1985.
49. **Keith, L. H.,** *Anal. Chem.,* 55, 2210, 1983.
50. **Luchtefeld, R. G.,** U.S. Food and Drug Administration Laboratory Information Bulletin (LIB) No. 3309, May 1989.

Chapter 8

TWO-DIMENSIONAL CAPILLARY GAS CHROMATOGRAPHY FOR RESIDUE ANALYSIS

Hans-Jürgen Stan

TABLE OF CONTENTS

I. INTRODUCTION

Pesticides are applied worldwide to increase food production and to protect crops during transport and storage. Trade between countries with differing pesticide regulations makes it necessary to analyze food on the home market for all known pesticides. Several hundred pesticides are allowed with varying maximum tolerances, depending on the type of food. For a reliable screening of pesticide residues, multimethods are generally required. The most suitable multimethods are based on capillary gas chromatography.

Compounds can be identified on the basis of retention time in a calibrated gas chromatographic system. The quality of information depends on the separation efficiency, which is related mainly to column length and selectivity of the stationary phase for pesticides. The information obtained from one column is commonly used for screening analysis. The absence of peaks in a chromatogram correlating to the retention times of calibrated pesticides indicates that the sample is free from the corresponding pesticide residues. The presence of peaks correlating to calibrated peaks makes a confirmatory analysis necessary. This can be achieved by running the same sample on another capillary column coated with a separation phase of different polarity. In terms of information theory, the correlation of retention data from different phases has the highest discriminating value if the stationary phases are as dissimilar as possible.[1,2] Our experience in routine pesticide residue analysis demonstrates, however, that a pair of capillary columns with the nonpolar methyl silicone (SE-30) or methyl silicone with 5% phenyl (SE-54) and the medium polar methyl phenyl silicone (OV-17) phases gives the necessary discriminating power.[3] Similar good results[4,5] are obtained when the second column is coated with methyl silicone having 14% cyanopropyl (OV-1701), which exhibit polarity very similar to OV-17. In this chapter, the practical aspects of applying two-dimensional gas chromatography to pesticide residue analysis rather than theoretical considerations are discussed. Therefore, the chapter mainly will reflect methods developed in the author's laboratory and the current state of the art.

II. REVIEW OF TWO-DIMENSIONAL GAS CHROMATOGRAPHY

A. SCOPE

The term "two-dimensional chromatography" was at first used in paper chromatography and later in thin-layer chromatography (TLC), in which a chromatographic development in one direction can be followed by a second, perpendicular one. Usually the eluent for each development is different.[2] Giddings[6] pointed to the tremendous increase in separation power, which is based on the fact that the overall peak capacity is roughly the product of the peak capacities of both dimensions. There are many meanings of the term two-dimensional or multidimensional chromatography. In the context of this chapter, two-dimensional gas chromatography will be understood as follows:[2]

Two columns of different selectivity are combined such that eluate fractions can be directly transferred from one column to another.

This definition does not include analytical methods where eluate fractions are trapped at the outlet of one column and reinjected into a second column.

Several review articles and book chapters dealing with two-dimensional gas chromatography have been published in the last decade. The first reviews of two-dimensional techniques in high-resolution gas chromatography were written by Bertsch.[7-9] Deans[10] and Schomburg[11] reviewed heart-cutting techniques, and recently Bertsch[2] summarized newer developments in multidimensional gas chromatography. The application of two-dimensional gas chromatography in pesticide residue analysis was described in detail by Mrowetz.[12] The principle and merits of the method were described by Stan[4,13] and more recently by Stan and Heil.[3]

B. TECHNIQUE OF TWO-DIMENSIONAL GAS CHROMATOGRAPHY

Two-dimensional gas chromatography was first used in process gas chromatography, where continuous analysis of the composition of a reaction mixture is necessary to control a reaction process. The analytical problem to be solved is usually the separation of compounds of lower concentrations from the main components. This is achieved by means of heartcutting, where only a part of the sample is analyzed on the second column. The minor components that are overlapped on the first column are transferred to the second one with only a portion of the main component. The efficiency of the second column is now sufficient to separate the minor component from the interfering compound so that all overlapped minor components eventually are separated and quantified on the second column.

When analyzing only a few volatile compounds in a complex mixture also containing low volatiles, back-flush was introduced to shorten analysis time and to protect the separation column from the thermal burden that arises from the high temperature necessary to elute the low volatiles. With back-flush, the direction of the carrier gas stream in the first column is reversed after the compounds of interest have been transferred to the second column.

In many laboratories these two techniques are applied to a combination of a shorter precolumn with limited separation power and a longer analytical column where the actual gas chromatographic analysis takes place. The first column can intentionally be overloaded to increase the amount of minor components to be transferred to the analytical column.[14]

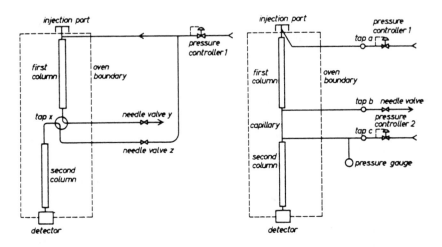

FIGURE 1. Principle of two-dimensional gas chromatography. Left: column switching with four-way valve; right: pneumatic switching according to Deans. (Reproduced from Deans, D. R., *Chromatographia, 1, 18, 1968.* With permission.)

A third technique, which is not really a two-dimensional technique, is analyzing the whole sample once on the first column and then in a second run over both columns coupled in series. This approach resembles selectivity tuning, as described by Sandra et al.[15,16]

C. VALVE SWITCHING VS. PNEUMATIC SWITCHING

Sample transfer by column switching can be achieved by two different techniques. The principles are depicted in the flow diagrams in Figure 1. The effluent stream of the first column enters a four-way valve that can be operated in two positions. In one position the effluent stream passes through a restricting valve to the open air or a monitoring detector, whereas in the other position the effluent stream enters the second column. In both positions the second column is rinsed with carrier gas. The elegant technique of pneumatic column switching without a valve in the effluent stream was introduced by Deans[17] and is shown on the right in Figure 1. The effluent stream from the first column is conducted to the second column or vented by means of pressure differences built up with the restricting needle valve and the carrier gas that additionally enters through tap c. Taps a and c are left open at all times during cutting separations; they are used for setting purposes only. The direction of the effluent stream is now controlled with tap b. With tap b closed, the two columns are used in series; with tap b open, the effluent stream is vented or directed to a monitoring detector.

D. EQUIPMENT FOR PNEUMATIC EFFLUENT TRANSFER

In Figure 2, three commercially available column switching systems are shown. They all operate with the Deans principle of pneumatic switching.

Two of them are add-on accessories from SGE and Chrompack that can be installed with various gas chromatographs. The MCS gas chromatograph from Gerstel is designed as an add-on device for the Hewlett-Packard HP 5890 gas chromatograph and is installed by the manufacturer. It was developed as a preparative gas chromatograph. A short precolumn is usually combined with a longer analytical column, and reactions are transferred to the second column by heartcutting. The pure fractions eluting from the second column are split and monitored with a flame ionization detector (FID) or other appropriate detector. The main part of the fraction, however, is collected in a cryogenic trap for further investigations off-line. The MCS gas chromatograph from Gerstel is fully under the control of sophisticated integrated chromatography software. A set of flow controllers is used to adjust the carrier gas as well as the auxiliary gas streams. This instrument appears to be very versatile for all kinds of two-dimensional gas chromatographic work.

The Sichromat 2 gas chromatograph from Siemens comes as an integrated system with a dual oven. The two columns can be operated with independent temperature programs. The connecting device between the two columns is the patented, so-called, "live" T-piece exhibiting a sophisticated design for optimized gas flow as shown encircled in Figure 3. Auxiliary gases are adjusted with high performance digital flow controllers. Together with the reading of the differential gas pressure over the Deans balance at the live T-piece, readjustment of the system after maintenance work can be easily achieved. The configuration of the system is described in detail in the appendix.

E. MODES OF OPERATION

In Figure 4, the three operation modes of pneumatic switching with the live T-piece from Siemens are demonstrated. The principle of Deans switching has already been described. The live T-piece, however, exhibits a special design feature. The two columns are connected by a narrow platinum capillary. By building up pressure differences between the two ends of the connecting capillary in the live T-piece using the auxiliary gas lines, the carrier gas can be directed as required. There are three possibilities for column switching:

1. Straight: The effluent from the first column is led directly to the detector. This switching position is used for screening analysis.
2. Cut: The effluent from the first column (in total, only a single peak or a whole group of peaks) is transferred to the second column. This switching mode is used for confirmatory analysis.
3. Back-flush: The compounds on the first column are flushed back through the open split valve. This switching mode is used to shorten the analysis time and to avoid interference of later eluting components from the first column with compounds having already passed the second column.

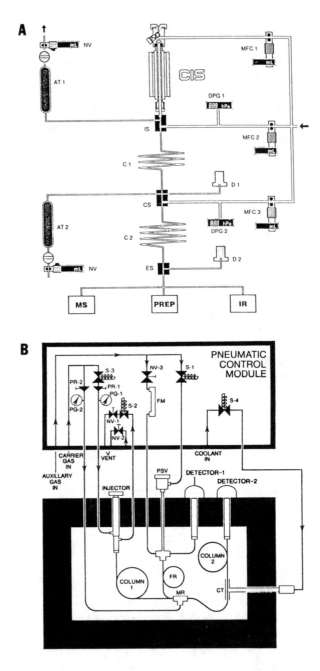

FIGURE 2. Commercial equipment of two-dimensional systems with pneumatic switching. (A) MCS from Gerstel, Mülheim, Germany; (B) PCS from SGE, Weiterstadt, Germany; (C) MUSIC from Chrompack, Frankfurt/Main, Germany. (Courtesy of the manufacturing companies.)

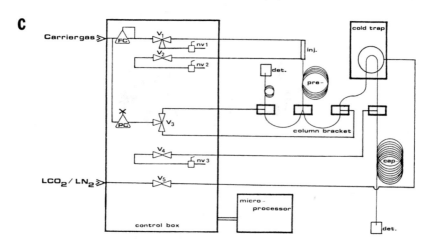

FIGURE 2 (continued).

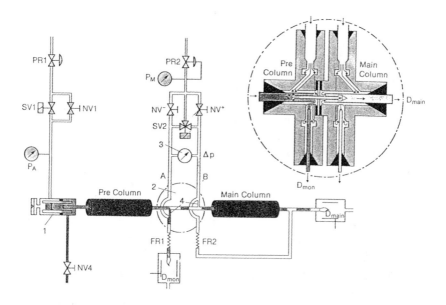

FIGURE 3. Two-dimensional gas chromatography (Siemens, Karlsrue, Germany). 1, Injector; 2, live-T-piece; 3, differential pressure manometer; 4, ring slot; NV, needle valve; SV, solenoid valve; PR, pressure regulator; FR, flow restrictor; P_A, inlet pressure; P_M, mean pressure; p, differential pressure; A, control line A; B, control line B; D_{mon}, monitor detector; D_{main}, main detector; *light,* control line A; *dark,* carrier gas line.

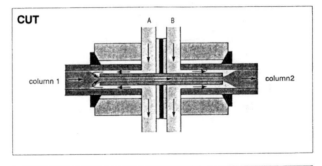

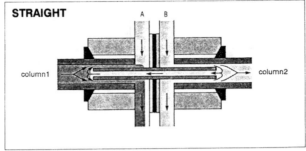

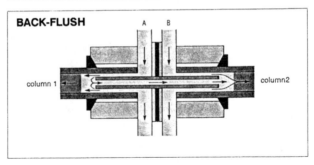

FIGURE 4. Operation modes in live chromatography.

The application of two-dimensional gas chromatography in process gas chromatography and with complex mixtures, such as crude oils, pursues the aim of separating a few target compounds from the bulk of uninteresting compounds or overlapping major components. The introduction of live chromatography with the Sichromat 2 made the precise cutting more convenient in that the chromatogram of the first column (precolumn) could be monitored by means of the monitor detector. The actual analysis was performed on the second column (analytical or main column) with the effluent stream directed to the main detector. Both detectors were of the same type, preferentially the FID.

III. APPLICATION TO PESTICIDE ANALYSIS

A. EQUIPMENT REQUIREMENTS

Our goal was to exploit all the information resulting from the tremendous increase in resolving power by combining two high-performance analytical columns. With a suitable pair of capillary columns, it should be possible to achieve the total separation of all pesticides amenable to gas chromatography. At the same time, pesticides should be separated from all interfering matrix compounds. Therefore, a 50-m capillary column with high separation efficiency was chosen as the first column. The effluent stream was directed to the detectors and the analysis performed one-dimensional like a normal screening analysis. Peaks that correlated in their retention times with calibrated pesticides were considered as possible pesticide residues. In a second confirmatory run, these peaks were transferred to the second column of different polarity. Each pesticide in this system has a linked pair of retention data. A pesticide residue is considered as present in a food sample if the peak recognized on the first column exhibits the expected retention data on the second column.

When analyzing food or other samples with biological matrices, the application of selective detectors is very important. In combination with an appropriate cleanup, the huge number of substances from the matrix appear transparent. Pesticides are small and relatively lipophilic compounds that contain at least one of the heteroatoms Cl, Br, I, S, P, or N. These heteroatoms seldom occur in small lipophilic compounds of biological origin. Therefore, selective detectors responding to these heteroatoms are well suited to pesticide residue analysis. The most prominent selective detectors have been, up to now, the electron capture (ECD), nitrogen-phosphorus thermionic (NPD), and flame photometric (FPD) detectors. Pesticides contain at least one heteroatom that can be taken advantage of for selective detection. Unfortunately, until recently, no selective detector was available for the determination of all kinds of pesticides. Chlorinated pesticides are easily detected with electron capture and nitrogen-containing and organophosphate pesticides with thermionic detection. The two detectors can be used in parallel with effluent splitting from a capillary column.[18] Parallel detection produces an additional increase in information due to the response ratio, which is an important indicative property of many pesticides. A considerable drawback in multiresidue analysis proved to be the lack of selectivity of the NPD with respect to nitrogen and phosphorus because of the greater number of pesticides containing either one or both of these elements in their molecules. A differentiation between the nitrogen and phosphorus response is, however, possible by means of the FPD, which can be tuned to phosphorus selectivity with the appropriate filter. Therefore, maximizing information about any chromatographed substance is achieved by splitting the effluent to the three selective detectors: ECD, NPD, and FPD.

As described, in two-dimensional gas chromatography with live switching, two detectors of the same type are installed as monitor detector after the first column and as main detector after the second column. Our approach with splitting to three selective detectors would make necessary the installation of six selective detectors connected to six integrator channels. We found this an inappropriate expenditure. Therefore, the effluent streams from the first as well as the second column were initially united and then split to the three selective detectors, as shown in Figure 5. The installation of the two columns, the live T-piece, and the splitters is also depicted in the lower part of Figure 5. Details of the equipment are described in the appendix.

In summary, pesticide residue analysis applying gas chromatography is based on retention data from separation phases of different polarity and the response to selective detectors. The presence of a pesticide in a sample is better confirmed the more retention data and response ratios correlate. Two-dimensional capillary gas chromatography is a means to make the correlation of retention data in samples of complex nature more reliable. Splitting of the effluent streams to three selective detectors makes the detection of all pesticides possible. The response ratios of the signals in the three detectors are very important additional data about the nature of a detected compound.

In Figure 6 parallel detection of pesticides with splitting to the three detectors is shown. A mixture of 13 pesticides was chosen representing the various chemical structures found with active substances. The mixture contains pesticides that respond only to the ECD, such as quintozen, aldrin, or dieldrin, and those that respond only to the NPD, such as ethiolate or atrazine. The other pesticides respond to two or all three detectors. Therefore, many pesticides eluting at similar retention times can be easily differentiated by their response behavior. It is not, however, possible to differentiate all of the more than 300 pesticides by chromatography on one column.

B. ANALYSIS OF CRITICAL PAIRS

The merits of the two-dimensional gas chromatographic system are demonstrated in Figure 7. Five pesticides in common use form two critical pairs or even groups of three, respectively, when analyzed on a 50-m capillary column coated with SE-54 as used for screening analysis.[3] Chlorpyrifos and parathion form a critical pair at a retention time (RT) of 28.52 and 28.57 min. Both pesticides are organophosphates and give a response with all three detectors. In the NPD and the FPD trace they show partial separation, but in the ECD trace there is only one single peak to be observed. After transfer to the second column, the two organophosphates are well separated. The methyl esters of these organophosphates also form a critical pair that is additionally overlapped by the fungicide vinclozolin. ECD and NPD respond to all three pesticides, with vinclozolin producing only a small peak in the NPD trace. The FPD responds also to the organophosphates, but not to vinclozolin, which does not contain phosphorus. After transfer to the second column, all three pesticides are well separated and exhibit the response characteristics described. In this example, no disturbance by mixing the compounds on the second column is to be expected. Therefore, the second run was performed just by connecting the two columns in series, which may be considered as a kind of selectivity tuning.[15,16]

C. ANALYTICAL PROCEDURE — CLEANUP

Before analyzing food samples by means of a gas chromatograph, the substances that should be detected have to be extracted from their biological matrix and enriched.

Using the multimethod S19,[19-21] most of the pesticides detectable by gas chromatography are recovered. After extraction with acetone and liquid/liquid partitioning with dichloromethane, the extract is cleaned up by gel permeation chromatography, followed by partition chromatography on a silica gel mini-column applying eluents with increasing polarity. Five fractions are produced in the original method,[19-21] but they can be combined to only three fractions to reduce analysis time.[22] In our laboratory, aldrin is added as an internal standard (ISTD) at a concentration of 0.1 ppm to the homogenized food sample before starting the cleanup procedure. Aldrin is recovered completely in the combined fractions 1 and 2. After having carried out the cleanup procedure on the combined fractions 3 and 4 as well as on fraction 5, an equivalent of 0.2 mg/kg of chlorthion is added as a chromatographic standard. This second standard is added in our laboratory to check for the proper operation of the detectors and for retention time reproducibility.

D. GAS CHROMATOGRAPHIC ANALYSIS

The gas chromatographic analysis starts as described with a screening run on the first column with parallel detection by three detectors: ECD, NPD, and FPD. The pesticide residues are detected by one or more detectors de-pending on their retention times and their responses. Any peak exhibiting a similar retention time and response comparable to a calibrated pesticide is suspected to be the specified pesticide. If there are no such peaks observed, the analysis ends at this point. Otherwise, a second gas chromatographic analysis is required to confirm the presence of the pesticide suspected to be in the sample. This can be executed by simply running the analysis over the two columns in sequence if no overlapping from later eluting compounds is expected. The transfer to the second column is initiated some time before the first suspect pesticide elutes from the first column. With many foods, this simple procedure works well because of the clarity of the chromatograms produced with the selective detectors (see examples of green pepper and broccoli as shown in Figures 8 and 9).

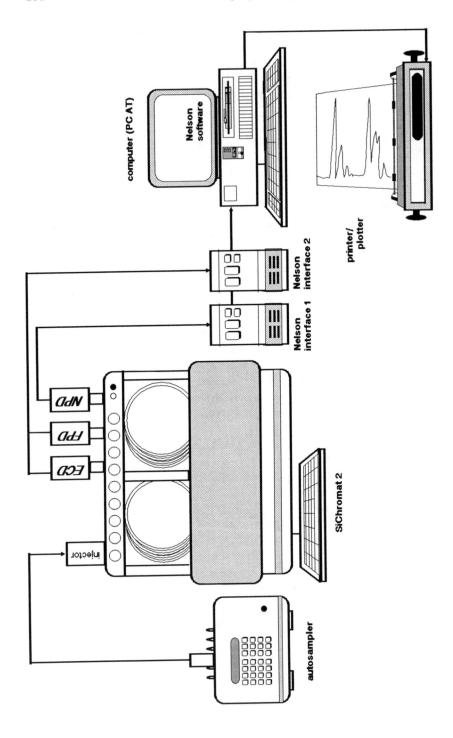

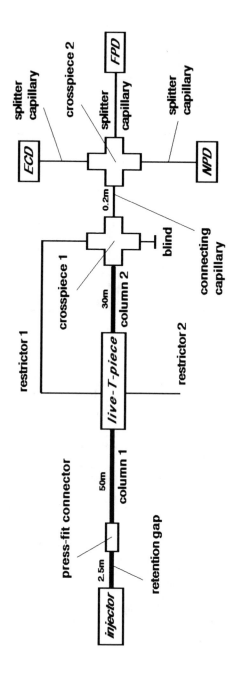

FIGURE 5. Two-dimensional gas chromatograph SiChromat 2 (Siemens) with three detectors and chromatography data processing from PE-Nelson (Perkin Elmer). Top: instruments; bottom: column configuration. (Reproduced from Stan, H.-J. and Heil, S., *Fresenius J. Anal. Chem.*, 339, 34, 1991. With permission.)

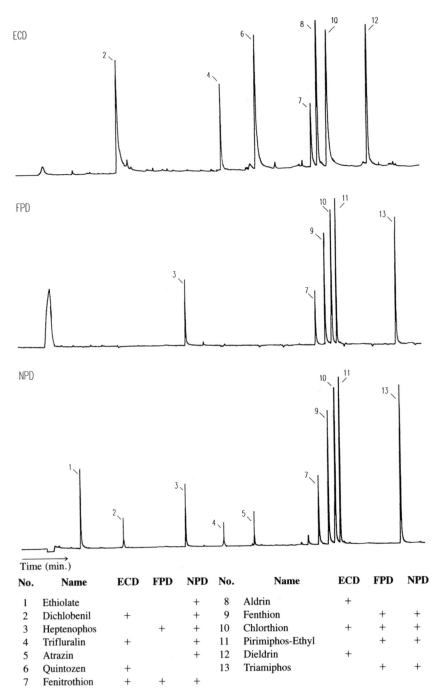

No.	Name	ECD	FPD	NPD	No.	Name	ECD	FPD	NPD
1	Ethiolate			+	8	Aldrin	+		
2	Dichlobenil	+		+	9	Fenthion		+	+
3	Heptenophos		+	+	10	Chlorthion	+	+	+
4	Trifluralin	+		+	11	Pirimiphos-Ethyl		+	+
5	Atrazin			+	12	Dieldrin	+		
6	Quintozen	+			13	Triamiphos		+	+
7	Fenitrothion	+	+	+					

FIGURE 6.   Parallel detection of pesticides by means of three selective detectors. Composition of mixture MIX1130 (6 ng/μl, Chlorthion 12 ng/μl).

To achieve better resolution and to avoid overlapping on the second column by peaks already separated on the first one, the heart-cutting technique may be applied for a confirmatory analysis. Defined fractions of the eluate from column 1 are transferred to column 2; overlapping of such fractions with later-eluting compounds from column 1 is prevented by subsequent back-flushing of the first column. An example is given with onions in Figure 10.

The resulting data are evaluated by means of calibration tables. Peaks must show the same retention time and relative responses in the screening run and in the confirmatory run as the calibrated pesticides. In this case, the presence of the suspected pesticide residues in the sample is considered as proven. As described in detail elsewhere,[22,23] the detection of more than 300 pesticides can be achieved when the cleanup is carried out according to multimethod S19 with final chromatographic separation on a silica gel mini-column into three fractions (combined fractions 1 + 2, 3 + 4, and 5).

The final step of an analysis is quantification. The concentration of the pesticide detected may be estimated by means of the internal standards added to the sample. According to this estimate, the corresponding test substances are dissolved at the appropriate concentrations. Then the sample and this test solution are injected directly following one another for quantitative analysis.

IV. APPLICATION TO REAL FOOD SAMPLES

A. FIRST EXAMPLE: GREEN PEPPER

The procedure described previously is illustrated with a real food sample analyzed in our laboratory, namely, green pepper imported from Egypt.

The fractions 1 and 2 received from the cleanup according to multimethod S19 were collected together, and this extract was injected onto the first column (screening run). The sample contained 0.1 mg/kg of aldrin as an ISTD, which was equivalent to 1 ng/μl. Chromatograms were obtained for all three detectors and the data were automatically compared with the calibration tables to check for pesticide residues.

In Figure 8, the screening analysis on the first column and the corresponding reports are given in the upper part. The pepper sample was suspected to be contaminated with parathion or chlorpyrifos at an RT of 28.50 min. All three detectors respond to both these pesticides, and they form a well-known critical pair on the first column. The endosulfan group of three metabolites was suspected to be present as well (RTs 31.35, 34.00, and 36.28 min, ECD +). An excerpt from the calibration table containing the pesticides found in fractions 1 and 2 is given in Table 1.

A second injection with transfer of the relevant peaks to the second column was necessary to differentiate between the two pesticides of the critical pair. The resulting chromatograms of this confirmatory analysis together with the

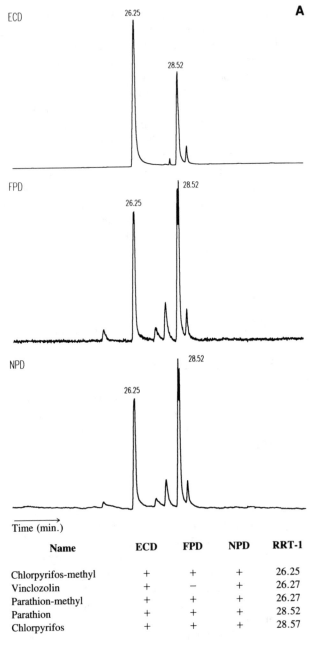

Name	ECD	FPD	NPD	RRT-1
Chlorpyrifos-methyl	+	+	+	26.25
Vinclozolin	+	−	+	26.27
Parathion-methyl	+	+	+	26.27
Parathion	+	+	+	28.52
Chlorpyrifos	+	+	+	28.57

FIGURE 7. Separation of critical pairs. (A) Chromatograms of the first column; (B) chromatograms after both columns.

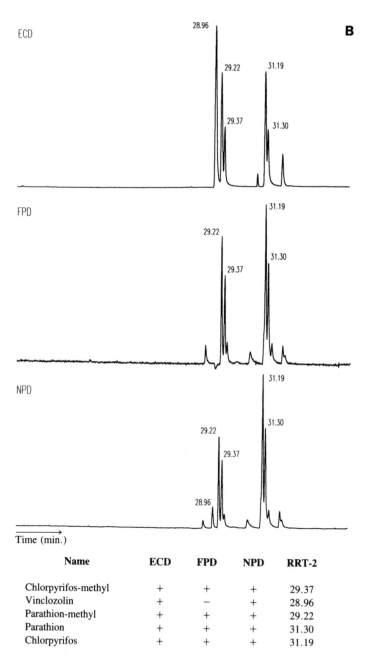

ECD

28.96
29.22 31.19
29.37 31.30

FPD

31.19
29.22
29.37 31.30

NPD

31.19
31.30
29.22
29.37
28.96

Time (min.)

Name	ECD	FPD	NPD	RRT-2
Chlorpyrifos-methyl	+	+	+	29.37
Vinclozolin	+	−	+	28.96
Parathion-methyl	+	+	+	29.22
Parathion	+	+	+	31.30
Chlorpyrifos	+	+	+	31.19

FIGURE 7B.

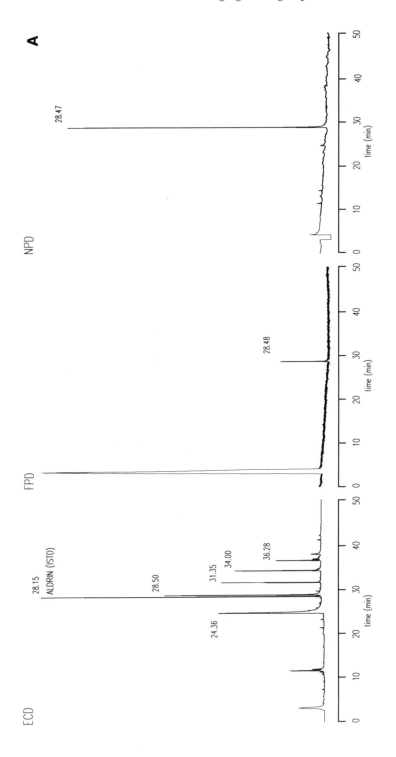

PEAK NUM	RET TIME	NAME	AREA
1	3.732		328091
2	11.207		117463
3	24.361		786110
4	28.153	ALDRIN	614180
5	28.495	PARATHION/CHLORPYRIFOS	398827
6	31.354	ENDOSULFAN-ALPHA	198307
7	34.003	ENDOSULFAN-BETA	224271
8	36.281	ENDOSULFAN-SULPHATE	103090

PEAK NUM	RET TIME	NAME	AREA
1	4.324		157129
2	28.485	PARATHION/CHLORPYRIFOS	31070

PEAK NUM	RET TIME	NAME	AREA
1	4.920		82958
2	28.470	PARATHION/CHLORPYRIFOS	142552

FIGURE 8. Analysis of a green pepper sample (fraction 1 + 2). (A) Screening; (B) confirmation. (Reproduced from Stan, H.-J. and Heil, S., *Fresenius J. Anal. Chem.*, 339, 34, 1991. With permission.)

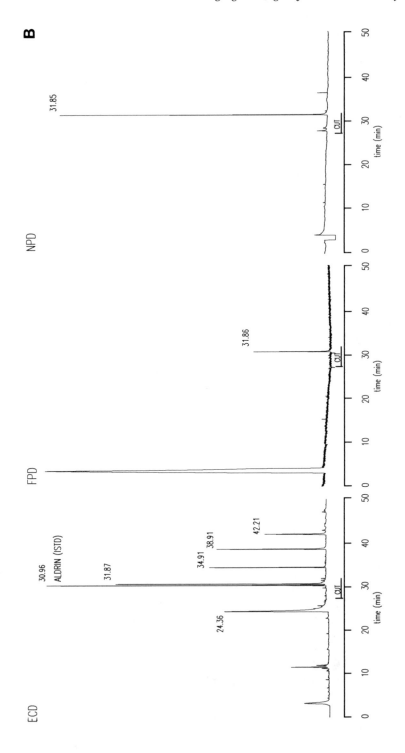

PEAK NUM	RET TIME	NAME	AREA
1	3.732		320193
2	24.361		1240277
3	30.963	ALDRIN	719370
4	31.874	CHLORPYRIFOS	568411
5	34.909	ENDOSULFAN-ALPHA	161198
6	38.911	ENDOSULFAN-BETA	179625
7	42.210	ENDOSULFAN-SULPHATE	103090

PEAK NUM	RET TIME	NAME	AREA
1	4.324		1494922
2	31.864	CHLORPYRIFOS	56795

PEAK NUM	RET TIME	NAME	AREA
1	4.920		33164
2	31.852	CHLORPYRIFOS	126148

FIGURE 8B.

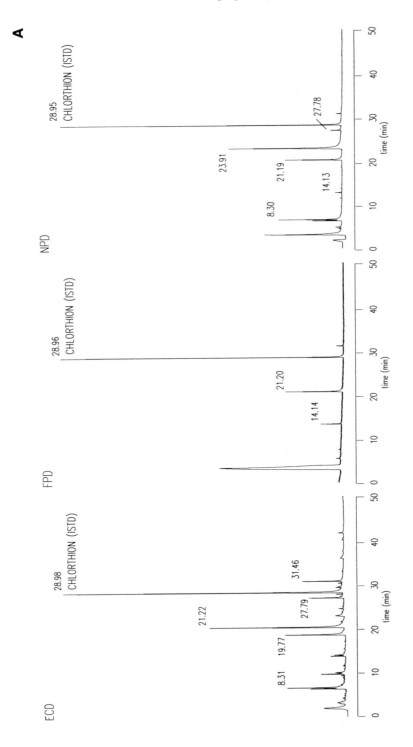

PEAK NUM	RET TIME	NAME	AREA
1	8.308	DICHLORVOS	312707
2	11.578		112620
3	19.766	BENFLURALIN	668717
4	21.221	DIMETHOATE	402223
5	27.792	BROMACIL	238277
6	28.976	CHLORTHION	2272620
7	31.456		485261

PEAK NUM	RET TIME	NAME	AREA
1	14.137	METHACRIFOS	29603
2	21.201	DIMETHOATE	96302
3	28.956	CHLORTHION	371013

PEAK NUM	RET TIME	NAME	AREA
1	8.290	DICHLORVOS	964501
2	14.130	METHACRIFOS	62454
3	21.189	DIMETHOATE	84551
4	23.910		2885772
5	27.780	BROMACIL	133556
6	28.946	CHLORTHION	936279

FIGURE 9. Analysis of a broccoli sample (fraction 3 + 4). (A) Screening; (B) confirmation.

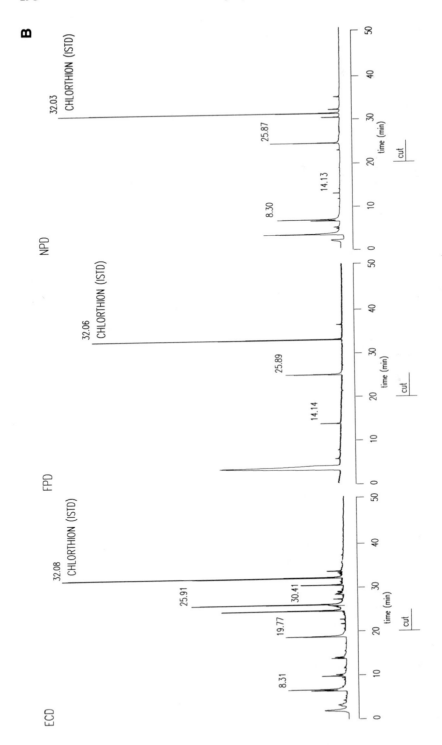

PEAK NUM	RET TIME	NAME	AREA
1	25.906	DIMETHOATE	436751
2	30.411		245447
3	32.077	CHLORTHION	2056584

PEAK NUM	RET TIME	NAME	AREA
1	25.886	DIMETHOATE	101156
2	32.056	CHLORTHION	511026

PEAK NUM	RET TIME	NAME	AREA
1	25.876	DIMETHOATE	95392
2	32.030	CHLORTHION	3499048

FIGURE 9B.

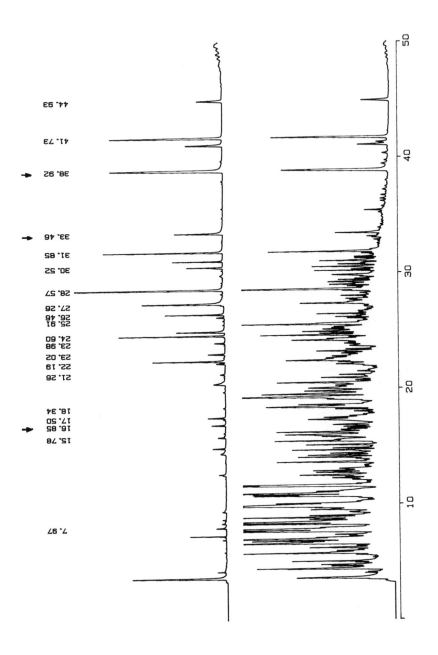

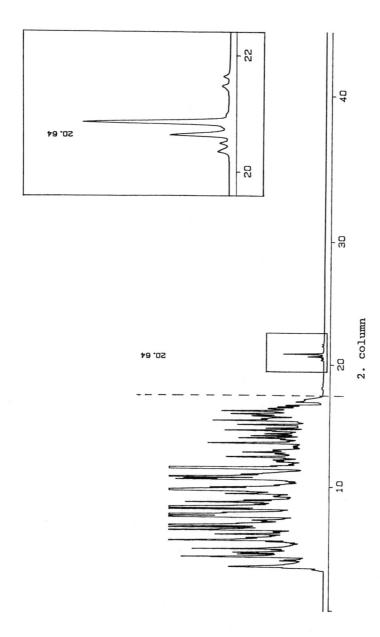

FIGURE 10. ECD-chromatograms of a screening run of an onion sample (fraction 3 + 4). Top: pesticide mixture; middle: onion spiked with pesticide mixture; bottom: transfer of propachlor to the second column with identification. (Reproduced from Stan, H.-J. and Christall, B., *Fresenius J. Anal. Chem.*, 339, 395, 1991. With permission.)

TABLE 1
Excerpt from the Calibration Table of Pesticides
Appearing in Combined Fractions 1 and 2

Pesticide	RRT-1	RRT-1+2	ECD	NPD	FPD
Dichlobenil	0.3660	0.4266	+	+	
Chlormephos	0.4363	0.4932	+	+	+
Nitrapyrin	0.4515	0.5180	+	+	
Etridiazole	0.4524	0.5108	+	+	
Chloroneb	0.5038	0.5673	+		
Tri-allate	0.8616	0.8826	+	+	
Chlorothalonil	0.8630	0.8991	+	+	
Bromocyclen	0.8799	0.9018	+		
Dichlofenthion	0.9146	0.9242	+	+	+
Vinclozolin	0.9296	0.9376	+	+	
Chlorpyrifos-methyl	0.9296	0.9471	+	+	+
Parathion-methyl	0.9296	0.9508	+	+	+
Tolclofos-methyl	0.9397	0.9582	+	+	+
Heptachlor	0.9421	0.9491	+		
Tridiphane	0.9541	0.9591	+		
Fenchlorphos	0.9573	0.9637	+	+	+
Fenitrothion	0.9800	0.9886	+	+	+
Dichlofluanid	0.9916	1.0033	+	+	
Aldrin	1.000	1.000	+		
Fenthion	1.0097	1.0185		+	+
Chlorpyrifos	1.0130	1.0280	+	+	+
Parathion	1.0130	1.0325	+	+	+
Chlorthal	1.0200	1.0172	+		
Nitrothal-isopropyl	1.0254	1.0055	+	+	
Bromophos-ethyl	1.1006	1.1035	+	+	+
DDE-*o,p′*	1.1030	1.1135	+		
Endosulfan-α	1.1159	1.1278	+		
Phoxim	1.1185	1.1318		+	+
Perthane	1.1915	1.1234	+		
Chlorthiophos I	1.1984	1.2242	+	+	+
Endosulfan-β	1.2090	1.2566	+		
Chlorthiophos II	1.2131	1.2443	+	+	+
DDD-*p,p′*	1.2220	1.2579	+		
Cyanofenphos	1.2789	1.3546	+	+	+
Endosulfan-sulfate	1.2880	1.3699	+		
DDT-*p,p′*	1.2920	1.3446	+		

Note: RRT-1: Relative retention time on the screening column — 50-m × 0.32-mm fused silica SE 54, 0.25 μm film, Nordion. RRT-1 + 2: Relative retention time on the screening and the confirmatory column — 30-m × 0.32-mm fused silica DB-17, 0.25 μm film, J & W. RRT of Aldrin = 1.000

corresponding reports are given in the lower half of Figure 8. Pepper has no complex matrix; therefore, the interesting peaks were transferred totally to the second column by switching to the second column at 27 min as indicated. The retention times and the response values were found to be identical to those of chlorpyrifos (RT 31.87 min). Parathion was rejected because no peak was found at the expected retention time of 31.99 min. The endosulfan group was additionally confirmed by RTs of 34.91, 38.91, and 42.21 min. These three peaks are only detected by ECD as expected.

Final quantification was carried out on the first column. Looking at the peak intensities from the screening, the concentrations of the confirmed pesticides were estimated to be approximately 0.15 mg/kg of chlorpyrifos and 0.35 mg/kg of endosulfan. A standard containing concentrations equivalent to 0.15 mg/kg of chlorpyrifos and 0.1 mg/kg of aldrin (ISTD for the ECD) as well as a mixture of 0.2 mg/kg each of α- and β-endosulfan and 0.1 mg/kg of endosulfan sulfate was prepared. In this example, the green pepper sample was found to be contaminated with 0.13 mg/kg of chlorpyrifos and 0.38 mg/kg of endosulfan determined as the sum of the three metabolites.

B. SECOND EXAMPLE: BROCCOLI

A second example was chosen to demonstrate the determination of a more polar pesticide, which was found in broccoli from Italy. In Figure 9, the three chromatograms of the combined fractions 3 and 4 from the cleanup according to multimethod S19 are demonstrated. In these fractions, medium polar pesticides are found as shown in an excerpt of the corresponding calibration in Table 2.

As usual in the ECD trace a few peaks were found, but they appear small compared to the chlorthion peak. Four pesticides were indicated in the report table of the ECD chromatogram. One of them also exhibited response to the FPD and NPD, indicating the presence of an organophosphate; therefore, dimethoate was a suspect pesticide residue. The confirmatory run with transfer of the effluent from the first to the second column at the indicated retention time of 20 min confirmed the presence of dimethoate, while the other suspects were eliminated. This is a good example for recognition of individual pesticides in critical pairs by their difference in detector response. Bendiocarb forms a critical pair with dimethoate on the first column and is poorly resolved on the second one.

C. THIRD EXAMPLE: ONIONS

The last example from our laboratory was chosen to demonstrate the capacity of two-dimensional GC with a food sample of very complex nature, namely, onions.[24,25] In the combined fractions 1 and 2, no matrix peaks are observed in the NPD and FPD trace and only a few in the ECD trace. In the combined fractions 3 and 4, negligible matrix peaks were found in the NPD

TABLE 2

Excerpt from the Calibration Table of Pesticides Appearing in Combined Fractions 3 and 4

Pesticide	RRT-1	RRT-1+2	ECD	NPD	FPD
Dichlorvos	0.2926	0.3400	+	+	+
Pebulate	0.4494	0.4944		+	
Methacrifos	0.4952	0.5512	+	+	+
Molinate	0.5254	0.5041		+	
Heptenophos	0.5773	0.6525	<	+	+
Atratone	0.7411	0.8078		+	
Demeton	0.7439	0.8069	<	+	+
Bendiocarb	0.7454	0.8376		+	
Dimethoate	0.7454	0.8385	+	+	+
Simazine	0.7539	0.8271	<	+	
Prometon	0.7563	0.8172		+	
Carbofuran	0.7587	0.8461		+	
Chlorbufam	0.7618	0.8209	<	<	
Terbutryn	0.9743	0.9806		+	
Ioxynil	0.9809	1.2063	+		
Pirimiphos-methyl	0.9729	0.9838		+	+
Pentanochlor	0.9848	0.9799	<	<	
Bromacil	0.9874	1.0071	+	+	
Dichlofluanid	0.9916	1.0033	+	+	
Dipropetryn	0.9950	1.1601		+	
Amidithion	0.9950	1.0124		+	+
Thiobencarb	0.9957	0.9995		+	
Malathion	0.9964	0.9990	+	+	+

Note: RRT-1: relative retention time on the screening column — 50-m × 0.32-mm fused silica SE 54, 0.25 μm film, Nordion. RRT-1 + 2: relative retention time on the screening and the confirmatory column — 30-m × 0.32-mm fused silica DB-17, 0.25 μm film, J & W. RRT of Aldrin = 1.000.

and FPD trace; the ECD chromatogram, however, was overcrowded with matrix peaks, as can be seen in the middle chromatogram of Figure 10. All pesticides that respond to NPD and FPD can easily be detected in onions. However, pesticides that respond only to the ECD and elute with retention times up to 31 min in the GC system described cannot be recognized in one-dimensional gas chromatography. This holds true also when analyzing on columns of different selectivity because of changing overlaps.

A mixture of pesticides currently in use and occurring in fractions 3 and 4 after cleanup was added to onions. The mixture of these 22 pesticides was analyzed together with the internal standard chlorthion on the first column (screening). Many of these pesticides produce signals with the ECD (see the upper part of Figure 10). The same mixture was added to an onion sample before cleanup was started.

Therefore, the middle chromatogram represents all matrix compounds from onions and the pesticides as detected with the ECD. Three of the pesticides can only be recognized by their ECD signal, namely, propachlor, chloropropylate, and bromopropylate. The corresponding peaks are indicated by arrows. Two of these pesticides, namely, chloropropylate and bromopropylate, elute late in the chromatogram so that they are not overlapped by matrix compounds. Propachlor, however, cannot be detected among the matrix compounds. The lower chromatogram demonstrates how to detect propachlor in onions by applying two-dimensional gas chromatography. A heart-cut was carried out precisely between 16.71 and 16.95 min, because propachlor was expected to elute at 16.85 min. At 16.97 min, back-flush was set into operation to avoid overlapping signals from later eluting matrix compounds. Six peaks were found after chromatography of this small fraction on the second column, with only two of them of noticeable size. The peak at 20.64 min is indicative for propachlor, while the other peak is produced by an unknown matrix compound.

V. ANALYSIS OF PCB CONGENERS

Another example of the application of two-dimensional gas chromatography was chosen not from pesticide residue analysis but from the very similar analysis of polychlorinated biphenyls (PCBs), which is of great importance in surveying seafish. PCBs are ubiquitous environmental contaminants that may be present as complex mixtures of as many as a total of 209 theoretically possible congeners. At present not all congeners that occur in commercial mixtures or environmental samples can be separated into individual peaks with the use of only one column. Separation, however, is desirable, in particular, for the exact quantitative determination of those congeners which are the most toxic. Duinker et al.[26] demonstrated the separation of a few such congeners from commercial mixtures. An example is given in Figure 11, where Clophen A 50 is first separated on a 25-m fused silica SE-54 column. Consecutive cuts were made to transfer critical pairs to the second column for separation. The second column was 30-m fused silica coated with OV-210. Back-flush was not necessary because both columns were connected to separate ECDs.

Other authors also applied two-dimensional gas chromatography to PCB congener analysis. Günther et al.[27] used a DB-5 capillary column (equivalent to SE-54) as the first column in order to separate according to volatility and, as a second column, he used SB-Smectic, a liquid crystalline phase. A mass

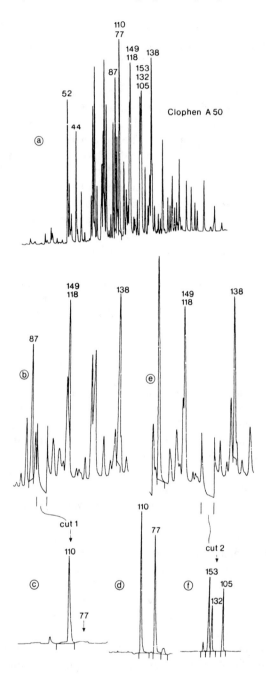

FIGURE 11. Separation of PCB congeners of Clophen A50. (Reprinted with permission from Duinker, J. C., Schulz, D. E., and Patrick, G., *Anal. Chem.*, 60, 478, 1988. Copyright 1988 American Chemical Society.)

spectrometer was connected to measure the major congeners, while an ECD was used to detect the minor components. The same column combination was used by Issaq et al.,[28] but they connected to two FIDs, while Sonchik[29] combined a methylsilicone capillary column with a DB-1701 column and used an FID as a monitor and an ECD as the main detector.

In the field of pesticide residue analysis, Tuinstra et al.[30] has already demonstrated two-dimensional gas chromatography as a sampling technique that enables the injection of larger sample volumes. Up to 400 μl of sample extract was injected by the hot splitless technique onto a short wide-bore column. After venting the solvent, the analysis of chlorinated pesticides took place on a fused silica analytical column with 0.22 mm I.D. The chromatograms exhibited good peak shape, and the detection limit could be decreased without further concentration of the extract.

VI. SAMPLE TRANSFER WITH INTERMEDIATE TRAPPING

Two-dimensional gas chromatography can be performed as described in this chapter by direct transfer of small or larger fractions from the first to the second column with the second column held at about the same temperature as the first one. The separation achieved on the first column is fully retained at the beginning of the chromatographic process on the second column. As can be seen from the examples from our laboratory shown in Figures 7 to 10, the additional residence time in the second column is very short, i.e., only 3 min on a 30-m capillary column for the peaks eluting within 35 min from the first column.

Another approach is transfer with intermediate trapping at the top of the second column in a narrow band. With pesticides, this can easily be achieved by holding the second column at 100°C. The pesticides are trapped in a narrow band, and no peak broadening was observed.[12] The analysis on the second column can be started at any time by applying an independent temperature program. In a few cases, a better separation of critical pairs was observed; in most cases, however, the elongation of analysis time is a drawback that is not offset by a corresponding better separation.[12] There is one important exception when using our two-dimensional approach unifying the effluent streams from both columns to one set of detectors. If, for instance, a series of fractions from a complex chromatogram has to be transferred from the first to the second column, then the fractions have to be trapped together at the top of the second column before back-flush of the first column may be started. Then the chromatography on the second column can be performed without possible interferences from later eluting compounds from the first column. We applied this method to onion and leek samples, which produce chromatograms of a very complex nature, and trapped five transfer fractions before starting the confirmatory analysis on the second column.[31]

VII. CONCLUSIONS

The technique of two-dimensional capillary gas chromatography applying pneumatic switching between two columns of different polarity with three selective detectors has been proven to allow the separation and identification of more than 300 pesticides on the basis of their retention times and three detector signals (see Tables 1 and 2).

Critical pairs of pesticides can additionally be distinguished by using the response ratio of the selective detectors characteristic for each substance.

The system configuration used allows an efficient, semiautomated pesticide residue analysis. It seems to be very promising for routine application in the laboratory.

Identification and quantification of all 305 pesticides is achieved by using linked retention data from both columns and the response ratios of the three detectors.[23] The analytical method presented has been proven to give reliable results with the analysis of food samples with matrices of varying complexity. It should be emphasized that this method enables both screening and confirmatory analysis using one single instrument. On the other hand, handling and maintaining the system described here, based on a Sichromat 2 gas chromatograph, is not as easy as an instrument equipped with only one column and one detector.

The two-dimensional gas chromatographic system, however, is versatile with respect to column selection and detector connection. It is unsurpassed by any other chromatographic system when looking at separation power. This makes it the best choice for combination with mass spectrometry. A combination of these two techniques may gain much popularity in pesticide residue analysis in the future because it can also be applied equally to automated screening analysis[32] and to confirmatory analysis with unequivocal pesticide identification.

VIII. APPENDIX

A. INSTRUMENTATION

The system configuration used and shown in Figure 5 consists of a double oven gas chromatograph SiChromat 2 (Siemens AG, Karlsruhe, Germany) equipped with an autosampler AS 200 (Siemens AG), one splitless injection port, and three selective detectors in parallel: an electron capture detector (ECD; Siemens AG) for compounds containing electron-capturing groups, a flame-photometric detector (FPD; Siemens AG) in phosphorus mode, and a flame-thermionic detector (NPD; Carlo Erba Strumentazione, Rodano, Italy) to detect nitrogen and phosphorus. To apply two-dimensional gas chromatography, a live T-piece (Siemens AG) is installed allowing valveless pneumatic column switching.

The gas chromatographic system is linked to an industry standard personal computer (PC-AT) by two intelligent multichannel interfaces (Nelson Analytical, Cupertino, CA, U.S.) connected to the analog output signals of the gas chromatograph. In the computer system, a special software package for the processing of chromatographic data is installed (Chromatography Software Model 2600, Nelson Analytical). The personal computer is equipped with 640 kilobyte RAM, a 40 megabyte hard disk, a single floppy disk drive (1.2 megabyte), and a hercules compatible graphic monitor. For printouts, a matrix printer is used.

B. COLUMN CONFIGURATION

Parallel detection using two-dimensional capillary gas chromatography cannot be carried out with the configuration as suggested by the manufacturer. Our realization of the parallel detection is shown in Figure 5 (bottom).

In the gas chromatograph, two columns of different polarity are installed, coupled by the live T-piece. The first column is a fused silica column 50-m × 0.32-mm I.D. SE-54, chemical-bonded, of 0.25 μm film thickness (Nordion Instruments, Helsinki, Finland). A chemical-bonded fused-silica column 30-m × 0.32-mm I.D. DB-17 with $D_f = 0.25$ μm (J & W Scientific, Folsom, CA,U.S.) is used as the second column. To avoid overloading the capillary columns with the matrix from food samples, a retention gap (fused silica, DPTMDS-deactivated: 2.5 m × 0.32 mm; SGE, Austin, TX, U.S.) is connected to the first column by means of a press-fit connector (0.32 mm/0.32 mm; ICT, Frankfurt, Germany). For two-dimensional operation, two restrictor capillaries (fused silica, DPTMDS-deactivated: 2 m × 0.31 mm; SGE) are necessary, connected to the live T-piece. Parallel detection is performed by splitting the effluent from the first as well as the second column to all three detectors by means of three fused-silica splitter capillaries (DPTMDS-deactivated: 0.45 m × 0.31 mm; SGE) and two crosspieces (Gerstel, Mülheim, Germany).

C. GAS CHROMATOGRAPHIC CONDITIONS

Hydrogen is used as carrier gas with $p_a = 1.05$ bar and $p_m = 0.6$ bar. The temperature of the splitless injection port is set at 220°C, and the split valve is closed for 60 s during injection. The injection volume is 1 μl (autosampler).

The ECD (280°C) runs with nitrogen as makeup gas (30 ml/min). The NPD is set to 300°C with hydrogen (35 ml/min) and air (350 ml/min) as fuel gas. The makeup gas is helium with a flow rate of 35 ml/min. The temperature of the FPD in phosphorus mode is 270°C. The fuel gases are hydrogen (72 ml/min) and air (65 ml/min). Nitrogen is used as makeup gas (20 ml/min).

Temperature programming conditions are 1 min at 100°C, with 25°/min to 150°C; after 2 min with 3°C/min to 205°C; and then with 9°C/min to 260°C, 20 min.

ACKNOWLEDGMENTS

The work was in part financially supported by a grant from Der Bundesminister für Wirtschaft in Bonn in the ERP program and a scholarship from the Hermann-Schlosser-Stiftung given to B. Christall.

REFERENCES

1. **Huber, J. F. K. and Smit, H. C.**, *Fresenius Z. Anal. Chem.*, 245, 84, 1969.
2. **Bertsch, W.**, in *Multidimensional Chromatography*, Cortes, H. J., Ed., Marcel Dekker, New York, 1990, 74.
3. **Stan, H.-J. and Heil, S.**, *Fresenius J. Anal. Chem.*, 339, 34, 1991.
4. **Stan, H.-J.**, *Lebensmittelchem. Gerichtl. Chem.*, 42, 31, 1988.
5. **Anderson, A. and Ohlin, B.**, *Var Föda*, 38 (Suppl. 2), 79, 1986.
6. **Giddings, J. C.**, *J. High Resolut. Chromatogr. Chromatogr. Commun.*, 10, 319, 1987.
7. **Bertsch, W.**, *J. High Resolut. Chromatogr. Chromatogr. Commun.*, 1, 85, 1978.
8. **Bertsch, W.**, *J. High Resolut. Chromatogr. Chromatogr. Commun.*, 1, 187, 1978.
9. **Bertsch, W.**, *J. High Resolut. Chromatogr. Chromatogr. Commun.*, 1, 289, 1978.
10. **Deans, D. R.**, *J. Chromatogr.*, 203, 19, 1981.
11. **Schomburg, G.**, in *Sample Introduction in Capillary Gas Chromatography*, Sandra, P., Ed., Huethig, Heidelberg, 1985, chap. 12.
12. **Mrowetz, D.**, Dr. Rer. Nat. thesis, Technical University, Berlin, Germany, 1985.
13. **Stan, H.-J.**, *CLB Chem. Labor. Betrieb.*, 35, 284, 1984.
14. **Schomburg, G.**, *Gas Chromatography — a Practical Course*, VCH Publishers, Weinheim, Germany, 1990, chap. 7.3.
15. **Sandra, P., David, F., Proot, M., Diricks, G., Verstappe, M., and Verzele, M.**, *J. High Resolut. Chromatogr. Chromatogr. Commun.*, 8, 782, 1985.
16. **Sandra, P. and David, F.**, in *Multidimensional Chromatography*, Cortes, H. J., Ed., Marcel Dekker, New York, 1990, chap. 4.
17. **Deans, D. R.**, *Chromatographia*, 1, 18, 1968.
18. **Stan, H.-J. and Goebel, H.**, *J. Chromatogr.*, 268, 55, 1983.
19. **Specht, W. and Tillkes, M.**, *Fresenius Z. Anal. Chem.*, 301, 300, 1980.
20. **Specht, W. and Tillkes, M.**, *Fresenius Z. Anal. Chem.*, 322, 443, 1985.
21. **Thier, H.-P. and Zeumer, H., Eds.**, Method S19, in *DFG Pesticide Commission Manual of Pesticide Residue Analysis*, VCH Publishers, Weinheim, Germany, 1987.
22. **Stan, H.-J. and Lipinski, J.**, *HP Pesticide Library Manual*, Hewlett Packard, Palo Alto, 1989.
23. **Stan, H.-J. and Heil, S.**, in preparation.
24. **Stan, H.-J. and Christall, B.**, *Fresenius J. Anal. Chem.*, 339, 395, 1991.
25. **Christall, B.**, Dr. Rer. Nat. thesis, Technical University, Berlin, Germany, 1992.
26. **Duinker, J. C., Schulz, D. E., and Patrick, G.**, *Anal. Chem.*, 60, 478, 1988.

27. **Günther, F. R., Chesler, S. N., and Rebbert, R. E.**, *J. High Resolut. Chromatogr. Chromatogr. Commun.*, 12, 812, 1989.
28. **Issaq, H. J., Fox, S. D., and Muschik, G. M.**, *J. Chromatogr. Sci.*, 27, 172, 1989.
29. **Sonchik, S. M.**, *J. Chromatogr. Sci.*, 24, 22, 1986.
30. **Tuinstra, L. G. M., Traag, W. A., van Munsteren, A. J., and van Hese, V.**, *J. Chromatogr.*, 395, 307, 1987.
31a. **Christall, B. and Stan, H.-J.**, *Lebensmittelchemie*, 45, 26, 1991.
31b. **Klaffenbach, P. and Stan, H.-J.**, *J. High Resolut. Chromatogr.*, 14, 754, 1991.

Chapter 9

HEADSPACE METHODS FOR DITHIOCARBAMATES*

Alan R. C. Hill

TABLE OF CONTENTS

I. INTRODUCTION

Dithiocarbamates are widely used in agriculture, but they also have important uses in inorganic analysis, in the rubber industry, and in medicine. Useful reviews of many aspects of the properties, uses, and analysis of the dithiocarbamates were provided by Thorn and Ludwig[1] and Raizman and Thompson.[2]

Most dithiocarbamates are produced by the reaction of carbon disulfide (CS_2) with an amine under alkaline conditions, which may be followed by oxidation, to form a disulfide (such as thiram, a thiuram disulfide), or precipitation, either as a heavy metal salt of defined chemical composition (such as ziram, a dialkyldithiocarbamate) or as a heavy metal salt of incompletely defined, polymeric nature (such as mancozeb, an ethylenebisdithiocarbamate).

$(CH_3)_2NCS.SSCS.N(CH_3)_2$
Thiram

$(CH_3)_2NCS.SZnSCS.N(CH_3)_2$
Ziram

$[-SCS.NHCH_2CH_2NHCS.S.Mn-]_x(Zn)_y$
Mancozeb

Ethylenethiourea (ETU)

In agriculture, a few dithiocarbamates are used for soil sterilization and others are used as bird and animal repellents, but the most widespread use has been as protectant fungicides. These fungicides are, or have been, used on practically all edible and ornamental crops for the control of mildews, rusts, blights, *Botrytis,* root rots, etc. Dithiocarbamates themselves are generally of low mammalian toxicity, but they are often applied to crops in high doses to achieve disease control, and thus the residues of the parent compounds at harvest can be significant. However, these pesticides have been a focus of attention for many years, mainly because of concern about the risks associated with ethylenethiourea (ETU), a metabolite and decomposition product of ethylenebisdithiocarbamates (EBDCs). Nonagricultural uses of dithiocarbamates can be important in pesticide residue analysis, because of the risk of sample contamination with elastomer vulcanization accelerators.

The physical and chemical properties of the more complex dithiocarbamates make analysis and specific identification difficult. The large number of different analytical methods published reflects the formidable problems presented by these compounds and their decomposition products. Most maximum residue limits for dithiocarbamates are expressed as CS_2 (rather than as the dithiocarbamates) in recognition of the widespread use of analytical methods that determine this moiety. Headspace methods, because of their speed and simplicity, are widely used for monitoring residues of dithiocarbamates. However, all methods for dithiocarbamate residue analysis require considerable care in sample preparations, the analysis, and the interpretation of the results.

II. GENERAL ANALYTICAL CONSIDERATIONS

Many of the factors affecting headspace analysis for dithiocarbamates are not specific to the technique; some apply to most methods, and these are considered in this section.

A. PHYSICAL AND CHEMICAL PROPERTIES

The dithiocarbamates used in agriculture are not volatile, are rather labile, and some are insoluble (without decomposition) in any solvent. The properties of those in most widespread use are given by Worthing and Hance.[3] The compounds of simple composition, such as thiram, can be extracted from crops with a solvent, but the more complex and insoluble compounds (e.g., mancozeb, metiram) can only be extracted after decomposition of the residues in some way.

The dithiocarbamate soil fumigants (e.g., dazomet, metam) decompose rapidly in soil to give methylisothiocyanate. Their formulations have been analyzed by decomposition to CS_2[4,5] (though not by headspace techniques), but their primary residues are not degradable to CS_2.

B. CONDITIONS FOR DECOMPOSITION TO CS_2

The decomposition of dithiocarbamates under acidic conditions usually leads (at least in part) to the formation of carbon disulfide and amine. This degradation is equivalent to the reverse of synthesis, but the reaction conditions markedly alter the yield of the products. The effects of the type of acid used, the reaction temperature, and the inclusion of reducing agents have been studied by many workers and, although most of this work did not relate specifically to headspace analysis, it is directly applicable.

Raizman and Thompson[2] pointed out that EBDCs may decompose under acid conditions by two routes. At high temperatures, 2 mol of carbon disulfide may be produced per mol of EBDC, but at lower reaction temperatures 1 mol each of CS_2, hydrogen sulfide (H_2S), and ETU may be produced. They also suggested that thiram does not give 2 mol of CS_2 on acid hydrolysis, but 1 mol each of CS_2, carbonyl sulfide (COS), and elemental sulfur. Some recent reports have reiterated the need for rapid heating and high temperatures to obtain quantitative degradation of dithiocarbamates to CS_2.[6,7] However, Keppel[8] demonstrated that good recoveries of CS_2, and little or no H_2S, were obtained from a range of dithiocarbamates if stannous chloride ($SnCl_2$) was included and if the HCl was added cold and then boiled, rather than adding the acid hot. Kiba et al.[9] refluxed dithiocarbamates in 2 M HCl with SnCl$_2$ and found that the concentration of the latter could be varied from 0.5 to 3% without adversely affecting the yield of CS$_2$, but they also found that EBDCs produced a small proportion of H$_2$S and COS (in addition to the CS$_2$). Details of the digestion conditions used for headspace methods are given in Section III.A, but most authors employed a mixture of HCl and SnCl$_2$, and good recoveries of CS$_2$ were obtained from the dithiocarbamates analyzed.

The inhibitory effect of copper compounds on the degradation of dithiocarbamates to CS$_2$ has been noted by many workers, and it can affect headspace analysis.[10,11] Lesage[12] attributed the effect to the low solubility and high stability of copper complexes. Raizman and Thompson[2] reviewed improvements to the digestion conditions of copper-containing dithiocarbamate formulations, including the use of sulfides, EDTA, and ethanol. Maini and Boni[13] assessed their solvent-layer method for the analysis of formulations containing copper oxychloride and found no apparent depression of CS$_2$ evolution under the digestion conditions employed (3% SnCl$_2$ in concentrated HCl at 80°C). Irth et al.[14] found that a 1:1 mixture of 10 mM EDTA and 10 mM potassium citrate could inhibit, but not reverse, the complexation of thiram with cupric ions. Residues in water samples were stabilized during up to 16 h of storage by addition of this reagent.

C. CALIBRATION STANDARDS

The CS$_2$ calibration solutions may be prepared by weight or volume, dispensing the pure liquid into a solvent, such as ethanol, acetone, or hexane. Thier and Zeumer[7] recommended that larger volumes than those suggested by Keppel[8,15] should be weighted and pipetted, and this should minimize evaporation errors. Because of the high volatility of CS$_2$, it is important to use flasks with tightly fitting stoppers, and the solutions are best prepared immediately before use.

Soluble dithiocarbamates, such as thiram, can be obtained in a pure state, and the preparation of calibration standards is straightforward, whereas for the more complex dithiocarbamates it is not. The latter are not normally available in a pure form, they cannot be dissolved without decomposition, and most are not particularly stable. If their purity is determined by a CS$_2$-generation technique, no independent method of calibration is involved. The standards can be prepared by dilution in powder, and this approach obviously has merit if the material used for calibration is a dust formulation. Gordon et al.[16] prepared dilutions of dithiocarbamates in attapulgite, which they claimed gave very stable standards. Howard and Yip[17] used dilutions in starch, as did Keppel,[8] who found them to be stable for at least 6 months when stored in dry, closed containers. McLeod and Ritcey[18] recommended preparation of standards in talc, by mixing in a blender, but they stated that the mixtures would be stable for only 24 h. Yao et al.[19] also used talc and stored the standards in a low-temperature refrigerator, but they noted that care is required to get a homogeneous mixture. Thier and Zeumer[7] also recommended talc dilutions, but others have suggested mixing with lactose,[20] sodium sulfate, starch, or sand.[21]

As many of the complex dithiocarbamates are formulated for use as wettable powders, it can be convenient to prepare standard suspensions of them. These should be prepared and agitated thoroughly immediately before use. Sasaki et al.[6] determined the stability of aqueous suspensions of maneb and zineb at 5°C. They found that zineb was much more stable than maneb, but nonetheless recommended that the suspensions should be prepared immediately before use. McGhie and Holland[22] used maneb formulations in aqueous suspension; using Tween 80 as a dispersing agent and sonication to homogenize the suspensions, a new standard was prepared every second day. Some workers have used aqueous solutions in EDTA,[24] which, in the case of EBDCs, represent partially degraded standards. As noted by Ripley,[25] most authors using EDTA solutions stress that the standards should be prepared daily. Ripley investigated the further decomposition in EDTA solution and showed that maneb had a half-life of about 2 d. Rangaswamy et al.[23] used 0.5 M sodium hydroxide to dissolve their standards of zineb but noted that these had to be prepared freshly because inconsistent results were obtained if the standards were stored for 3 to 4 d in the refrigerator.

D. SAMPLE PROCESSING

Residues of dithiocarbamates on plants are frequently distributed very unevenly. Because of the labile nature of the residues, satisfactory subsampling for accurate dithiocarbamate analysis is difficult, although the problem is not very widely acknowledged. Yip et al.[26] were concerned by differences of 10 to 20 mg/kg between duplicate determinations of lettuce samples that contained 40 to 50 mg/kg, which were ascribed to subsampling problems. In the author's experience, this appears to demonstrate fairly good agreement between subsamples. Few data have been published on the effects of sample processing on dithiocarbamate residues, although several recent authors have cautioned against excessive tissue disruption. Howard and Yip[17] investigated the breakdown of maneb and zineb in fresh, refrigerated, or frozen kale that had been chopped prior to the addition of the fungicides. Up to 25% of maneb was lost immediately after mixing, but very litle zineb was lost at this stage. A further 5% of the maneb and about 10% of the zineb was lost after 3 d in the refrigerator. There was only a slow decline during 3 d in the freezer (after the initial losses incurred at mixing). The losses of field-incurred residues could have been greater than those observed by addition of the pesticides after chopping, but this was not tested. McLeod and Ritcey[18] noted that blending of leafy samples gave low recoveries of extractable dithiocarbamates. Nonetheless, many other workers (most of whom did not use headspace techniques) have reported residue data from samples that were thoroughly comminuted before analysis.

Thier and Zeumer[7] suggested cutting wedge-shaped subsamples from large fruit and vegetables, but that samples such as lettuce should have the leaves detached and mixed. They recommended that subsamples should be analyzed without delay after preparation, and they cautioned that samples should not be finely comminuted because high losses of dithiocarbamates may result. Others[10,20] have made similar recommendations. Experience at the author's laboratory indicates that samples (of perishable commodities, at least) should be analyzed as soon as possible after receipt and that all subsampling should involve a minimum of cutting. We have found that the simplest way to take subsamples from most fruits and vegetables is to remove segments and place them in the digestion bottles with a minimum of further tissue disruption. As segments may not provide fully representative subsamples, it may be necessary to perform more replicate analyses than if the samples could be thoroughly comminuted and mixed.

Where residues of the complex dithiocarbamates are apparently detected deep inside the tissue of crops or in processed foods, it seems likely that metabolites capable of producing CS_2 are being detected. Marshall[27] demonstrated that a number of EBDC metabolites (not ETU) could degrade with the liberation of CS_2 when heated in aqueous buffers, and this may also occur under the conditions used for headspace analysis. There is a conflict between the need to avoid premature degradation of dithiocarbamate residues and the need to comminute samples, such as apples, so that the digestion of residues to CS_2 is complete. Further work is needed to optimize methods for sample processing for dithiocarbamate analysis, whatever determinative technique is used.

E. NATURAL SOURCES OF CS_2

There are sources of CS_2 other than dithiocarbamates. Certain plant materials produce small amounts of CS_2, either naturally or under the reaction conditions used for digestion. Improvements in the sensitivity of detectors and regulatory changes toward lower maximum residue limits (MRLs) have made it important to take into account CS_2 of natural origin. Most information published on natural sources of CS_2 is not related to dithiocarbamate analysis, but if CS_2 (or a precursor that can be decomposed to it) is present in untreated plant materials, it may be interpreted erroneously as representing residues of dithiocarbamates. After digestion for headspace analysis, some vegetables (unspecified) were reported to produce up to 0.08 mg/kg of CS_2.[20] Ott and Gunther[24] stated that background interference is to be expected from fruit peels, especially banana. Carbon disulfide was identified as a natural volatile component of shiitake mushrooms by Chen and Ho.[30] It has also been reported at low concentrations (in the microgram per liter range) in the natural headspace volatiles of some wines by Leppänen et al.[31] and Spedding et al.[32] and in those of raw grain spirit by Ronkainen et al.[33] Heikes[34] determined the CS_2 content of a wide range of table-ready foods, but the only samples found to contain it were of certain cereal products (up to 0.24 mg/kg in corn chips). The source of this CS_2 was not identified.

Onions and brassicas are most frequently claimed to be sources of natural CS_2, but Sasaki et al.[6] found no CS_2 produced by digestion of oranges, tomatoes, cucumbers, potatoes, or onions. Carson and Wong[28] found no CS_2 among the natural volatiles of onions, and, although Schreyen et al.[29] reported CS_2 among the natural volatiles of leeks, they indicated that it originated from the use of the compound as a solvent. In the author's laboratory, little CS_2 has been found to be produced from untreated onions or leeks during headspace analysis for dithiocarbamates. The occurrence of CS_2 as a natural volatile component of cabbages was noted by Buttery et al.,[35] although Gordon et al.,[16] in determining dithiocarbamates, did not report any CS_2 from cabbages or broccoli. Ripley[25] recorded 0.15 mg/kg of zineb in untreated control cabbages. McLeod and McCully[36] found that H_2S, not CS_2, was produced by untreated cabbages in headspace analysis for dithiocarbamates, but it is possible that they would not have detected the low levels which can sometimes occur. Uno et al.[37] noted that untreated cabbage and rape produced CS_2 up to 0.1 mg/kg after $HCl/SnCl_2$ digestion. Thier and Zeumer[7] also noted that high blank values may be obtained with rape seed, swedes, cauliflowers, and Savoy cabbages. In the U.K., a high proportion of brussels sprouts samples analyzed for dithiocarbamate residues by a headspace method appeared to produce CS_2.[37a] A subsequent study in the author's laboratory has shown that this particular commodity is capable of producing CS_2 (identity confirmed by coupled gas chromatography/mass spectrometry (GC/MS)) at levels up to 2 mg/kg from samples untreated with dithiocarbamates, particularly if the sprouts are allowed to age for a few days after picking. Other brassicas generally seem to produce less CS_2, although this may be dependent upon the variety. Carbonyl sulfide is also to be found in the headspace of some types of untreated plant material digested for analysis.

F. CONTAMINATION

Recently, a problem of contamination in the dithiocarbamate headspace analysis of melons was traced to handling of the fruit with latex rubber gloves.[38a] The levels recorded were quite low (generally less than 0.5 mg/kg), but, presumably, commodities smaller than melons could receive higher levels of contamination. Ordinary household gloves may be formulated with dithiocarbamates as vulcanization accelerators at about 1% by weight.[38] The potential for transfer of these compounds to samples (and to analysts' hands) is, therefore, high and it is recommended that gloves used for handling samples, or at any stage in the analysis, should be checked as a source of contamination. Samples may have been handled with rubber gloves or may have been in contact with rubber objects before arrival at the laboratory, and this could make it difficult to be certain that residues are derived from the use of agricultural pesticides.

A very wide range of dithiocarbamates and related compounds may be used as vulcanization accelerators in the production of a wide range of elastomers. Some (such as thiram) are also used as agricultural pesticides and others (such as ETU) occur as metabolites of agricultural pesticides. Franta[39] provided a comprehensive guide to vulcanization accelerators and their uses. Fortunately, CS_2-generating vulcanization accelerators are not used in the manufacture of silicone rubber septa of the type normally used to seal headspace analysis bottles, but it is worth noting that other elastomers are used in certain analytical vial caps, and ETU has been found to be present in some of these.[40] The potential CS_2 contamination from septa should be assessed if silicone rubber is not used.

G. CONFIRMATION OF IDENTITY AND QUANTITY

Given the potential for sample contamination, the natural sources of CS_2, and the possibility of interference from other volatile sulfur compounds (see Section III.B), it is important to consider qualitative and quantitative confirmation when unexpected residues are found. The labile nature of residues may make it necessary to carry out such confirmation quickly. Because of the highly uneven distribution of dithiocarbamate residues and the rapid nature of headspace analysis, the simplest confirmation is to carry out replicate analyses to establish the generality of the finding. Confirmation that the component measured is CS_2 is straightforward by mass spectrometry. However, for additional confirmation of residues, an alternative technique that does not rely on CS_2 generation is required. Determinations of the amine, other degradation products, or, in a few cases, the intact dithiocarbamates are possible but are outside the scope of this review. It should be remembered that the many difficulties that beset analysis via CS_2 may also apply to alternative methods.

Analysis of dithiocarbamates as CS_2 provides only a measure of total dithiocarbamates, which may include metabolites or other decomposition products (but not ETU) that are not dithiocarbamates as such. The distinction between residues of individual dithiocarbamates and/or their metabolites has been the goal of many workers, but, although this can be achieved in some cases, the complete identification of residues of the complex dithiocarbamates may remain impossible. However, Steinwandter[21] attempted partial characterization of dithiocarbamates by a headspace method. He found that decomposition of the water-insoluble dithiocarbamates was much slower than the water-soluble ones and proposed a differentiation of the residues on this basis, although no results on real samples were presented to support this.

III. HEADSPACE METHODS

A. DIGESTION TECHNIQUES AND CS₂ DETERMINATION METHODS

The headspace technique, as used for dithiocarbamates, is very different from most pesticide analyses. The principle is very simple: the dithiocarbamates are decomposed to CS_2 in a sealed bottle fitted with a septum, and aliquots of the headspace vapors are withdrawn for analysis, usually by GC with electron capture (ECD) or flame photometric (FPD) detectors. The decomposition is usually carried out in a heated water bath.

To calibrate dithiocarbamate headspace analysis, it is not necessary to know the amount of CS_2 injected or its partition between the vapor and liquid phases, although the distribution of CS_2 is expected to follow Henry's law (i.e., the partial pressure of the CS_2 in the vapor above the liquid should be directly proportional to the CS_2 concentration in the liquid). The linear calibrations produced by the CS_2 analysis system indicate that it conforms to expectations. The practice of stripping volatiles from the headspace to concentrate them prior to GC analysis, often used in flavor analysis, is generally too labor intensive for use in dithiocarbamate analysis and is not necessary with modern, sensitive detectors.

The pioneering work of McLeod and McCully[36] in the development of headspace analysis for dithiocarbamates has been refined by a number of workers, but the general approach remains unchanged. McLeod and McCully digested 1 g subsamples of crop in 250-ml glass bottles at temperatures up to 90°C, which they considered to be the safe temperature limit. Even at this temperature, they obtained poor recovery from EBDCs when using sulfuric acid, whereas digestion with 10 ml of 1.5% $SnCl_2$ in 4 M of HCl gave recoveries of CS_2 from ferbam, thiram, ziram, zineb, maneb, and nabam from lettuce, cucumber, carrot, apple, cabbage, and strawberries mostly in the range 70 to 100% (recoveries of nabam from lettuce and cucumber were lower). They investigated digestion temperatures and found that the optimum for headspace concentration of CS_2 was 60 to 80°C. They recommended digestion at 60°C for 30 min because the yield of CS_2 and the reaction rate were similar to those at 80°C, and there was less condensation of water in the syringe at the lower temperature when sampling the headspace. However, they found it essential to agitate the contents of the bottles to ensure a complete and rapid equilibration between the liquid and gas phases. Headspace volumes of 25 µl were analyzed on an SE-30 packed column at 60°C, using an FPD. Most headspace methods for dithiocarbamates that have been published subsequently are based on this technique. McLeod and Ritcey[18] described essentially the same method and recommended analysis of headspace within 30 min of the completion of digestion.

A headspace method has been subjected to collaborative study.[10] The digestion was carried out in 50 ml of 5 M HCl containing 1.5% $SnCl_2$ at

80°C for 1 h. Again, 250-ml bottles were used but 50-g subsamples were digested. The need to shake the contents of the bottles thoroughly at intervals during digestion was stressed. Average recovery of zineb, mancozeb, maneb, and thiram added to lettuce at about 0.2 to 10 mg/kg was in the range 71 to 94%, except that recovery of mancozeb at 0.2 mg/kg averaged 60%. For GC, OV-1 was recommended as a stationary phase, with a column temperature of 60°C. It was recommended that the saturation level of the FPD should be checked at intervals.

Steinwandter[21] proposed the use of a headspace method to differentiate water-soluble and water-insoluble dithiocarbamates. He used 250-ml bottles, with 150 ml of 3 *M* HCl and 1.2 g of $SnCl_2$ and performed the digestion at 40 to 80°C in a drying oven. The CS_2 determination was performed by GC-ECD.

Doroshenko and Pokhil'chenko[41] used 50 ml of 1.2 *M* HCl containing 1.5% $SnCl_2$ in 200- to 300-ml bottles to digest 50-g subsamples at 80°C (for not less than 1 h) to determine zineb and metiram residues in sugar beet, apples, grapes, onions, wheat grain, and potato tops and tubers. Headspace aliquots of 0.5 ml or more were injected with a syringe preheated to 50 to 60°C, onto an OV-1 or OV-17 column for GC-ECD. Recoveries were in the range 84 to 95%, and the limit of determination was 0.01 mg/kg. The method was compared with a colorimetric technique for CS_2, and a very close correlation was found with crops field-treated with a dithiocarbamate.

Yao et al.[19] digested 100-g subsamples in 600-ml bottles with 0.1 g of ascorbic acid and 150 ml of 4 *M* HCl containing 2% $SnCl_2$ at 70°C for 32 h. They checked for leaks by immersing the bottles in the water at 70°C to detect escaping bubbles. The bottles were cooled before injection of aliquots of headspace onto an SE-30 column at 60°C for determination by GC-FPD. Recovery of mancozeb, zineb, and thiram added to tomatoes, cucumbers, Chinese cabbages, and hyacinth beans was in the range 80 to 102%, and the minimum detectable level was 0.01 mg/kg.

Ott and Gunther[24] described a unique, colorimetric headspace method for rapid screening of dithiocarbamate residues in the field. They used 125-ml vials in which 30-g subsamples were digested with 0.5 g of $SnCl_2$ and about 100 ml of 4 *M* HCl at 100°C for 1 h. Then, with the vial still in the water bath, two 10-ml volumes of headspace were removed and discarded. After a further 5 min, 9 ml of headspace was withdrawn into a syringe containing 1 ml of the colorimetric reagent, and the contents of the syringe were swirled for 1 min before the absorbance of the solution was determined.

McGhie and Holland[22] digested maneb in 8-oz medicine bottles, using 25 ml of 5 *M* HCl containing 1.5% $SnCl_2$ at 70°C for 45 min.

Van Haver and Gordts[42] compared results using a modified McLeod and McCully[36] headspace method and the Keppel[15] colorimetric method for the analysis of lettuce. The correlation factor was r = 0.9903. The headspace analysis of 20-g subsamples was done in 120-ml bottles with 25 ml of concentrated HCl and 1 g of $SnCl_2$, digested overnight in a water bath at 80°C. Aliquots of the headspace were injected onto an OV-1 column at 75°C for GC-FPD.

Another method[20] recommends that 20-g subsamples are digested with 100 ml of 3 M HCl containing 1.5% $SnCl_2$ at 70°C for 2 h and that the bottles should be cooled to room temperature before injection. It is recommended that 25- to 300-μl aliquots of headspace are injected onto a Tenax GC column at 100°C for GC-FPD or GC-ECD. The method has been applied to apples, beans, broccoli, brussels sprouts, cabbages, cauliflowers, cucumbers, endives, leeks, lettuce, mushrooms, onions, pears, radishes, spinach, strawberries, and tomatoes. Average recoveries of EBDCs were 95%, of thiram were 87%, and of other dimethyldithiocarbamates were 97%. The limit of determination is quoted as 0.01 to 0.1 mg/kg, depending upon whether CS_2 of natural origin is produced by the commodity.

Spiegelenberg et al.[43] digested 20 g subsamples with 25 ml 4M HCl and 1 g of $SnCl_2$ for 4 h at 85 to 90°C. Gas chromatographic analysis was performed with OV-101 or Chromosorb 105 columns and an FPD. They obtained recoveries of thiram and zineb from apples, leeks, turnips, lettuce, and shallots in the range of 89 to 102%. The detection limit was 0.05 mg/kg except in the case of leeks, for which it was 0.2 mg/kg.

Delventhal[44] digested 30- to 50-g subsamples in 250-ml bottles with 100 ml of water, 40 ml of 4 M HCl, and 1 g $SnCl_2$ at 90°C for 2 h. Headspace aliquots of 150 to 200 μl were injected onto a mixed-phase column of DC-200 and QF-1 at 50°C for GC-ECD. Recovery of sodium diethyldithiocarbamate from lettuce, peppers, apples, and red cabbage was in the range 83 to 104%.

Hill and Edmunds[11] reviewed headspace methods for dithiocarbamates and described some of the practical problems involved. However, the claim that most plant materials do not affect the partition of CS_2 is now known to be incorrect, and, at the author's laboratory, calibration of the analyses with untreated crop samples is routinely carried out whenever possible. The analysis of high-sugar fruit, oily seeds, and nuts is particularly affected by changes in the partition of the CS_2, and the sensitivity of the analysis may be significantly lowered. Nawar[45] demonstrated that the presence of sugar, gelatin, or glycerol in the aqueous phase had marked effects on the distribution of volatiles in the headspace, and that these effects varied from one volatile chemical to another. Nawar's work was not related to dithiocarbamate analysis, but most headspace analyses may be expected to show these effects. The use of thiophene as an internal standard provides a considerable improvement in both efficiency and reliability of the assay. For the analysis of 50-g subsamples in 250-ml bottles, the addition of 2.5 mg of thiophene is suitable for calibration of CS_2 residues in the range of 0.1 to 5 mg/kg. If calibration outside this range is required, the quantity of thiophene can be adjusted accordingly. Although the relative partition of CS_2 and thiophene is not completely identical in all crop digests to that in water, the ratio in most digests is sufficiently similar to permit the use of a common crop sample for calibration. For crops in which the partition ratio differs, an untreated sample of the commodity (or a sample that has been shown not to contain residues) can be used for calibration. The internal standard also acts as a leakage check: if little or no thiophene is present, it is probably not worth measuring the CS_2. An internal standard also eliminates much of the variation introduced by the manual injection technique. A minor disadvantage of thiophene is that it is not quite as stable as CS_2 under the digestion conditions, and its relative response may decrease with time, albeit usually very slowly. For this reason, the headspace injections are normally made within an hour of the completion of the digestion. Alternative internal standards have been sought among the low molecular weight thiophenes and thiols, but those compounds that are sufficiently volatile appear to be less stable than thiophene, tend to interfere with the CS_2, or are interfered with by natural products. Thiophene appears to be produced only in very small quantities from most crops, although the methylthiophenes and dimethylthiophenes may be produced in greater quantities.

Solvent layer analyses, used by a few workers, resemble headspace techniques, but the place of the usual vapor phase is taken by a solvent overlaying the digestion mixture. One assessment of a solvent-layer procedure[10] indicated that it was relatively insensitive and subject to interference problems from the solvent; a conventional headspace technique was recommended instead. However, Maini and Boni[13] attempted to use published headspace methods[10,18] for the analysis of dithiocarbamate dusts in workroom air and found it difficult to ensure that the headspace CS_2 concentration remained within the dynamic range of the FPD. They, therefore, developed an isooctane-layer modification in which aliquots of the solvent layer could be removed and diluted as required for GC-ECD.

Large, wide-necked bottles should be used for the digestion so that large subsamples of the plant material can be introduced with the minimum of tissue disruption. However, it can be difficult to seal wide-necked bottles. The bottles used at the author's laboratory are 250-ml Schott & Mainz Duran (Fisons BTF-682-090M), the necks of which have a drip-resistant plastic pouring ring, which supports the septum without distortion when the cap is screwed on. The septum used should be of silicone rubber, because no CS_2-generating compounds are used in its manufacture, unless an alternative type is known not to produce interference. It is useful to drill an off-center sampling hole in the bottle cap so that septa can be reused. Septa can be reused for at least several analyses if they are baked in an oven at 60°C for several hours before use to avoid cross contamination.

The use of large bottles makes automation of the headspace technique difficult, but commercially available headspace analyzers can be used when small vials are employed for samples such as grain.

B. GAS CHROMATOGRAPHIC DETERMINATION OF CS$_2$

The manual injection technique used for headspace analysis is rather different from that used for more conventional GC, particularly if aliquots are removed from the headspace while it is hot. The water bath, or other heater, (preferably in a fume cupboard) should be close to the GC instrument. Use of gas-tight syringes is essential, but it is desirable to keep the syringe hot. The syringe should be rinsed with hot, distilled water, both to avoid carryover contamination and to create a meniscus at the plunger tip to help maintain the seal. The syringe should be rinsed several times with the headspace, without removing it from the bottle, and then the system should be allowed to equilibrate for a few seconds before withdrawing the syringe. It is important to make the injection quickly to avoid losses, although if thiophene is used as an internal standard, this is less critical.

The GC of the CS$_2$ produced in headspace analysis is not particularly demanding.[10] McGhie and Holland[22] investigated the influence of solvents, used for preparing the CS$_2$ calibration solutions, and of GC columns used in the headspace analysis of dithiocarbamate residues. Use of a mixed phase (1% OV-17 and 4% QF-1) column gave quenching of the sulfur response of the standards, by the coeluting hexane solvent, and a consequent overestimation of the dithiocarbamate residues. They recommended that either the CS$_2$ calibration solution be made up in acetone or octane, or an SE-30 column be used. The headspace vapors injected include fairly large quantities of oxygen, water, and hydrochloric acid, which tend to damage silicone stationary phases; this is a good reason for using cheap, robust, packed columns. With an appropriate choice of carrier gas flow rate and column temperature, it is possible to separate CS$_2$ from COS, H$_2$S, methanethiol, and thiophene in a run time of 1 min. Such short retention times provide narrow peaks and very good sensitivity. Although bonded-phase capillary columns have been used in headspace analysis at the author's laboratory, they are quickly damaged. Porous polymer packings and solid adsorbents are more stable to the headspace gases than silicones but require higher column temperatures to avoid long retention times and broad peaks (which would reduce sensitivity markedly when using the FPD). Porous-layer open-tubular columns may offer a convenient solution when the chromatographic resolution of packed columns is inadequate. The volumes of headspace injected may be from a few microliters to volumes up to 1 ml. The larger volumes may seem surprising to analysts accustomed to injecting only 1 to 5 μl of solvent extracts, but these solvent volumes may occupy an even larger volume upon vaporization.

The CS$_2$ detection is usually by use of the FPD or ECD, but other detectors, including mass spectrometers, can be employed. The ECD is highly responsive to some of the other compounds likely to be present in the head-space and usually has a smaller dynamic range[20] for CS_2 than is expected for other compounds. The problem with the FPD is that the sulfur response is an approximately square function of the quantity of CS_2 injected, so that the effective dynamic range of the detector is quite small. The limit of deter-mination and the saturation level of the FPD or ECD should be checked frequently[10] and, although the volume of headspace injected can be varied over a wide range to achieve the sensitivity required, it is better to ensure that the calibration is carried out over similar volume ranges. Retention times and peak widths should be kept to the minimum for good sensitivity, but they may be deliberately increased in order to reduce sensitivity. The use of the FPD exaggerates the apparent efficiency of the chromatographic column.

Because of the limited dynamic range of the detectors normally used, the headspace technique is rarely used for analysis of dithiocarbamate formula-tions — the subsamples analyzed would be limited to a few milligrams. However, cooling the bottles to room temperature and using smaller injection volumes would assist with this problem. The analysis of spray tank mixes may be more practical by headspace methods than other CS_2-generation tech-niques, particularly using the Ott and Gunther method.[24]

Apart from CS_2, the compounds produced from plant substrates during headspace analysis for dithiocarbamate residues have not been studied to any extent. At the author's laboratory, GC/MS of the headspace vapors from various digested plant substrates has shown that many are qualitatively similar. Common to most of them is the occurrence of H_2S, methanethiol, and various furans and furfurals; however, brassicas, onions, and related commodities tend to produce methylthiophene, dimethylthiophene, and also small quan-tities of alkyl and alkenyl sulfides. Most commodities do not appear to produce significant quantities of thiophene, which is fortunate because of its use as an internal standard. Typical total ion mass chromatograms obtained from headspace analysis of brussels sprouts and onions are presented in Figure 1.

The chromatograms in Figure 1 were produced using a JEOL DX300 double-focusing, magnetic sector mass spectrometer, with on-column injec-tion into a 1-m × 0.53-mm retention gap of uncoated, deactivated silica tubing coupled to a 25-m × 0.25-mm CP-Sil 5 column. The column was operated isothermally at 40°C for 1 min and then programmed at 10°C/min to 80°C; the helium carrier gas flow rate was approximately 1 ml/min. The silicones were from the stationary phase.

Onions tend not to produce very much COS, but this commodity can produce significant quantities of *n*-propanethiol and *n*-propenethiol. The CS_2 in the brussels sprouts was equivalent to 0.5 mg/kg. By comparison with a sample digested in water (i.e., no HCl and $SnCl_2$), it was evident that most of the dimethyl disulfide present in the brussels sprouts had been degraded by the normal digestion, but this compound may elute with CS_2 on some columns and could be confused with it. Many of the compounds found in the headspace from digested crops are also found in either the raw or cooked commodities. Interference with the CS_2 by other sulfur compounds is not common, but its occurrence is readily checked by mass spectrometry. The possible production of thiophene (if it is used as an internal standard) from the substrate may be checked by omitting it from a duplicate analysis.

IV. POSSIBLE FUTURE DEVELOPMENTS

Most methods for the analysis of dithiocarbamates have drawbacks. Those factors that are important in headspace analysis are dealt with above, but further developments are required to simplify the interpretation of results.

Improved methods of subsampling and sample storage are required for dithiocarbamate analysis. A comminution technique is needed to enable the complete degradation of residues (which may be of metabolites) within the flesh of commodities such as apples, without losses prior to the digestion process. The development of methods for the quantitative degradation of dithiocarbamates in the presence of high concentrations of copper compounds is desirable, but, if this is achieved, it may be possible to use the extraordinarily high stability of the copper-dithiocarbamate complexes as a means of stabilizing residues during sample processing or storage.

An ideal internal standard for calibration of headspace analysis of dithiocarbamates would be $^{13}CS_2$, which implies the use of mass spectrometric detection. It normally occurs in $[^{12}C]CS_2$ at about 1% abundance, but, if this is taken into account, its use could improve the reliability of the calibration because the partition would always be identical to that of the $^{12}CS_2$. This approach has been investigated at the author's laboratory with promising results, and the internal standard is not prohibitively expensive because of the very small quantities required for each sample.

The occurrence of CS_2 of natural origin, which is produced from some crops under the conditions of reaction, needs to be investigated further. It is possible that it could be distinguished from CS_2 of dithiocarbamate origin by employing different digestion conditions. It may be possible to distinguish residues of thiram from EBDCs in a similar fashion.

V. CONCLUSIONS

Headspace analysis for dithiocarbamates is widely used for monitoring residues in crops and foods. It is a simple and rapid technique, particularly suited to the screening of large numbers of samples. It is evident from the wide range of GC systems employed, and the differing times, conditions, and containers for digestion that have been used successfully, that headspace methods provide a flexible and robust approach to dithiocarbamate analysis.

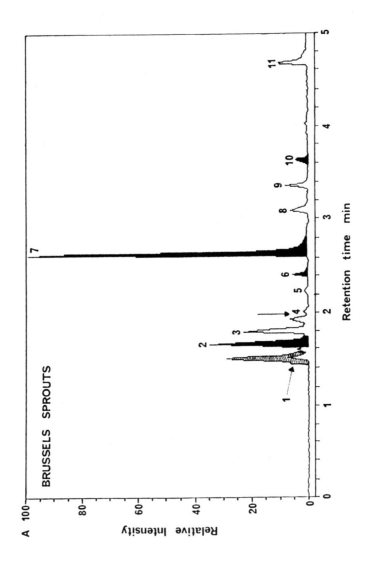

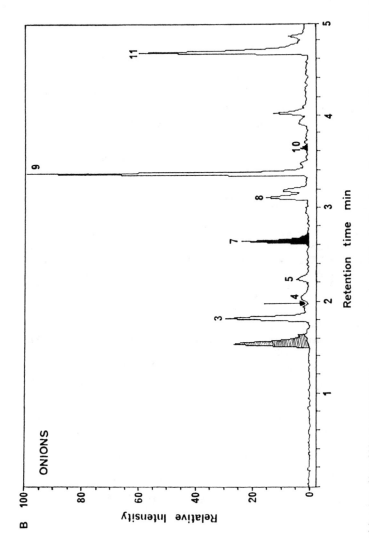

FIGURE 1. Total ion (*m/z* 60 to 200) electron-impact mass chromatograms of 20 μl of headspace from digests of 50-g samples of (A) brussels sprouts and (B) onions, untreated with dithiocarbamates. Identification: 1, COS; 2, CS₂; 3, methylfuran; 4, acetic acid; 5, dimethylfuran; 6, dimethyl disulfide; 7, methylthiophene; 8, furfural; 9, silicone; 10, dimethylthiophene; 11, methylfurfural. The sulfur compounds are indicated in black. The peak corresponding to H₂S (*m/z* 33) and methanethiol (*m/z* 48) is shown superimposed (shaded) on that of the coeluting COS. HCl is coeluted with methylfuran. The vertical arrow indicates the elution time of thiophene.

Most of the drawbacks to the technique are related to the chemical and physical properties of the dithiocarbamates and are common to many of the alternative analytical techniques. The technique does not differentiate between different types of dithiocarbamates, but, in the absence of a convenient and reliable alternative, headspace methods are likely to remain an important tool for monitoring residues of dithiocarbamates in foods for the foreseeable future.

ACKNOWLEDGMENTS

I am very grateful to all colleagues who have contributed to the development of headspace methods at this laboratory. The mass spectrometry referred to in this review was carried out by J. P. G. Wilkins; the assessment of internal standards was done by D. J. Mason; the assessment of subsampling techniques and natural sources of CS_2 were done by E. Patel.

REFERENCES

1. **Thorn, G. D. and Ludwig, R. A.,** *The Dithiocarbamates and Related Compounds,* Elsevier, Amsterdam, 1962.
2. **Raizman, P. and Thompson, Q. E.,** in *The Analytical Chemistry of Sulfur and its Compounds,* Part 2, Karchmer, J. H., Ed., Wiley-Interscience, New York, 1972, 620.
3. **Worthing, C. R. and Hance, R. J.,** *The Pesticide Manual,* 9th ed., British Crop Protection Council, Farnham, 1991.
4. **Stansbury, H. A.,** in *Analytical Methods for Pesticides and Plant Growth Regulators,* Vol. 3, Zweig, G., Ed., Academic Press, New York, 1964, 119.
5. **Gray, R. A.,** in *Analytical Methods for Pesticides and Plant Growth Regulators,* Vol. 3, Zweig, G., Ed., Academic Press, New York, 1964, 177.
6. **Sasaki, K., Takeda, M., and Uchiyama, M.,** *J. Food Hyg. Soc. Jpn.,* 17, 72, 1976.
7. **Thier, H.-P. and Zeumer, H.,** Eds., *Manual of Pesticide Residue Analysis,* Vol. 1, Dtsch. Forschungsgemeinschaft, Weinheim, Germany, 1987, 353.
8. **Keppel, G. E.,** *J. Assoc. Off. Anal. Chem.,* 54, 528, 1971.
9. **Kiba, N., Sawada, Y., and Furusawa, M.,** *Talanta,* 29, 416, 1982.
10. Report by the Panel on Determination of Dithiocarbamate Residues, *Analyst (London),* 106, 782, 1981.
11. **Hill, A. R. C. and Edmunds, J. W.,** *Anal. Proc.,* 19, 433, 1982.
12. **Lesage, S.,** *J. Agric. Food Chem.,* 28, 787, 1980.
13. **Maini, P. and Boni, R.,** *Bull. Environ. Contam. Toxicol.,* 37, 931, 1986.
14. **Irth, H., de Jong, G. J., Brinkman, U. A. Th., and Frei, R. W.,** *J. Chromatogr.,* 370, 439, 1986.
15. **Keppel, G. E.,** *J. Assoc. Off. Anal. Chem.,* 52, 162, 1969.
16. **Gordon, C. F., Schuckert, R. J., and Bornak, W. E.,** *J. Assoc. Off. Agric. Chem.,* 50, 1102, 1967.
17. **Howard, S. F. and Yip, G.,** *J. Assoc. Off. Anal. Chem.,* 54, 1371, 1971.
18. **McLeod, H. A. and Ritcey, W. R.,** Eds., *Analytical Methods for Pesticide Residues in Foods,* Department of National Health and Welfare, Ottawa, Canada, 1973.

19. **Yao, J., Zheng, Y., Shuzhen, J., Wang, Zh., and Zhao, F.,** *Sci. Agric. Sinica,* 22, 81, 1989.
20. *Analytical Methods for Residues of Pesticides,* Part 2, 5th ed., Ministry of Welfare, Health and Cultural Affairs, Rijswijk, Netherlands, 1988, 81.
21. **Steinwandter, H.,** *Fresenius Z. Anal. Chem.,* 321, 375, 1985.
22. **McGhie, T. K. and Holland, P. T.,** *Analyst (London),* 112, 1075, 1987.
23. **Rangaswamy, J. R., Poornima, P., and Majumder, S. K.,** *J. Assoc. Off. Anal. Chem.,* 54, 1120, 1971.
24. **Ott, D. E. and Gunther, F. A.,** *J. Assoc. Off. Anal. Chem.,* 65, 909, 1982.
25. **Ripley, B. D.,** *Bull. Environ. Contam. Toxicol.,* 22, 182, 1979.
26. **Yip, G., Onley, J. H., and Howard, S. F.,** *J. Assoc. Off. Anal. Chem.,* 54, 1373, 1971.
27. **Marshall, W. D.,** *J. Agric. Food Chem.,* 25, 357, 1977.
28. **Carson, J. F. and Wong, F. F.,** *J. Agric. Food Chem.,* 9, 140, 1961.
29. **Schreyen, L., Dirinck, P., Van Wassenhove, F., and Schamp, N.,** *J. Agric. Food Chem.,* 24, 1147, 1976.
30. **Chen, C.-C. and Ho, C.-T.,** *J. Agric. Food Chem.,* 34, 830, 1986.
31. **Leppänen, O. A., Denslow, J., and Ronkainen, P. P.,** *J. Agric. Food Chem.,* 28, 359, 1980.
32. **Spedding, D. J., Eschenbruch, R., and McGregor, P. J.,** *Food Technol. Australia,* 35, 22, 1983.
33. **Ronkainen, P., Denslow, J., and Leppänen, O.,** *J. Chromatgr. Sci.,* 11, 384, 1973.
34. **Heikes, D. L.,** *J. Assoc. Off. Anal. Chem.,* 70, 215, 1987.
35. **Buttery, R. G., Guadagni, D. G., Ling, L. C., Seifert, R. M., and Lipton, W.,** *J. Agric. Food Chem.,* 24, 829, 1976.
36. **McLeod, H. A. and McCully, K. A.,** *J. Assoc. Off. Anal. Chem.,* 52, 1226, 1969.
37. **Uno, M., Okada, T., Nozawa, M., and Tanigawa, K.,** *J. Food Hyg. Soc. Jpn.,* 23, 474, 1982.
37a. **Lindsay, D. A., Scottish Office,** Agriculture and Fisheries Department, Edinburgh, U.K., personal communication, 1990.
38. Natural Rubber Technical Information Sheet L18, Malaysian Rubber Producers' Association, Hertford, U.K., 1977.
38a. **Reynolds, S. L.,** Ministry of Agriculture, Fisheries and Food, Norwich, U.K., personal communication, 1990.
39. **Franta, I., Ed.,** *Elastomers and Rubber Compounding Materials,* Elsevier, Amsterdam, 1989.
40. **Pattinson, S. J. and Wilkins, J. P. G.,** *Analyst (London),* 114, 429, 1989.
41. **Doroshenko, N. D. and Pokhil'chenko, I. N.,** *Agrokhimiya,* 9, 116, 1989.
42. **Van Haver, W. and Gordts, L.,** *Z. Lebensm. Unters. Forsch.,* 165, 28, 1977.
43. **Spiegelenberg, W., Wanningen, H., and Perquin, L.,** *De Ware(n) — Chemicus,* 9, 33, 1979.
44. **Delventhal, J.,** *Lebensm. Gerichtl. Chem.,* 38, 30, 1984.
45. **Nawar, W. W.,** *J. Agric. Food Chem.,* 19, 1057, 1971.

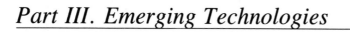

Part III. Emerging Technologies

Chapter 10

FIBER OPTIC SPECTROSCOPY

Alan E. Grey

TABLE OF CONTENTS

I. INTRODUCTION

While this chapter is titled "Fiber Optic Spectroscopy", recent advances in piezoelectric and surface acoustic wave (SAW) devices will also be covered. These technologies, along with fiber optic spectroscopy, are still in their infancy, and are just beginning to emerge as viable techniques for the detection and/or monitoring of biological active substances.

Each of these devices have certain advantages and disadvantages that will be covered later in this chapter. However, all three have in common the ability to be miniaturized, the ability to analyze a substance *in situ* without the need for sample collection, high sensitivity to the analyte of interest, and they are quite inexpensive when compared to standard analytical instrumentation. Because of these advantages and other advantages specific to the device in question, interest is continually increasing for their development as analytical sensors.

II. PIEZOELECTRIC SENSORS

Piezoelectric sensors can best be described as microbalances. The phenomenon of piezoelectricity was first postulated by Raleigh in 1885; however, in 1959, the potential for using piezoelectric devices as analytical sensors was realized by Sauerbrey[1] through the development of a frequency to mass relationship. Sauerbrey's equation is

$$XF = -2.3 \times 10^6 F^2 XW/A$$

where XF is the change in frequency due to the coating on the crystal (hertz), F is the resonant frequency of the crystal (megahertz), XW is the weight of the coating deposited on the crystal (grams), and A is the area of the crystal that is coated (centimeter squared).

This equation predicts that for a 9 MHz crystal, the mass sensitivity would be approximately 400 Hz/μg. The detection limit is estimated to be approximately 10^{-12} g. It is, therefore, apparent that the vibrating quartz crystal can be an extremely sensitive weight indicator.

One of the most active research groups in developing piezoelectric sensors is headed by Dr. G. G. Guilbault at the University of New Orleans. This group has been involved in the development of a sensor for organophosphorus compounds that would be applicable to many of the pesticides and to nerve agents used in chemical warfare. The following is a brief summary of the development process and its rationale.

Initial attempts at developing a detector for organophosphorus compounds consisted of coating a piezoelectric crystal with solid transition metal salts.[2] Solutions of mercury(II) bromide, chloride, and iodide were prepared in ether at concentrations of 0.01 *M*, and the piezoelectric crystals were coated with salts by dipping them into the desired solution. The weight of the coating was determined by using a semimicro analytical balance. In these experiments, it was found that a 14-MHz crystal was superior to a 9-MHz crystal. Using diisopropylmethyl phosphonate (DIMP) as the model compound, a frequency change of 400 cycles was obtained when in contact with 0.7 mm of the DIMP. Whereas, with a 9-MHz crystal, a frequency change of 250 cycles was obtained when subjected to the same conditions. Water vapor and oxygen gave a minor response and were easily removed under vacuum. The DIMP, however, was removed only with difficulty, and the crystal never attained its base frequency. This would indicate that the DIMP was held on the crystal by chemisorption.

For optimum utility of a piezoelectric sensor, or any other type of sensor for that matter, the reaction of the analyte on the crystal surface should be reversible. Studying transition metal halides, Scheide and Guilbault[3] found that $FeCl_3$ is very reactive toward DIMP. It was determined that two processes were being observed. The DIMP was initially combining with the $FeCl_3$ by chemisorption forming the complex $Fe(DIMP)_xCl_3$ until all active sites on the crystal had been covered. Additional DIMP could then be adsorbed onto the crystal by physisorption, and could be readily removed by high vacuum.

To test this hypothesis, a crystal was coated with the $Fe(DIMP)_xCl_3$ complex and subjected to the pesticide paraoxon (*O,O*-diethyl-*o-p*-nitrophenyl phosphate). The response was extremely poor. It was postulated that steric factors were a major reason for the lack of physisorption of the paraoxon onto the DIMP complex. Therefore, it was reasoned that if the structure of the substrate was similar to that of the adsorbent, or capable of weak interaction with the adsorbent, then a good surface for physisorption would be prepared. To test this theory, a detector was built using a $FeCl_3$-paraoxon complex as the substrate on the piezoelectric crystal. Using this crystal, paraoxon could be detected down to approximately 10 ppm in air.

A study of potentially interfering air contaminants, using car exhaust, showed that SO_2 was the only gas that presented even a slight interference. However, this was not considered a problem since the SO_2 could either be removed from the air sample, or tested separately and the frequency change due to it subtracted from the total.

Studies to increase the sensitivity of the piezoelectric detector below the 10-ppm limit previously obtained using the $FeCl_3$ complexes were continued.[4]

Essentially all organophosphorus pesticides are toxic since they contain either a phosphoryl or a thiophosphoryl group, and are cholinesterase inhibitors. Oximes have been shown to react with organophosphorus cholinesterase inhibitors in solution, and are used as antidotes for organophosphorus poisoning.[5] It was, therefore, of interest to investigate oximes as coating on piezoelectric crystals to capture the organophosphorus compounds.

From a study of the literature, two oximes of proven reactivity toward organophosphorus compounds were evaluated as substrates. These oximes were 2-pyridylaldoxime methiodide and isonitrilobenzoylacetone. The first was soon discarded when it was found to be too volatile for use in a flowing gas stream detector. The isonitrilobenzoylacetone, however, was quite stable when used as the sodium salt. When used alone, the reaction between the organophosphorus compounds and the oxime was not reversible. To correct this situation, the cobalt complex of the oxime was prepared. This complex was dissolved in methylene chloride and applied to the piezoelectric crystal. The resulting detector was stable, selective, sensitive, and completely reversible. A further refinement was made by adding a small amount of the pesticide to the methylene chloride solution of the cobalt-oxime complex. This increased the sensitivity of the detector. It also increased the lifetime of the detector from approximately 1 to 3 weeks. It is postulated that the increased sensitivity is due to the added sorption effect of the organophosphorus compound, and the longer lifetime is due to a decrease in volatility of the surface coating.

The response of the detector to parathion, dimethyldichlorovinyl phosphonate (DDVP), and DIMP was determined. Parathion, which is a stronger cholinesterase inhibitor than DDVP, showed much more interaction with the detector than did DDVP, and recovery time was longer. However, with both pesticides an essentially linear response was obtained from 1 to 20 ppb concentration in air. The curvature of the plots at high concentration was felt to be due to different absorption mechanisms as the active sites on the crystal are filled. At these low concentrations, the detector gave no response to DIMP; this would indicate that the nonpesticide organophosphonates would not be interfering compounds.

Work continued to find other coatings for the piezoelectric crystal in attempts to improve sensitivity, obtain faster response times, and increase the lifetime of the sensor. Epstein et al.[6] found that 1-*n*-dodecyl-3-(hydroximinomethyl) pyridinium iodide (3-PAD) is an effective nucleophilic reagent for the hydrolysis of organophosphates. Epstein's work also showed that at high pH and through addition of a surfactant, the hydrolysis reaction between organophosphates and 3-PAD in solution is accelerated.

Tomita and Guilbault[7] were the first to apply this reaction in the solid state. The optimum coating for the piezoelectric crystal was found to be a mixture of 56% Triton X-100, 13% NaOH, and 31% 3-PAD.

As in previous studies, DIMP was used as the model compound. Using the above coating, the frequency change of the crystal was 612 Hz for 15 ppm DIMP, and 63 Hz for 1 ppb DIMP. The frequency change of 63 Hz suggests that detection at sub-parts per billion levels might be possible.

The lifetime of the coating was also improved. Even after a month of use, the loss of sensitivity was only 12%. One of the major factors limiting lifetime is the loss of reactant due to evaporation. Initial attempts to reduce evaporation by using a polymeric binder were unsuccessful. It is hypothesized that the polymer covers or blocks the active sites on the crystal.

Another coating for piezoelectric crystals was based on the work of Wagner-Jauregg et al.,[8] who found that Cu(II) catalyzes the hydrolysis of diisopropylphosphonate. The overall hydrolysis reaction occurs in two steps. In the first step, the copper complex binds the phosphorus ester reversibly. In the second step, the adduct-product is irreversibly broken down by hydrolysis. The property of the first reaction is the one of most interest for detector application. It was suggested that the second reaction would probably not take place because of the low water content of the air.

Based on this chemistry, Guilbault et al.[9] prepared copper complexes of Cu(butyrate)$_2$-ethylenediamine, Cu(butyrate)$_2$-diethylene-diamine, and Cu(butyrate)$_2$-2-ethylenediamine. The different complexes were then mixed in a polymer binder. It is believed that the polymer-bound copper complexes adhered to the crystal surface by electrostatic attraction, but this did not produce long-term stability.

The polymer-Cu^{2+} diamine substrate was found to be the best coating. However, without long-lasting adhesion to the crystal electrodes, the coating had little practical value. This problem was solved by first coating the crystal with poly(hexadecylmethacrylate), a highly viscous liquid. The polymer-copper complex could then be sprayed onto the crystal and the excess wiped off. This procedure was quite reproducible. The response of this coating to DIMP was found to be 2.6 (ΔF[Hz]/ppb DIMP/μg coating). This is superior to the response of 1.9 found in the previous study using 3-PAD.

Additional studies were pursued using the copper complex.[10] In these studies, a (tetramethylenediamine) copper(II) chloride complex was prepared from CuCl$_2$·2H$_2$O and tetramethylenediamine. This complex combined with either a hydrophilic polymer, poly(vinylpyrolidone), or a hydrophobic polymer, poly(vinylbenzyl) chloride. These polymer-bound complexes (or chelates) were then evaluated using DIMP as the model compound for the organophosphorus pesticides.

The piezoelectric crystal coated with the copper chelate-hydrophilic polymer showed a linear response to DIMP from 0 to 30 ppb. Above 30 ppb, a saturation effect was observed with a corresponding increase in response and recovery time, and a decrease in reproducibility. Interference by water was also noted. Thus, dried air would be necessary for this coating.

The copper chelate-hydrophobic polymer showed a linear response to DIMP from 0 to 20 ppb. Above 20 ppb, the response to DIMP continued to increase in a nonlinear fashion, and no saturation was observed. Water vapor up to 80% relative humidity caused no serious effect, nor did other potentially interfering compounds.

After 25 days, the detector with the hydrophilic polymer showed a 20% loss in sensitivity, while the detector with the hydrophobic polymer showed only an 8% loss.

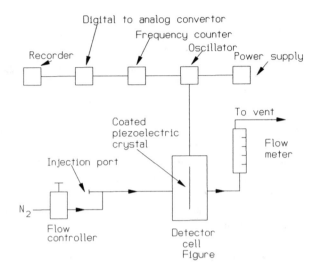

FIGURE 1. Schematic of a piezoelectric detector. (Reproduced from Guilbault, G. G., Luong, J. H. T., and Pruzak-Sochaczewski, E., *Bio/Technol.*, 7, 349, 1989. With permission. © 1989 BIO/TECHNOLOGY.)

One of the last studies in this series was to evaluate different electrode materials,[11] still using DIMP as the model compound. The electrodes studied were composed of gold, silver, or nickel, and used on an uncoated 9-MHz piezoelectric crystal.

The nickel-plated crystal showed the highest sensitivity to DIMP, but the reproducibility was poor, and the response and recovery times were too long. Silver-plated crystals gave better results, but recovery times were still high, and nitrogen dioxide can irreversibly adsorb on the material. The gold-plated crystal showed the fastest response and recovery time. In addition, the gold-plated crystal was essentially interference free for all potentially interfering compounds tested. From practical consideration, this study recommended that gold-plated crystals be used.

A general schematic of the instrumentation used in the preceding experiments is shown in Figure 1.

A. BIOSENSORS

All of the sensor development work reported above could be classified as generic. The reactants immobilized on the piezoelectric crystals were applicable for all organophosphorus pesticides, even though some variation in reaction kinetics may be noted. Interest was now turned to biosensors as a means of developing specificity for a given pesticide.

Increased interest in biosensors is due to the recent advances in biotechnology and the need for selectively determining a specific analyte. In general, biosensors consist of an immobilized biologically active material on a substrate that can detect the reaction between the material and the analyte of interest. The biological components may be enzymes, antibodies, organelles, tissues, whole cells, etc. An excellent review of biosensors as applied to piezoelectric detectors, fiber optics, surface acoustic wave (SAW) devices, and field effect transistors is given by Luong et al.[12]

Most biosensors are developed for use in an aqueous medium. One of the first reported uses of a biosensor for insecticides in the gas phase was by Ngeh-Ngwainbi et al.[13] In this study, the piezoelectric crystal was coated with antiparathion antibodies.

While good results were obtained, the study brought out several problems that need to be resolved before these detectors can become practical reality. These questions include: the nature of the antigen/antibody interaction on the crystal, the association and dissociation rate constants, the antibody binding affinities in the gas phase, and the need for an independent method of confirming the antigen/antibody interactions on the crystals.

One of the more critical aspects of biosensor development is the bonding of the protein material to the surface of the detector. Guilbault et al.[14] have reviewed the methods currently in use, and these are summarized below. They fall into four main categories: adsorption, gluteraldehyde cross-linking of the protein, chemical bonding of the protein onto an electrode surface, and polymeric adhesion to an electrode surface.

In the chemical immobilization process, the quartz surface of the detector was silylated with glycidoxypropyltrimethoxysilane. The material was oxidized with periodate to convert the terminal epoxide group to an aldehyde. The terminal aldehyde could then react with a free primary amine on the protein material to form a stable linkage. This method was used by Roederer and Bastiaans[15] to immobilize goat antihuman IgG onto a quartz surface.

Another method involved surface adhesion, as used by Rice,[16] in which the crystal was coated with 2-hydroxy-3-dimethyl-1,4-amino butane. An antibody was then adsorbed onto this layer during an incubation period.

Krube and Gotoh[17] used gluteraldehyde to bond the antibodies. The quartz surface was first silylated with 3-aminopropyltrimethoxy silane, then treated with the gluteraldehyde followed by the antibody. The gluteraldehyde formed stable linkages between the antibody and the quartz crystal.

The optimum method found by Guilbault's group was reported by D'Souza et al.[18] This method was to precoat the crystal with polyethyleneimine, then chemically attach the antibody to the quartz crystal using gluteraldehyde. This procedure gave the best balance between sensitivity and stability.

A further review of antibody immobilization methodologies is given by Prusak-Sochaczewski et al.[19]

III. SURFACE ACOUSTIC WAVE (SAW) DEVICES

A SAW device is essentially the same as a piezoelectric sensor. The primary difference is in the frequency of vibration of the crystal. While a piezoelectric crystal vibrates in the 9- to 14-MHz range, a crystal used in a SAW device can vibrate in the 100+-MHz range. Therefore, theoretically a SAW device would be much more sensitive than a piezoelectric detector.

This increase in sensitivity also presents problems. The coatings on a SAW device crystal are normally measured in angstroms rather than microns. The coating thickness is reported in terms of the frequency shift (i.e., kHz) when the coating is applied to the crystal. When a polymer is applied to the SAW crystal, the elastic properties and shear modulus of the coating have an effect on the response of the crystal. Another major factor is temperature. A SAW device is very temperature sensitive. It is estimated that for best results, the temperature during measurements should be held to ±0.2°C.

To date, SAW devices are being developed primarily for airborne gases, including organic vapors. No references could be found where SAW devices were being developed for the detection and/or monitoring of pesticides. A good general review of SAW devices can be found in *Chemical Sensors and Microinstrumentation.*[20]

IV. FIBER OPTIC SENSORS

The development of chemical sensors based on fiber optic spectroscopy is probably being pursued in more laboratories than any other sensing technique. One of the reasons for this high level of interest is the variety of techniques that can be used for detection. Some of these techniques are absorption, refractive index, phase change, optical rotation, and fluorescence. In general, any chemical or physical reaction that will perturbate the light transmission through the optical fiber can be used as the basis for a fiber optic detector.

The advances in fiber optic sensors have been rapid. This interest is probably due to the advantages fiber optic sensors have over other types of analytical instrumentation. Some of these advantages are their immunity to electromagnetic interferences, their insensitivity to surface contamination, and the physical separation of the sample from the instrumentation. In the latter case, remote sensing has been accomplished at a distance of over 1 km from the detector electronics.[21] Another major advantage is the ability to multiplex the system so that more than one analyte can be determined using the same electronics. An overview of chemically selective fiber optic sensors is given by Angel.[21]

The majority of fiber optic sensors utilize quartz fibers because of their availability and durability in hostile environments. Typically, quartz fibers have a usable spectral range from approximately 400 to 1100 nm. However,

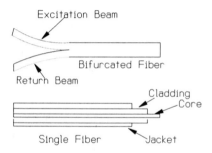

FIGURE 2. Commonly used optical fibers.

fibers are now being developed that will transmit light further into the infrared region, which will expand the utility of fiber optic sensors.

The operation of a fiber optic sensor is based on the transmission of light through the fiber and the measurement of the change in light intensity caused by either a chemical or physical reaction with the analyte.

A single optical fiber and a bifurcated fiber bundle are shown in Figure 2. Each fiber is composed of a quartz core, a cladding, and a jacket. If the refractive index of the cladding is less than that of the core, light will be reflected internally and travel through the fiber with minimal loss. If the refractive index of the cladding is greater than that of the fiber, light can escape from the core.

In general, the optical fibers used in chemical sensors can range in size from approximately 10 to 1000 μm in diameter. Normally, fibers of 200-μm diameter are used.

One of the first papers to report the concept of using fiber optics for sensor development was by Vali and Shorthill.[22] The early concept was based on the change in light transmission through the fiber caused by a physical property such as pressure.

While no fiber optic sensors have been developed for pesticides per se, the techniques used for the development of other biosensors are in most cases directly applicable to pesticides. There is considerable scope for these applications since antibodies for a large number of pesticides of interest to this book are commercially available.

One of the first sensors that could be classified as a fiber optic biosensor was reported by Andrade et al.[23] They describe the detection of protein using the evanescent wave. The evanescent wave excited the protein, and produced a fluorescence emission from the tryptophane in the protein.

A unique fiber optic biosensor for glucose has been developed by Meadows and Schultz.[24] For this detection, fluorescence energy transfer between dextran labeled with fluorescein isothiocyanate and a glucose receptor protein (Concanavalin A) labeled with rhodamine is used. When dextran, a glucose analog, bonds to the Concanavalin A, the fluorescein transfers its energy to the rhodamine. When glucose is added to the solution, it replaces the dextran, and the fluorescein emission intensity increases because energy transfer diminishes. The fluorescence increase is directly related to the glucose concentration in the solution. It should be noted that the solution volume used in this detector was only approximately 1 μl.

Angel and co-workers[25,26] have reported two biosensors under development in their laboratory. In one sensor for penicillin, a polymer membrane is attached to the distal tip of the optical fiber. The membrane contains covalently bound fluorescein and serves to trap the enzyme penicillinase. In the presence of penicillin, the enzyme catalyzes the cleavage of the β-lactam on the penicillin molecule to produce the acidic penicilloic acid. The fluorescein detects this micro-pH change with a detectable change in the fluorescent signal.

For the other type of biosensor under development by Angel, an antibody is attached to a short length of unclad fiber at the distal end. The antibody is then saturated with an antigen labeled with a fluorescent probe. In the presence of the antigen to be detected, there is competition for the binding sites on the antibody between the tagged and untagged antigen. As the tagged antigen is displaced, there is a decrease in the fluorescent signal. The decrease in fluorescent intensity is, therefore, related to the concentration of the antigen in the environment.

Recent work in our laboratory has been devoted to developing a fiber optic sensor based on an antibody-antigen type reaction. Current efforts involve solving the problem of attaching the antibody to the optical fiber. In general, an antibody has two active sites. One at the F_a region, and the other at the F_b region of the antibody. The third region (F_c) is unreactive to the antigen. Methods of antibody attachment, as described in the piezoelectric section, are nonspecific as to the point of attachment. Therefore, the spatial orientation of the antibody on the optical fiber may render a certain percentage of the antibodies unavailable for antigen attachment. One technique, which we are evaluating to circumvent this problem, is to first silylate the fiber with 3-aminopropyltrimethoxy silane and then chemically attach Protein A from *Staphylococcus aureus*. Protein A has the characteristic of bonding specifically to the nonreactive F_c region of the antibody. This should produce a fiber on which the active (F_a, F_b) sites of the antibody are oriented away from the fiber and available for antigen interaction.

Another method of accomplishing antibody orientation was reported by Bright et al.[27] In their procedure, they cleaved the antibody and attached the fragments to a quartz surface. The F_a and F_b halves of an antibody are held together by disulfide bonds. Cleaving these bonds produces two halves with a relatively high concentration of sulfhydryl groups along one surface, which is oriented away from the active region of the antibody. The quartz surface is silylated with 3-glycidoxypropyltrimethoxy silane. The epoxide group is

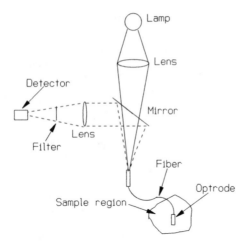

FIGURE 3. General presentation of a fiber optic sensor. (From Angel, S. M., *Spectroscopy,* 2(4), 38, 1987. With permission.)

broken and the terminal hydroxy group is reacted with tresyl chloride. The resulting product then reacts selectively with the sulfhydryl groups on the antibody fragment. An additional interesting aspect of this paper was the long-term stability of the antibody, which was up to 4 months, and the ability to regenerate the detector.

A general schematic of a fiber optic sensor is shown in Figure 3.

V. CONCLUSIONS

Piezoelectric, surface acoustic wave, and fiber optic sensors are still in their infancy. Many problems still need resolution, such as long-term stability, reversibility, and a more detailed understanding of the reaction kinetics, mechanisms, and mass transfer that occur at the sensor surface. The potential for these sensors, with their high sensitivity, small size, and portability, makes the necessary effort very worthwhile.

By presenting the historical development of a piezoelectric device for the detection of organophosphonates and the techniques used for the development of fiber optic biosenors, it is the hope of the author to generate interest in the research community for further development of these much needed sensors. The potential for development is only limited by the imagination of the researchers involved.

REFERENCES

1. **Sauerbrey, G. Z.**, *Z. Physik,* 155, 206, 1959.
2. **West, P. W. and MacDonald, A. M. G.**, *Anal. Chim. Acta,* 39, 260, 1967.
3. **Scheide, E. P. and Guilbault, G. G.**, *Anal. Chem.,* 44(11), 1764, 1972.
4. **West, P. W. and MacDonald, A. M. G.**, *Anal. Chim. Acta,* 73, 383, 1974.
5. **Ellin, R. I. and Wills, J. H.**, *J. Pharm. Sci.,* 44, 995, 1964.
6. **Epstein, J., Kaminski, J. J., Bodor, N., Enever, R., Sowa, J., and Higuchi, T.**, *J. Org. Chem.,* 43, 2816, 1978.
7. **Tomita, Y. and Guilbault, G. G.**, *Anal. Chem.,* 52, 1484, 1980.
8. **Wagner-Jauregg, T., Hackley, B. C., Lies, T. A., Owen, O. O., and Proper, R. J.**, *J. Am. Chem. Soc.,* 77, 922, 1955.
9. **Guilbault, G. G., Affoiter, J., and Tomita, Y.**, *Anal. Chem.,* 53, 2057, 1981.
10. **Guilbault, G. G. and Kristoff, J.**, *Anal. Chem.,* 57, 1754, 1985.
11. **MacDonald, A. M. G., Pardue, H. L., Townshend, A., and Clerc, J. T.**, *Anal. Chim. Acta,* 149, 337, 1983.
12. **Luong, J. H. T., Mulchandani, A., and Guilbault, G. G.**, *Tibtech,* 6, 310, 1988.
13. **Ngeh-Ngwainbi, J., Foley, P. H., Kuan, S. S., and Guilbault, G. G.**, *J. Am. Chem. Soc.,* 108(18), 5444, 1986.
14. **Guilbault, G. G., Luong, J. H. T., and Pruzak-Sochaczewski, E.**, *Bio/Technol.,* 7, 349, 1989.
15. **Roederer, J. E. and Bastiaans, G. J.**, *Anal. Chem.,* 7, 2333, 1983.
16. **Rice, J. T.**, U.S. Patent 4,236,893, 1980.
17. **Krube, I. and Gotoh, M.**, *Immunosensors, NATO Advanced Research Workshop on Analytical Uses of Immobilized Biological Compounds,* Reidel Press, Amsterdam, 1987, 187.
18. **D'Souza, S. A., Melo, J. S., Deshpande, A., and Nadkarni, G.**, *Biotechnol. Lett.,* 8, 643, 1986.
19. **Prusak-Sochaczewski, E., Luong, J. H. T., and Guilbault, G. G.**, *Enzyme Microbiol. Technol.,* 12, 173, 1990.
20. **Wohltjen, H., Ballantine, D. S., Jr., and Jarvis, N. L.**, in *Chemical Sensors and Microinstrumentation,* ACS Symposium Series No. 403, Murray, R. W., Dessy, R. E., Heineman, W. R., Janata, J., and Seitz, W. R., Eds., American Chemical Society, Washington, D.C., 1989, 157.
21. **Angel, S. M.**, *Spectroscopy,* 2, 39, 1987.
22. **Vali, V. and Shorthill, R. W.**, *Appl. Opt.,* 15, 1099, 1976.
23. **Andrade, J. D., Vanwagenen, R. A., Gregonis, D. E., Newby, K., and Lin, J. N.**, *IEEE Trans. Electron Devices,* 32(7), 1175, 1985.
24. **Meadows, D. and Schultz, J. S.**, *Talanta,* 35(2), 145, 1988.
25. **Angel, S. M., Daley, P. F., and Kulp, T. J.**, *Proc. Electrochem. Soc.,* 87(9), 484, 1990.
26. **Fuh, M. S., Burgess, L. W., and Christian, G. D.**, *Anal. Chem.,* 60, 433, 1988.
27. **Bright, F. V., Betts, T. A., and Litwiler, K. S.**, *Anal. Chem.,* 63, 1065, 1990.

Chapter 11

ENZYME-LINKED COMPETITIVE IMMUNOASSAY

Patricia A. Nugent

TABLE OF CONTENTS

I. INTRODUCTION

The use of immunoassays as an analytical tool in the detection of pesticides has been rapidly increasing. The enzyme-linked immunoadsorbant assay (ELISA) has been found to be especially useful as a residue analysis method.[1-5] Immunochemical assays for pesticides are by no means a replacement for traditional analytical techniques such as high-performance-liquid or gas-liquid chromatography. Immunoassays are especially useful to quantify residues when chromatographic methods are difficult and time-consuming. With the growing concern of regulatory agencies over the effects of agricultural chemicals in our environment, monitoring studies have increased significantly, both in size and actual numbers of studies. Immunoassays are now being looked at as a critical analytical tool in dealing with the increased volumes of residue samples. The procedures offer advantages of decreased sample processing, high compound specificity, and significant increase in the number of samples that can be analyzed compared to traditional analytical techniques. Depending upon the direction of the initial antibody development, the assay can be class specific (i.e., able to screen for a certain class of compounds such as the *s*-triazine herbicides) or compound specific (i.e., having a high specificity for the compounds of interest).

In this chapter, a detailed review of the basic immunology principles involved in the development of competitive enzyme-linked immunoassay will not be discussed. Van Emon et al.[6] give a very complete overview of the immunoassay terminology and basic principles in designing immunochemical assays.

II. DEVELOPMENTAL STEPS IN COMPETITIVE IMMUNOASSAY

A. HAPTEN DESIGN, SYNTHESIS, AND CONJUGATION

Most compounds of environmental interest are of low molecular weight. In order to induce an immune response, a compound must be linked to a large molecular weight protein carrier. The compound may already have a functional group that can be conjugated to the carrier, but frequently a derivative of the compound (hapten) must be synthesized in order for conjugation to occur.[7] This step in the assay development demands that careful structural analysis precedes the hapten synthesis and subsequent conjugation. At this point in assay development, one can also make choices in the hapten synthesis that will eventually lead to class specific antibodies or compound specific antibodies. Generally, antibody specificity is highest for the part of the molecule that is farthest from the carrier. If one then chooses to conjugate farthest from a functional group common to several compounds, the chances of developing a class-specific compound are much greater. The importance of conjugation position was demonstrated with producing antibodies to maleic hydrazine.[8] Immunization with O-conjugates of maleic hydrazine produced hapten specific antibodies, but immunization with N-conjugates developed antibodies both to the hapten and to maleic hydrazine. In another study done to produce antibodies to *s*-triazine herbicides, the degree of binding was dependent not only on the position of attachment of the spacer arm but also on the length.[9] In addition to the consideration of antibody specificity when choosing a hapten, one must also consider the ease of synthesis and the stability of the resulting hapten-protein conjugate.[10] Before conjugation, the haptens must also be pure and structurally characterized. Usually this is established by UV absorbance.

Once the hapten selection and synthesis have been completed, conjugation to the carrier molecule must take place. Generally, the functional group of the hapten determines the methods used to covalently bond to the protein. Several of these methods have been described previously.[6,10] Common protein carriers include bovine serum albumin (BSA), thyroglobulin (THY), ovalbumin (OA), human serum albumin (HSA), and rabbit serum albumin (RSA).

B. ANTIBODY PRODUCTION

The resulting conjugate is then injected into an animal host to induce an immune response. Rabbits and goats are commonly used for the production of polyclonal antibodies, and mice are used for the production of monoclonal antibodies. There is no one optimal method of immunization since immunologic response is species dependent, and individual animals within a species can produce a different group of antibodies. Polyclonal sera will contain a mixture of antibodies that will react to various portions of the conjugate. Immunizing mice for monoclonal antibodies will produce antibodies that will be specific for one epitope of the antigen. Both monoclonal[11-13] and polyclonal[4,11,14-16] antibodies have proven effective for pesticide analysis, but more assays have been reported using polyclonal antibodies. Polyclonal antibodies can be made more rapidly and inexpensively than monoclonal and can have comparable sensitivity limits, which can explain why polyclonal is more often the antibody of choice. After immunization, polyclonal antibodies can be ready for method development in a minimum of 3 months. Rabbits are easy to care for and will produce moderate amounts of antibody. The binding specificity of polyclonal antibodies is a result of a mixture of antibodies, each with different specificities. The amount of antibodies present in that mixture changes depending on the antibody production of the individual clones. The exact mixture will also change from one animal to the next. Polyclonal antisera, therefore, will have a specificity that changes over time and between animals.[17]

Monoclonal antibody production will take a minimum of 6 months to 1 year. Monoclonal antibody production, as described by Kohler and Milstein in 1975,[18] made it possible to develop cell lines that will produce a single antibody indefinitely. The animals of choice are mice, due to the availability of myeloma cell lines as fusion partners. After the immunization of several mice, they are bled to monitor antibody response. Shortly after the final boost, an animal is selected from the group on the basis of its serum antibody titer and the affinity and specificity of that titer. Cell fusion, selection of the hybrids, and subsequent establishment and scale-up of the hybrids is a time-consuming and meticulous task. General procedures for producing monoclonal antibodies have been described previously.[19,20]

Even though there is a considerably greater investment in producing monoclonal antibodies, there are two advantages that are difficult to discount. First, each monoclonal antibody is a uniform reagent with a single defined affinity and specificity. This allows for comparison of immunoassay results between different laboratories. Second, antibodies can be made in unlimited quantities as long as the hybridoma line is properly maintained in cell culture or storage. When considering an antibody for use as a reagent for trace analysis, monoclonals appear to offer a more specific and easily standardized reagent.

C. ASSAY DEVELOPMENT
1. Assay Formats
Although different immunoassay methods vary depending on the compound or compounds they are capable of detecting, they are all physical assays based on the laws of mass action. The enzyme-linked competitive immunoassay (cELISA) is one of the most common formats for pesticide immunoassays.[10] Two common variations of this assay consist of either (1) immobilizing antibody or (2) immobilizing conjugate on a solid-phase support such as a 96-well microtiter plate. The coating conjugate used to coat the microtiter plate must differ from the immunogen to prohibit binding from antibodies that recognize the carrier protein.[3] The format using immobilized conjugate is based upon the competition between the immobilized conjugate and an unknown amount of analyte and a known amount of antibody in the coated well. With no analyte in solution, a maximum amount of antibody binds to the plate. A high concentration of analyte blocks the binding of the antibody to the plate. Into a precoated well, equal volumes of antibody and analyte are added. The mixture is incubated for a period of time, and the solution phase is washed away. The antibody that is bound to the plate is left behind and must now be measured. If the antibody was conjugated to an enzyme before reacting with the analyte, then the final step is to add substrate and measure absorbance of the resulting color. If the enzyme was not conjugated to the antibody first, then a second antibody with an enzyme tag is added. This second antibody is usually commercially prepared and will react with species-specific antibodies, such as rabbit or mouse antibodies. After incubation, the unreacted materials are washed away, and substrate is added to measure enzyme activity (Figure 1). In the second format, the antibody is immobilized on the plate. Equal volumes of sample analyte and enzyme-

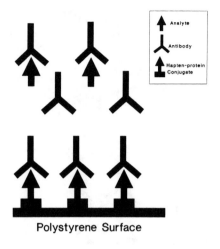

Polystyrene Surface

A

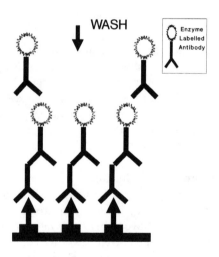

WASH

B

FIGURE 1.   Indirect ELISA. (A) The hapten-protein conjugate is coated to the polystyrene surface. Sample analyte and antibody are added and react. Unreacted antibody will then bind with antigen on the polystyrene surface. (B) Unreacted compounds are washed away. Enzyme-labeled antibody is then added to bind with first antibody. (C) Excess antibody is washed away. The addition of the substrate results in the formation of a colored product.

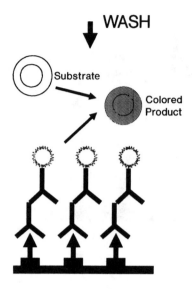

FIGURE 1C.

labeled analyte are added to the well and compete for the antibody binding sites. After incubation, the solution phase is washed away, substrate added, and the enzyme activity is measured by measuring the absorbance of the resulting color (Figure 2). In both formats, one can see that labeling either antibody or antigen with an enzyme reduces the steps in the assay. Not all antibodies or antigens can be easily labeled though, so these two shorter formats cannot always be used.

2. Enzyme Label and Data Reduction

A wide variety of enzymes can be used, such as peroxidase, alkaline phosphates, and urease, and they are all commercially available. Usually a chromogenic substrate is employed, and color intensity is measured with a spectrophotomer. With the competitive ELISA, the color intensity is inversely proportional to the concentration of the analyte. Usually with ELISA, sigmoidal standard curves are obtained.[6] The selection of a method for handling standard curve data can be made with commercially available curve-fitting programs. Programs such as 4-parameter logistic equations or log-logit data transformations can be used to linearize the standard curve. Data reduction techniques will not be covered in detail here but have been discussed.[21,22] A typical standard curve for a monoclonal antibody to the insecticide chlorpyrifos can be seen in Figure 3. This curve was generated with a commercial software package using a 4-parameter fit. Regression parameters, such as the R-squared value, and the concentrations at 10 and 90% of the optical density range are also calculated with the generation of the curve.

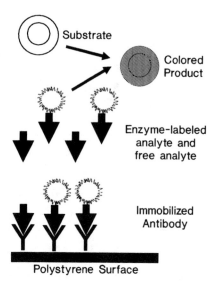

FIGURE 2. Direct ELISA. Antibody is immobilized on a polystyrene surface. Competition for binding sites on the coated surface takes place when free analyte (sample) and enzyme-labeled analyte are added to the well. The excess is washed away. The addition of a substrate results in the formation of a colored product.

3. Assay Optimization

Once a format is decided on, steps can be taken to further define the sensitivity of the antibody in the assay system. Sensitivity can be defined as the smallest amount of measurable analyte that is reliably not zero.[21] In immunoassays, sensitivity can be modified by several variables. Initially, when developing the assay format, the choice of enzyme label can optimize the assay sensitivity. The use of fluorogenic substrates in place of colorimetric substrates can increase sensitivity.[31] Recently, a commercial system to monitor a fluorescent ELISA has entered the commercial market. Another important variable is the choice and concentration of the coating conjugate used in the assay. The work of Wie and Hammock[32] illustrates that by having a library of antigens to compare, one can improve the sensitivity of the assay.

Also of importance in optimization is the choice of materials, such as microtiter plates, and the choice of a plate reader. There are a large variety of microtiter plates available. Some are more suitable for protein binding and others would be more appropriate for binding of an antibody. Harrison and Hammock[33] have shown that the well-location-dependent biases found in plate readers occur often enough to be of significant concern. The importance of routine testing of the plate reader needs to be understood by the user and monitored to control and prevent these sources of error.

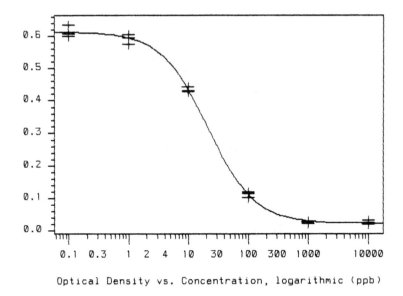

Optical Density vs. Concentration, logarithmic (ppb)

FIGURE 3. A standard curve for the determination of chlorpyrifos by competitive ELISA. Standards were run in triplicate.

4. Sample Preparation

The amount of work necessary to prepare a field sample for an immu- noassay will greatly depend on the individual antigen-antibody complex. For some assays, it can be as easy as buffering a sample and adding it directly to the assay. Because of the properties of antibodies, the effects of pH, organic solvents, and environmental conditions (i.e., temperature) are important con- siderations.

Pesticides are frequently extracted with organic solvents. Immunoassays are tolerant of solvents, but each system must be tested to determine the optimum percent of solvent tolerated in the assay and the most appropriate solvent. Since immunoassays are run most commonly in an aqueous solution, water soluble solvents are the first choice. In the development of an immu- noassay for chlorpyrifos in our laboratory, a comparison was made of several solvents diluted with buffer at different percentages. A 10% dilution of 2- propanol was found to give the greatest dynamic range and the best sensitivity for the assay.

The use of solid-phase extraction systems for sample cleanup in traditional pesticide analysis can also be adapted to sample preparation for immunoassay systems. The work of Stanker et al.[12] describes the sample preparation to determine permethrin in meats. A sample was extracted in an acetonitrile/ water solution, partitioned against hexane, partially purified on an alumina column, and analyzed with a cELISA. The work of Li et al.[34] demonstrated the use of ELISA coupled with solid-phase extraction to yield a sensitive and selective analytical method for molinate in water. After molinate water samples were loaded onto C_{18} cartridges, they were eluted with methanol for ELISA analysis.

D. ASSAY VALIDATION

The validation of an immunoassay method has not been developed to the point of having an official validation protocol, but that issue is being addressed by groups such as the Association of Official Analytical Chemists (AOAC) and the U.S. Environmental Protection Agency (EPA). The validation has several critical elements that should be addressed. One of the first areas an analytical chemist might examine is the comparison of the immunoassay to a traditional analytical method, such as gas chromatography (GC). The comparison must take into consideration several areas of difference. Immunoassays are run in aqueous systems and GC usually requires analysis using organic solvents. The data reduction methods usually are different. The calibration curves are most commonly sigmoidal for immunoassay, and data are obtained from the linear portion of the curve. Data from a GC method will be taken from a standard plot within the linear dynamic range.

The materials, instruments, and reagents used should be evaluated as part of an intralaboratory validation. Microtiter plates need to be reevaluated with each new lot received. The work of Harrison et al.[35] has shown with the molinate study that interwell variability within plates was the largest single contributor to total assay variability. The microplate readers must also be optimized and possible sources of error recognized. The antibody must be tested thoroughly for cross reactivity with other compounds, metabolites, and analogs that are similar in structure and could potentially be found in a sample being analyzed for the compound of interest. Cross-reactivity assays will help to define the specificity of the antibody. The specificity of an assay and the antibody used is defined as the ability to produce a measurable response only for the analyte of interest. Cross reactivity is defined as the dose of the analyte of interest divided by the dose of the competitor at 50% displacement. With complete cross-reactivity data, the selectivity of the assay can be well defined.

As with any analytical method, a statistically significant number of recoveries and samples must be run to prove the ruggedness of the method. Matrix effects must be addressed as well as biotransformation and variability of field samples.

III. APPLICATIONS

The cELISA has many different applications as an analytical method for pesticides. For in-house use, the ELISA can be a screening tool for studies having large amounts of samples to analyze with confirmation of positives by GC or GC/MS. If the antibody and the assay are selective and sensitive enough, the ELISA can be used to accurately quantify pesticides in place of

TABLE 1
Examples of Enzyme-Linked Competitive Immunoassays

Compound	Antibody	Detection limit	Ref.
Aldicarb	Polyclonal	15.6 ng	23
Methyl 2-Benzimidazolecarbamate[a]	Polyclonal	1 ng/g	24
Endosulfan	Polyclonal	3 ng/ml	25
Fenpropimorph	Polyclonal	13 pg/ml	26
Maleic hydrazide	Monoclonal	0.11 ppm	8
Norflurazon	Polyclonal	1 ng/ml	27
Paraquat	Polyclonal	0.1 ng/ml	28
Paroxon	Polyclonal	28 pg/ml	29
Pyrethroids	Monoclonal	50 ppb	12
s-Triazines	Polyclonal	Low to sub-ppb	30

[a]   Degradation product of benomyl.

traditional analytical procedures that are more labor intensive. Table 1 shows some of the competitive ELISA methods that have been developed to analyze pesticides.

The potential to apply the ELISA to a rapid field test kit is present with most assays. With the increased emphasis on environmental monitoring by our government, the ability to have an easily run, portable, and accurate field kit for pesticides will be very cost-effective. Field kits can be used in evaluating contamination sites as well as monitoring worker safety. Regulatory requirements for handling issues such as worker safety could be easily monitored with field test kits.

IV. CONCLUSIONS

The competitive ELISA, as an emerging technology, needs to continue to be optimized and standardized to gain wider acceptance and use as an analytical tool. Efforts on the part of groups such as the AOAC and the EPA to develop a validation protocol for immunoassays will help to standardize the methodology. More communication needs to take place among those scientists developing antibodies and methods so that time and energy are not wasted duplicating these efforts. That time and energy can be better spent streamlining and optimizing the assays already in use.

REFERENCES

1. **Hammock, B. D., Gee, S. J., Cheung, P. Y. K., Miyamato, T., Goodrow, M. H., Van Emon, J., and Seiber, J. N.,** in *Pesticide Science and Biotechnology,* Greenhalgh, R. and Roberts, T. R., Eds., Blackwell Scientific, Oxford, 1987, 309.

2. **Harrison, R. O., Gee, S. J., and Hammock, B. D.,** in *Biotechnology in Crop Protection,* ACS Symposium Series, No. 379, Hedin, P., Menn, J. J., and Hollingworth, R. M., Eds., ACS Publications, Washington, D.C., 1988, 316.

3. **Schwalbe-Fehl, M.,** *Int. J. Environ. Anal. Chem.,* 26, 295, 1986.

4. **Mumma, R. O. and Brady, J. F.,** in *Pesticide Science and Biotechnology,* Greenhalgh, R. and Roberts, T. R., Eds., Blackwell Scientific, Oxford, 1987, 341.

5. **Vanderlaan, M., Van Emon, J., Watkins, B., and Stanker, L.,** in *Pesticide Science and Biotechnology,* Greenhalgh, R. and Roberts, T. R., Eds., Blackwell Scientific, Oxford, 1987, 597.

6. **Van Emon, J., Seiber, J. N., and Hammock, B. D.,** in *Analytical Methods for Pesticides and Plant Growth Regulators: Advanced Analytical Techniques,* Vol. 17, Sherma, J., Ed., Academic Press, New York, 1989, 217.

7. **Van Emon, J. M., Seiber, J. N., and Hammock, B. D.,** in *Bioregulators for Pest Control,* ACS Symposium Series No. 276, Hedin, P. A., Ed., American Chemical Society, Washington, D.C., 1985, 307.

8. **Harrison, R. O., Brimfield, A. A., and Nelson, J. O.,** *J. Agric. Food Chem.,* 37, 958, 1989.

9. **Goodrow, M. H., Harrison, R. O., and Hammock, B. D.,** *J. Agric. Food Chem.,* 38, 990, 1990.

10. **Jung, F., Gee, S. J., Harrison, R. O., Goodrow, M. H., Karu, A.E., Braun, A. L., Li, Q. X., and Hammock, B. D.,** *Pestic. Sci.,* 26, 303, 1989.

11. **Deschamps, R. J. A., Hall, J. C., and McDermott, M. R.,** *J. Agric. Food Chem.,* 38, 1881, 1990.

12. **Stanker, L. H., Bigbee, C., Van Emon, J., Watkins, B., Jensen, R. H., Morris, C., and Vanderlaan, M.,** *J. Agric. Food Chem.,* 37, 834, 1989.

13. **Schlaeppi, J., Fory, W., and Ramsteiner, K.,** in *Immunomethods for Environmental Analysis,* ACS Symposium Series No. 442, Van Emon, J. M. and Mumma, R. O., Eds., American Chemical Society, Washington, D.C., 1990, 199.

14. **Gee, S. J., Miyamoto, T., Goodrow, M. H., Buster, D., and Hammock, B. D.,** *J. Agric. Food Chem.,* 36, 863, 1988.

15. **Koppatschek, F. K., Liebl, R. A., Kriz, A. L., and Melhado, L. L.,** *J. Agric. Food Chem.,* 36, 1519, 1990.

16. **Mei, J. V., Yin, C.-M., and Carpino, L. A.,** in *Immunomethods for Environmental Analysis,* ACS Symposium Series No. 442, Van Emon, J. M. and Mumma, R. O., Eds., American Chemical Society, Washington, D.C., 1990, 140.

17. **Vanderlaan, M., Watkins, B. E., and Stanker, L.,** *Environ. Sci. Technol.,* 22, 247, 1988.

18. **Kohler, G. and Milstein, C.,** *Nature (London),* 256, 495, 1975.

19. **Zola, H.,** *Monoclonal Antibodies: A Manual of Techniques,* CRC Press, Boca Raton, FL, 1987, 13.

20. **Gooding, J. W.,** *Monoclonal Antibodies; Principles and Practice,* Academic Press, London, 1986, 241.

21. **Feldkamp, C. S. and Smith, S. W.,** in *Immunoassay; A Practice Guide,* Chan, D. W., Ed., Academic Press, Orlando, FL, 1987, 49.

22. **Rodbard, D.,** in *Principles of Competitive Protein-Binding Assays,* Adell, W. D. and Daughaday, W. M., Eds., Lippincott, Philadelphia, 1971, 204.

23. **Brady, J. F., Fleeker, J. R., Wilson, R. A., and Mumma, R. O.,** in *Biological Monitoring for Pesticide Exposure: Measurement, Estimation, and Risk Reduction,* ACS Symposium Series No. 382, Wang, R. G. M., Franklin, C. A., Honeycutt, R. C., and Reinert, J. C., Eds., American Chemical Society, Washington, D.C., 1989, 262.

24. **Bushway, R. J. and Savage, S. A.,** *Food Chem.,* 35, 51, 1990.

25. **Dreher, R. M. and Podratzki, B.,** *J. Agric. Food Chem.,* 36, 1072, 1988.

26. **Jung, F., Meyer, H. H. P., and Hamm, R. T.,** *J. Agric. Food Chem.,* 37, 1183, 1989.

27. **Riggle, B. and Dunbar, B.,** *J. Agric. Food Chem.,* 38, 1922, 1990.

28. **Van Emon, J., Hammock, B., and Seiber, J. N.,** *Anal. Chem.,* 58, 1866, 1986.

29. **Hunter, K. W. and Lenz, D. E.,** *Life Sci.,* 30, 355, 1982.

30. **Harrison, R. O., Goodrow, M. H., and Hammock, B. D.,** *J. Agric. Food Chem.,* 39, 122, 1991.

31. **Shalev, A., Greenberg, A. H., and McAlpine, P. J.,** *J. Immunol. Methods,* 38, 125, 1980.

32. **Wie, S. I. and Hammock, B. D.,** *J. Agric. Food Chem.,* 32, 1294, 1984.

33. **Harrison, R. O. and Hammock, B. D.,** *J. Assoc. Off. Anal. Chem.,* 71, 981, 1988.

34. **Li, Q. X., Gee, S. J., McChesney, M. M., Hammock, B. D., and Seiber, J. N.,** *Anal. Chem.,* 61, 819, 1989.

35. **Harrison, R. O., Braun, A. L., Gee, S. J., O'Brien, D. J., and Hammock, B. D.,** *Food Agric. Immunol.,* 1, 37, 1989.

Chapter 12

THE USE OF ION TRAP MASS SPECTROMETRY FOR MULTIRESIDUE PESTICIDE ANALYSIS

Gregory C. Mattern and Joseph D. Rosen

TABLE OF CONTENTS

I. INTRODUCTION

The limited financial resources available to regulatory agencies mandates the use of a multipesticide residue method that determines as many pesticides as possible in one analysis. Multiresidue methods currently used are inefficient because they rely on packed column gas chromatography and several element-specific detectors needed to determine various groups of pesticides. A single, selective detector monitoring the eluates of a gas chromatography capillary column would greatly improve these analyses.

Most regulatory agencies employ the Luke extraction procedure,[1,2] which is capable of extracting at least 234 pesticides and their metabolites.[3] This method requires no column chromatography cleanup step because interferences from matrix substituents are minimized by the use of element selective detectors such as the Hall electrolytic conductivity detector (for organohalogen or organonitrogen pesticides) and the flame photometric detector (for organophosphate pesticides). Packed column gas chromatography columns of various polarities are used to separate the pesticides. The major problem with these methods is that an analyst determining all the possible pesticide residues extracted by the Luke procedure needs to make several separate determinations for each extract and that some sample constituents give detector responses at the same retention time as some pesticides (false positives). Pesticide residues are confirmed by using either different polarity columns or by a separate mass spectrometric determination.

Overall, we have approached methodology improvements as follows:

1. Retain the Luke extraction procedure (with minor modification) for fruit and vegetable analyses because of its simplicity and its ability to extract so many pesticides and pesticide metabolites.
2. Substitute capillary columns for packed columns in order to improve resolution, retention time precision (extremely important for rejection of false positives), and to shorten gas chromatography analysis time.
3. Employ cold on-column injection and a precolumn (a portion of which is disposed of after every five determinations) in order to prevent buildup of nonvolatile matrix components at the head of the capillary column, while at the same time improving the chromatography of polar pesticides and minimizing decomposition of some thermally labile pesticides.
4. Substitute a single specific detection system (based on mass) for the detection systems based on element specificity.

Mass spectrometry in the electron ionization (EI) mode is often used in pesticide analysis to confirm residues of pesticides in foods.[3] EI produces various ions that give structural information characteristic for each compound. Mass spectrometry can also be used to quantify pesticides in complex matrices because of its high specificity (based on mass) and excellent sensitivity, especially when selected ion monitoring (SIM) techniques are used. SIM allows the monitoring of only those ions known to be present in the mass spectrum of a particular analyte instead of scanning its entire mass range. This results in much enhanced intensities for those ions of interest, which, in turn, translates into greater sensitivity. While this technique works very well when there are only a few analytes, it would be overwhelmed in a multiresidue analysis where a tremendous number of different ions would have to be monitored. Nevertheless, some impressive results have been reported. West et al.[4] used an inexpensive mass selective detector (MSD) to determine as low as 10 ppb of trifluralin, benefin, ethalfluralin, and isopropalin in soil. Stan[5] described the detection of 18 pesticides in green peppers at the 10-ppb level with an MSD in the EI and SIM modes.

Another inexpensive mass spectrometer detection system, the ion trap detector (ITD), has also proven useful for trace analyses. Eichelberger and Budde[6] reported on the use of a GC/MS-ITD system for the quantitative determination of several environmentally significant chemicals, including many polyaromatic hydrocarbons. The ITD was found to be sensitive and precise, and linearity was obtained over a wide concentration range. Pereira et al.[7] used an ITD to determine several herbicides in surface and groundwaters. Full-scan EI mass spectra were obtained for 60 pg of the herbicides. An ion trap detector was also used to confirm and quantify chlorophenoxy herbicides in air and water.[8] The ion trap detector, because of its unique design, is very sensitive in the full-scan mode. The ability of the ion trap to perform more efficient ion storage and analysis is based on its mass selective instability operation in which only RF voltages are used. This high efficiency is also due to the absence of a physically separate analyzer, like that found in quadrupole and other mass spectrometers. Operating the ion trap detector in the full-scan mode allows for the simultaneous determination of an unlimited number of pesticides. Full-scan mass spectra, which give conclusive evidence for the presence and identity of a compound, are obtained. Sensitivity is further enhanced by the use of mass chromatography, which is a computer-generated chromatogram of a particular ion within the mass range of the analysis. This greatly improves signal-to-noise responses by blocking out the chemical noise generated from the coeluants. In addition, and of no small importance, the ITD can be purchased equipped for chemical ionization (CI).

In CI, ions are formed by reaction of the analyte with an ionized reaction gas such as methane or isobutane. Since ions are formed at a much lower energy than in EI, extensive fragmentation is usually avoided, and more intense ions at higher mass are produced.[9] The higher the mass at which an analysis is carried out, the lesser the chances for interferences from ions derived from the sample matrix. McLafferty[10] compiled a list of the most abundant ions found in a file of 4000 EI mass spectra. Using this data for the three most abundant ions in each spectrum, we have calculated that the relative probabilities of interfering masses in the 40 to 99, 100 to 159, 160

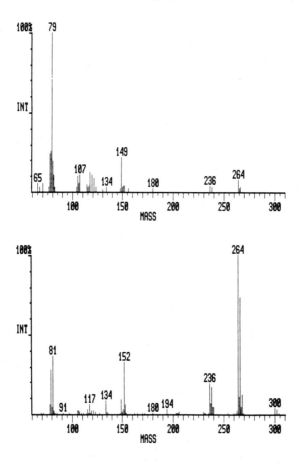

FIGURE 1. Comparison of electron ionization (upper trace) and methane chemical ionization (lower trace) mass spectra of captan. (Reprinted with permission from Mattern, G. C., Singer, G. M., Louis, J., Robson, M., and Rosen, J. D., *J. Agric. Food Chem.*, 38, 402, 1990. Copyright 1990 American Chemical Society.)

to 219, 220 to 279, and 280 to 339 mass ranges are 187:83:49:13:1. As a practical example, the EI and methane CI spectra of captan are compared in Figure 1. Thus, use of EI forces one into monitoring *m/z* 79, while CI allows one to monitor *m/z* 264. Figure 2 compares the results obtained from both EI and CI mass chromatography of an extract from peaches with an incurred captan residue of 0.4 ppm. The signal-to-noise ratio advantage of the CI mass chromatogram over the EI mass chromatogram is quite obvious.

In this report, we provide methods and results for the multiresidue determination of pesticides in fruits, vegetables, and surface water using gas chromatography/chemical ionization mass spectrometry (GC/CIMS) with an ion trap detector.

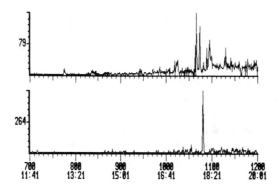

FIGURE 2. Comparison of mass chromatograms obtained from a peach sample with an incurred residue of 0.4 ppm of captan under electron ionization (upper trace) and methane chemical ionization (lower trace) conditions. (Reprinted with permission from Mattern, G. C., Singer, G. M., Louis, J., Robson, M., and Rosen, J. D., *J. Agric. Food Chem.*, 38, 402, 1990. Copyright 1990 American Chemical Society.)

II. EXPERIMENTAL

A. INSTRUMENTATION

A Varian Model 3400 gas chromatograph (Varian Associates, Walnut Creek, CA) interfaced to a Finnigan MAT ion trap detector in the chemical ionization mode (Finnigan MAT, San Jose, CA) and controlled by an IBM PC/AT was used. The analyses and quantifications were performed with Finnigan ion trap software version 3.15. A splitless on-column injector held at 50°C was used, and a 2-m × 0.53-mm I.D. deactivated fused silica precolumn was fitted between the injector and capillary column. Approximately 45 cm of the precolumn was removed after every 5 injections because of the accumulation of nonvolatile matrix components. A new 2-m precolumn was installed after every 20 injections. For apple, peach, potato, and tomato analyses, a 30-m × 0.25-mm I.D. (J & W, Rancho Cordova, CA) DB-1 fused silica capillary column (1 μm film thickness) was held at 50°C for 2 min and then temperature programmed from 50 to 300°C at 15°C/min. For the bean, corn, lettuce, pepper, and spinach analyses, the same column was held at 50°C for 1 min and then temperature programmed from 50 to 280°C at 15°C/min. For the determination of linuron in potatoes, a 15-m DB-1 column was temperature programmed from 50 to 300°C at 15°C/min. Carrier gas (He) velocity was 25 cm/s, and the injection volume was 1 μl. A 15-cm syringe needle was used for on-column injections. The mass spectrometer was operated in the chemical ionization mode using methane reagent gas at a source pressure that gave a 10:1 ratio for m/z 17 to 16, or isobutane reagent gas at a source pressure that gave a 2:1 ratio for m/z 43 to 57. Methane reagent gas was used for the apple, peach, potato, and tomato analyses; and isobutane reagent gas was used for the bean, corn, lettuce, pepper, and spinach analyses. The filament voltage and current were 70 eV and 80 μA, respectively. The electron multiplier gain was 10^5. The scan range was 70 to 420 amu at 1 s per scan for those analyses where methane was used, and 100 to 420 amu at 1 s per scan for those analyses with isobutane reagent gas. The transfer line and manifold temperatures were 250 and 220°C, respectively.

For the surface water samples, a 15-m × 0.25-mm I.D. (J & W, Rancho Cordova, CA) DB-5 fused-silica capillary column (1-μm film thickness) was temperature programmed from 80 to 260°C at 20°C/min. The mass spectrometer was operated in the chemical ionization mode using isobutane reagent gas at a source pressure that gave a 2:1 ratio for m/z 43 to 57. The scan range was 100 to 400 amu at 1 s per scan.

B. PREPARATION OF CALIBRATION CURVES

Calibration curves for the crop and surface water analyses were prepared using standard solutions diluted from stock solutions containing 100 μg/ml and 50 μg/μl of each pesticide in dichloromethane. Internal standard stock solutions contained 25 mg/ml of the internal standards. The standard solutions were analyzed two to three times at each concentration level. Calibration Files/Pesticide Libraries were generated by storing each pesticide name, retention time, and mass spectrum from one of the chromatograms into the file. This was performed by selecting "Internal Standard Calibration" in the "Quantitation" section of the "Chromatograms" programs. Response factors (area of pesticide/area internal standard) were obtained by using the "Auto Calibration" task of the main "Quantitation" program. The peak areas were obtained from the mass chromatograms of each pesticide quantification ion. Linear calibration curves were then generated by plotting response factors vs. the amount (nanograms per microliter) of pesticide. The ions used for quantification and confirmation are given in Table 1.

C. SAMPLE EXTRACTION

The Luke procedure,[2] with slight modifications, was followed[11] for fruit and vegetable analyses. Corn extracts had to be cleaned up prior to GC/CIMS analysis using a Florisil column solid-phase extraction procedure.[12] Fruit and vegetable extracts were concentrated to approximately 2 ml; 40 μl of internal standard solution containing 2.5 mg/ml each of *p*-bromonitrobenzene and fluorene was added; and the final volume was adjusted to 4 ml. Of this solution, 1 μl was injected into the GC/MS system.

For surface water analyses, the pesticides were extracted on a 30- × 1-cm I.D. glass Chromaflex column filled with 5 g each of XAD-2 and XAD-7 resins. After the column was conditioned, 2 l of water was pumped through the column at 75 to 100 ml/min using a stainless steel tank pressurized with nitrogen. The absorbed pesticides were then eluted from the column with methylene chloride. The eluate was collected, dried over anhydrous sodium

TABLE 1
Mass Spectral Data for Pesticides Discussed in This Chapter

Pesticide	Molecular weight	Quantitation ion	Other ions
Acephate	183	184(50)	143(100)
Alachlor	269	238	240(67), 270(30)
Atrazine	215	216	218(33)
Azinphos-methyl	317	160(80)	132(100), 318(5)
Butylate	217	218	156(10)
Captafol	347	312(100)	314(100), 278(40)
Captan	299	264	266(67), 152(50), 236(20)
Carbaryl	201	202(80)	145(100)
Carbofuran	221	222	165(40)
Chlordimeform	196	197	199(33)
Chlorothalonil	264	267	265(70), 269(20)
Chlorpyrifos	349	352(100)	350(100), 354(33)
Cyanazine	240	241	214(50)
Cypermethrin	415	191	193(67), 208(70)
Diazinon	304	305	
Dichloran	206	207	209(67), 211(11)
Diclofop-methyl	340	281	283(67), 341(20)
Dimethoate	229	230(80)	199(100)
Ethalfluralin	333	334	
Fenamiphos	303	304	
Folpet	295	260	262(67), 148(40)
Fonofos	246	247	
Isofenphos	345	245	287(30)
Linuron	248	249	251(67), 253(11)
Methamidophos	141	142	
Methyl parathion	263	264	
Metolachlor	283	284	286(67), 252(20)
Metribuzin	214	215	198(5)
1-Naphthol	144	145	
Oxadiazon	344	345	347(67),349(11)
Parathion	291	292	
Pendimethalin	281	282	264(50), 212(15)
Permethrin	390	183	365(25), 355(20)
o-Phenylphenol	170	171	
Phosmet	317	318(70)	160(100), 286(30)
Pronamide	255	256	258(67), 260(11)
Simazine	201	202	204(33)
Terbufos	288	103	233(100), 245(40), 199(30)
Terbutryne	241	242	
Tetrahydrophthalimide	151	152	
Trifluralin	335	336	

Note: All spectral information are from isobutane CI spectra. Values in parentheses are the percent intensity relative to the base peak. Quantitation ion is base peak unless otherwise noted.

sulfate, and concentrated to 0.5 ml in a Kuderna-Danish concentrator on a steam bath after the addition of 25 μl of the 0.4 mg/ml fluorene internal standard stock solution.[13]

D. DETERMINATION AND QUANTIFICATION

After analysis by GC/CIMS, the computer searched for each pesticide at the appropriate mass chromatographic retention time. This was performed by running the "Auto Quantitation" task of the "Quantitation" program. If a mass chromatographic peak was present in the retention time window, the computer compared the CI mass spectrum of that peak to the known spectrum of the pesticide stored in the calibration file. Any pesticides that were found were then quantified and listed by the computer. This entire process only takes approximately 3 min to perform for 40 to 50 pesticides. Confirmation of pesticide residues was performed by looking at the full-scale mass spectrum (usually after computer assisted background subtraction) and by plotting multiple mass chromatograms for the pesticides.

E. RECOVERY STUDIES

Recovery studies were performed at the 0.5-ppm level 3 or more times for each pesticide in residue-free crop samples. Fortification was performed prior to the extraction by adding the appropriate amount of the pesticide stock solution to 100 g of chopped sample. Apples, peaches, potatoes, and tomatoes were spiked with the stock solution containing acephate, azinphos-methyl, captafol, captan, carbaryl, carbofuran, chlorothalonil, dichloran, dimethoate, methamidophos, 1-naphthol, permethrin, phosmet, and tetrahydrophthalimide. Recovery studies were also performed for linuron in potatoes. Beans, corn, lettuce, peppers, and spinach were spiked with the stock solution containing acephate, alachlor, atrazine, azinphos-methyl, captan, carbaryl, carbofuran, chlordimeform, chlorothalonil, cypermethrin, dichloran, diclofop-methyl, dimethoate, ethalfluralin, folpet, methamidophos, methyl parathion, metolachlor, oxadiazon, parathion, permethrin, o-phenylphenol, pronamide, terbutryne, tetrahydrophthalimide, and trifluralin. In river water recovery studies for 18 pesticides (alachlor, atrazine, butylate, carbaryl, carbofuran, chlorpyriphos, cyanazine, diazinon, fenamiphos, fonofos, isofenphos, linuron, metolachlor, metribuzin, parathion, pendimethalin, simazine, and terbufos) were performed at the 1-ppb level; and for captan and chlorothalonil, at the 5-ppb level. The appropriate volume of pesticide stock solution was added to 2 l of river water previously shown to be pesticide free. The pesticide concentration (nanograms per microliter) in the crops and water were obtained by the instrument software, where the relative responses of each pesticide were applied to the calibration curves.

F. SENSITIVITY DETERMINATIONS

Pesticide-free apple, peach, potato, and tomato extracts were fortified with 32 μl of a 100-μg/ml stock solution. The average final filtrate volume from the acetone blending of the Luke extraction was approximately 250 ml. Assuming 100% extraction efficiency, these solutions represent extracts of crops spiked at the 0.1-ppm level. Beans, corn, lettuce, peppers, and spinach samples were fortified prior to extraction at the 0.05-, 0.10-, and 0.25-ppm level. Absolute limits of detection for all pesticides analyzed in standard solutions were below 0.1 ng/μl, except for captan and chlorothalonil, which could not be detected below 0.5 ng/μl. Pesticide-free water samples were spiked at the 0.05-, 0.005-, and 0.0005-ppb level in order to determine limits of detection. For the 0.005- and 0.0005-ppb spiked samples, the final extracts were concentrated to 100 μl. For a pesticide to be considered detectable at a certain level, it had to give a signal-to-noise ratio of at least three with the mass chromatographic peak being distinguishable from background and/or sample components.

III. RESULTS AND DISCUSSION

The crops analyzed in these studies were chosen because they are of economic importance to New Jersey agriculture, and they comprise a large segment of the dietary intake of fruits and vegetables. The pesticides were chosen because they are suspected oncogens, mutagens, or teratogens and/or are widely used on the nine crops of interest, or they are known to be subject to water runoff. A total of 48 out of 225 crop samples contained one or more pesticide residues, and there was a total of 74 positive findings. All but 3 of the 74 positives were below Environmental Protection Agency (EPA) tolerances. Of the 31 surface water samples analyzed, 29 contained residues of one or more pesticide, with the levels ranging from trace (<0.025) ppb to 5.48 ppb.[12] Only two pesticides, carbaryl and metolachlor, were present in average concentrations of more than 0.1 ppb.

Average recoveries of pesticides from apples, peaches, potatoes, and tomatoes at the 0.5-ppm level were between 73 and 120%, with the average coefficient of variation of 11% for the 56 analyses.[11] Average recoveries of pesticides at the 0.5-ppm level from beans, corn, lettuce, peppers, and spinach were between 70 and 123%, with an average coefficient of variation of 12%.[12] Recoveries of pesticides from surface water ranged from 75 to 113%, with an average coefficient of variation of 9%.[13]

Methane was originally chosen as the reagent gas because the Food and Drug Administration (FDA) has an extensive compendium of methane chemical ionizaton spectra for pesticides. The gentler ionization afforded by isobutane, in general, results in a lower number of fragment ions and a more intense protonated molecular ion that translates into improved sensitivity. The improved sensitivity was estimated to be approximately twofold. With a few exceptions, the pesticides determined in apples, peaches, potatoes, and tomatoes using methane reagent gas were detectable at the 0.1-ppm level. Estimated limits of detection using isobutane reagent gas in these four crops

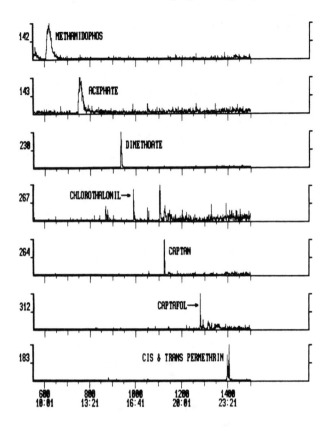

FIGURE 3. Mass chromatograms of several pesticides added to a tomato extract at the 0.1-ppm level. (Reprinted with permission from Mattern, G. C., Singer, G. M, Louis, J., Robson, M., and Rosen, J. D., *J. Agric. Food Chem.*, 38, 402, 1990. Copyright 1990 American Chemical Society.)

were 0.05 ppm or lower. Most of the limits of detection of pesticides in beans, corn, lettuce, peppers, and spinach were either 0.05 or 0.10 ppm, but some were 0.25 ppm or higher. Corn and spinach were the most difficult crops to analyze because they gave the dirtiest extracts, resulting in higher limits of detection for certain pesticides. Interferences in corn extracts necessitated the incorporation of a Florisil cleanup step, which was not completely successful because some pesticides could not be recovered from the Florisil. In spite of these shortcomings, all limits of detection were lower than the tolerance limits of the pesticides set in the crops by the EPA, except for permethrin in corn (limit of detection = 0.25 ppm), and chlorothalonil, captan, and captafol, which could not be recovered from corn using the Florisil cleanup procedure. Figure 3 shows the mass chromatograms of some pesticides at the 0.1-ppm level in tomatoes that were analyzed using isobutane reagent gas. Some of the mass chromatograms obtained from analysis of 0.05-ppm spiked peppers are shown in Figure 4.

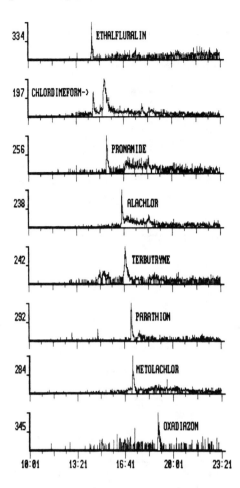

FIGURE 4. Mass chromatograms of several pesticides added to peppers at the 0.05-ppm level before extraction. (Reprinted with permission from Mattern, G. C., Low, C. H., Louis, J., and Rosen, J. D., *J. Agric. Food Chem.*, 39, 700, 1991. Copyright 1990 American Chemical Society.)

Limits of detection in surface water[13] were less than 0.05 ppb, except for captan and chlorothalonil, which had a limit of detection of 1 ppb. Alachlor, atrazine, butylate, carbofuran, chlorpyrifos, diazinon, fonofos, isofenphos, metolachlor, metribuzin, parathion, and simazine were all detectable at the 0.005-ppb level. The limit of detection of diazinon was less than 0.0005 ppb. The mass chromatograms of some pesticides in a spiked water sample at the 0.005-ppb level are illustrated in Figure 5.

It is important that the 3.15 version of the ITD software be used instead of the newer 4.0 version because the former gives better sensitivity for trace analyses. The 4.0 version has an Automatic Reaction Control (ARC) program for CI that controls ionization and reaction times as a function of total ion

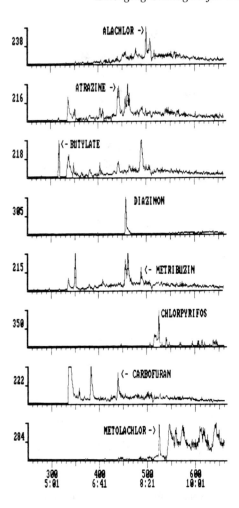

FIGURE 5. Mass chromatograms of several pesticides spiked into river water at the 0.005-ppb level. (From Mattern, G. C., Louis, J. B., and Rosen, J. D., *J. Assoc. Off. Anal. Chem.*, 74, 982, 1991. With permission.)

count in order to minimize saturation and space charging. In contrast, the 3.15 version has fixed ionization and reaction times that cannot decrease during an analysis. The lower sensitivity of CI-MS using the ARC program is believed to be caused by the prolonged decrease in the ionization and reaction times when major sample components enter the mass spectrometer.

The ion trap detector is a single, selective detector that has excellent sensitivity in the full-scan mode, making selected ion monitoring unnecessary. Thus, there is no limit on the number of pesticides that can be determined in an analysis, as would be the case if SIM were needed to enhance sensitivity. The major advantages of this method over presently used methods are (1) the substitution of cold on-column injection, capillary-column gas chromatography for hot injector packed-column gas chromatography to permit the analysis of polar as well as some thermally labile pesticides, such as linuron; (2) the use of a detector that in most cases gives immediate confirmation by the presence of multiple ions at a precise retention time; and (3) the substitution of a single specific detector (based on mass) for the array of specific detectors currently in use. These advantages are vividly demonstrated by considering the number of analyses (using current methodology) required for the 19 pesticides for which dietary oncogenic risk has been estimated by the EPA.[14] A flame photometric detector would be needed to determine acephate, azinphos-methyl, and parathion; a Hall detector in the halogen mode would be needed to determine alachlor, captafol, captan, chlordimeform, chlorothalonil, cypermethrin, diclofop-methyl, ethalfluralin, folpet, metolachlor, oxadiazon, permethrin, and pronamide; a Hall detector in the nitrogen mode would be required for the analysis of terbutryne; o-phenylphenol would require a flame ionization detector or HPLC; and linuron would have to be analyzed by still another procedure of HPLC. Two additional determinations at a higher column temperature would be required because azinphos-methyl, cypermethrin, and permethrin have such long retention times on packed columns.

IV. CONCLUSIONS

The multiresidue determination of trace levels of pesticides in fruits, vegetables, and surface water can be performed rapidly and accurately using gas chromatography/chemical ionization mass spectrometry with an ion trap detector. The cost of a gas chromatograph/chemical ionization ion trap detector including a computer system is much less than the cost of conventional mass spectrometers and compares favorably to the cost of the several specific detectors currently in use. These instruments are easily operated and maintained, and are suitable for these types of trace determinations.

ACKNOWLEDGMENTS

This work was supported by funds from the New Jersey Department of Environmental Protection and the New Jersey Agricultural Experiment Station (NJAES Publication Number F-10568-1-91).

REFERENCES

1. **Luke, M. A., Froberg, J. E., Doose, G. M., and Masumoto, H. T.,** *J. Assoc. Off. Anal. Chem.,* 64, 1187, 1981.
2. *Changes in Official Methods of Analysis,* 1st Suppl., 14th ed., Association of Official Analytical Chemists, Arlington, VA, 1985, sect. 29.A01-29.A04.
3. **Luke, M. A., Masumoto, H. T., Cairns, T., and Hundley, H. K.,** *J. Assoc. Off. Anal. Chem.,* 71, 415, 1988.
4. **West, S. D., Weston, J. H., and Day, E. W.,** *J. Assoc. Off. Anal. Chem.,* 71, 1082, 1988.
5. **Stan, H.-J.,** *J. Chromatogr.,* 467, 85, 1989.
6. **Eichelberger, J. W. and Budde, W. L.,** *Biomed. Environ. Mass Spectrom.,* 14, 357, 1987.
7. **Pereira, W. E., Rostad, C. E., and Leiker, T. J.,** *Anal. Chim. Acta,* 228, 69, 1990.
8. **de Beer, R. P., Sandmann, E. R., and Van Dyk, L. P.,** *Analyst (London),* 114, 1641, 1989.
9. **Munson, M. S. B. and Field, F. H.,** *J. Am. Chem. Soc.,* 88, 2621, 1966.
10. **McLafferty, F. W.,** *Mass Spectral Correlations,* Advances in Chemistry Series 40, American Chemical Society, Washington, D.C., 1963, 1.
11. **Mattern, G. C., Singer, G. M., Louis, J., Robson, M., and Rosen, J. D.,** *J. Agric. Food Chem.,* 38, 402, 1990.
12. **Mattern, G. C., Liu, C.-H., Louis, J. B., and Rosen, J. D.,** *J. Agric. Food Chem.,* 39, 700, 1991.
13. **Mattern, G. C., Louis, J. B., and Rosen, J. D.,** *J. Assoc. Off. Anal. Chem.,* 74, 982, 1991.
14. National Research Council, *Regulating Pesticides in Food: The Delaney Paradox,* National Academy Press, Washington, D.C., 1987, 45.

Chapter 13

HYPHENATED METHODS FOR PESTICIDE RESIDUE ANALYSIS

Damià Barceló

TABLE OF CONTENTS

I. INTRODUCTION

The term hyphenated methods was defined by Hirschfeld[1] as the marriage of two separate analytical techniques via appropriate interfaces. By combining several analytical techniques, it is possible to pool their virtues. The motivation for the development of these systems derives from a number of converging reasons. To accomplish general qualitative tasks more easily, a high discriminating power in one of the instruments can be combined with a high separating power in the other. Other possibilities include the complementarity of the qualitation and quantification performance of individual techniques being combined, e.g., gas chromatography (GC) is an excellent quantification and poor qualitation technique and is very well matched with infrared spectroscopy (IR) or mass spectrometry (MS), which offer good qualitation and poor quantification. Other examples are liquid chromatography with mass spectrometry (LC/MS) or supercritical fluid extraction with supercritical fluid chromatography (SFE/SFC).

The most common hyphenated methods used in pesticide residue analysis will include either chromatography, gas (GC) or liquid (LC) chromatography, or SFC and MS. In GC/MS, the selectivity is improved, compared to MS alone, by the physical separation of the components of a mixture by chromatography prior to mass analysis. The selectivity can be enhanced by the use of two or more analytical techniques in tandem, such as GC/MS/MS. In this sense it was shown that as the number of analytical stages of analysis is increased, the absolute levels of signal and noise decrease, dominating the chemical noise over the electrical noise (as is almost the case in MS) and resulting in an overall improvement in signal-to-noise ratio and detection limits.[2] An excellent review article[3] and two books[4,5] have been published recently concerning the use of MS methods for pesticide analysis. The use of other emerging hyphenated techniques such as directly linked GC/ Fourier Transform (FT)-IR/MS;[6] multidimensional GC and micro-LC/GC;[7] SFE coupled to GC, LC, and SFC;[8] and on-line pre- and postcolumn technology in LC[9,10] has also been reported. All of these techniques will be discussed in this chapter.

II. GAS CHROMATOGRAPHY/MASS SPECTROMETRY (GC/MS)

GC/MS is widely used by environmental laboratories involved in pesticide residue analysis. The most common practice in this context is to perform GC/ MS in the electron impact (EI) mode with library search for the unequivocal identification of the pesticide or with a second injection to check for coelution with an authenthic standard of the pesticide of interest. Thus, GC/MS with EI is commonly used to confirm the presence of organophosphorus pesticides,[3,11,12] triazines,[13,14] pyrethroids,[15] and paraquat and diquat after dehydrogenation.[16] Use of GC/MS with negative chemical ionization (NCI) is a selective approach, particularly suitable for pesticides containing electron-withdrawing groups (e.g., chlorine, nitro, etc.), which can resonance stabilize negative charges. The main advantages of NCI are its high selectivity and sensitivity to organochlorine and organophosphorus pesticides in environmental samples. GC/NCI-MS has been used to confirm the presence of organophosphorus compounds[3,14,17-19] and triazines.[20] The use of GC/PCI-MS has also been applied to organophosphorus pesticides[18] and triazines.[14,21] GC/MS with positive chemical ionization (PCI) and NCI and selected ion monitoring (SIM) of 2 to 3 characteristic fragment ions of each analyte allows the unequivocal identification of organophosphorus pesticide residues in different environmental matrices.[18]

As examples showing the differences between the various ionization modes in GC/MS, Figure 1A and B illustrates the EI- and PCI-MS spectra of simazine, and Figure 1C and D shows the PCI- and NCI-GC/MS spectra of cyanazine. Similar sensitivity is achieved when GC/MS spectra are obtained either with EI or PCI conditions except for cyanazine exhibiting detection limits in PCI two orders of magnitude poorer than in the EI mode, whereas in NCI the sensitivity was similar to EI. Detection limits at 100 pg using selected ion monitoring techniques can be provided by these GC/MS techniques. Although EI will be the common mode used for characterization of chlorotriazines in environmental samples,[14] GC/MS with PCI offers the advantage of providing higher selectivity, due to the formation of higher mass ions that can be easily used as diagnostic ions for monitoring purposes. In the case of cyanazine, the better signal obtained in NCI compared to PCI can be attributed to the lower gas-phase acidity of the molecule as a consequence of the CN group,[22] which can easily stabilize the negative charge by electron capture.

The EI spectra of cyanazine and simazine are characterized by base peaks that correspond to $[M - CH_3]^+$ (m/z 225), and $[M]^{\cdot +}$ (m/z 201), respectively. Another diagnostic ion of both chlorotriazines corresponds to the loss of $[C_2H_5NH]^+$ (m/z 44). The spectra from Figure 1 have been obtained under quadrupole MS, which are the common bench-tops used. However, GC-ion trap detectors are commonly used for pesticide confirmatory analysis, and in this latter case, the relative abundance of the different ions of chlorotriazines differs considerably[23] with regard to the data indicated in this chapter and previously reported.[14]

Under PCI conditions, the mass spectra of simazine and cyanazine showed $[M - Cl]^+$ ($m/z = 166$), and $[M + H]^+$ (m/z 241) as base peaks, respectively. The formation of $[M - Cl]^+$ ions agrees with the general fragmentation expected for chlorinated compounds under PCI conditions, e.g., aromatic chlorinated pesticides.[24] Other important ions in GC/MS with PCI were $[M + H]^+$ (m/z 202) for simazine and the adduct ion formed with the reagent gas, $[M + C_2H_5]^+$ at m/z 230 and 269 for simazine and cyanazine, respectively.

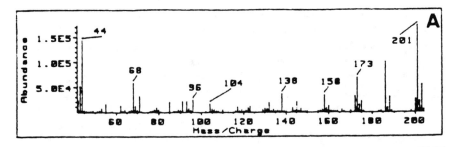

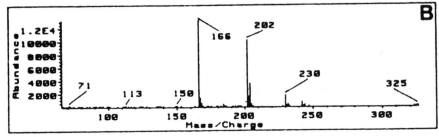

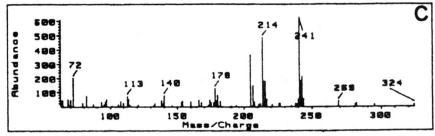

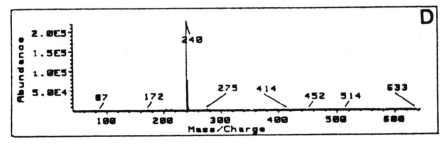

FIGURE 1. GC/MS spectra of simazine using EI (A) and PCI (B) and of cyanazine with PCI (C) and NCI (D). Amount injected of each pesticide onto the system: 5 to 10 ng. (Reprinted from Durand, G. and Barceló, D., *Anal. Chim. Acta,* 243, 259, 1991. With permission.)

The application of GC/MS in residue analysis is demonstrated in Figure 2, which illustrates the total ion current (TIC) and selected ion chromatograms for GC/MS with EI (A), PCI (B), and NCI (C) of the same soil extract containing 800 ng/g and 1 μg/g of fenitrothion and simazine, respectively. The ions monitored under GC/MS with EI (Figure 2A) corresponded to *m/z* values of 186 [M − CH$_3$]$^+$ and 201 [M]$^{\cdot+}$ for simazine and 260 [M − OH]$^+$ and 277 [M]$^{\cdot+}$ for fenitrothion. When GC/MS with PCI was employed, only the ions corresponding to simazine were monitored, with *m/z* values at 166 [M − Cl]$^+$, 202 [M + H]$^+$, and 230 [M + C$_2$H$_5$]$^+$ (see Figure 2B). Under these conditions, [M + H]$^+$ ion of fenitrothion was also detected at *m/z* 278 but not the adducts with the reagent gas. The amount of fenitrothion detected corresponded to 2 to 3 ng injected on column, which is fairly close to detection limits. As a consequence, only the base peak ion was observed, which makes the method unsuitable for confirmation purposes, because two or three diagnostic ions are also required. It is clear, therefore, that GC/MS with PCI is not recommended for screening most of the organophosphorus pesticides in environmental matrices.[14] In Figure 2C, the total ion current in NCI and the ions corresponding to fenitrothion at *m/z* values of 168 [SC$_7$H$_6$NO$_2$]$^-$ and 277 [M]$^{\cdot-}$ are illustrated from the same soil sample. Simazine, present in this soil, was not detected under NCI conditions, thus indicating the poor sensitivity of most of the chlorotriazines under NCI conditions.

The need to use GC/MS in environmental pesticide analyses for confirmation purposes arises from the enormous variety of pesticides currently used. Selectivity and sensitivity can be further enhanced by using various ionization modes. However, only a few libraries of standard mass spectra are available for PCI- and NCI-GC/MS, so each laboratory must build their own. This is one of the chief drawbacks of this approach and arises from the fact that instrumental parameters, such as the source temperature and reagent gas pressure, have a critical influence on the mass spectra.

III. GAS CHROMATOGRAPHY/MASS SPECTROMETRY/MASS SPECTROMETRY (GC/MS/MS)

GC/MS in the EI (also with PCI and/or NCI) mode and using selected ion monitoring (SIM) are, in general, the universal techniques for monitoring pesticides in environmental samples. However, in many instances, the specificity of the analysis is often compromised by using SIM. In this sense, information is lost in a GC/MS approach since in order to achieve a better limit of detection (LOD), only one or a few selected ions are monitored. The use of GC/MS/MS enables analysis of pesticides and their metabolites at trace levels in the presence of many coeluting compounds with minimal sample preparation. Such instrumentation includes, in most of the cases, triple quadrupole instruments combined with or without GC, since MS/MS is already a hyphenated technique and combines separation and identification as well. In addition, spectral information is gained by using the different modes of operation of GC/MS/MS, e.g., the use of daughter scan mode where a parent ion characteristic of the analyte is selected in the first mass analyzer, fragmenting it by collisionally activated dissociation (CAD) in the collision cell,

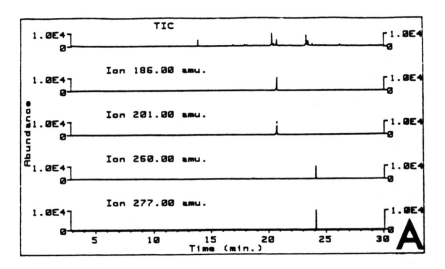

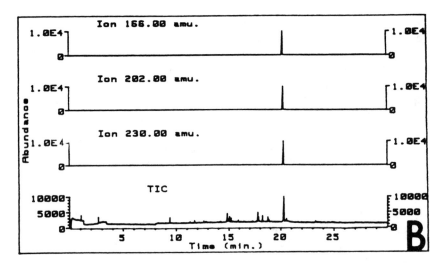

FIGURE 2. Total ion current (TIC) and selected ion chromatograms using GC/MS with EI (A), PCI (B), and NCI (C) of the same soil extract containing 800 ng/g of fenitrothion and 1 μg/g of simazine. Ions monitored by GC/MS with EI (Figure 1A) corresponded to *m/z* values of 186 and 201 for simazine and 260 and 277 for fenitrothion. In GC/MS with PCI (B), simazine ions were monitored at *m/z* values at 166, 202, and 230; and in GC/MS with NCI (C), fenitrothion ions were monitored at *m/z* values of 168 and 277. A 15 m × 0.15 mm I.D. fused-silica capillary GC column coated with chemically bonded cyanopropylphenyl DB-225 was used in (A) and (B). In Figure 1C, a 30 m × 0.25 mm I.D. bonded polyphenylmethylsiloxane RSL 300 capillary column was used. (Reprinted from Durand, G. and Barceló, D., *Anal. Chim. Acta*, 243, 259, 1991. With permission.)

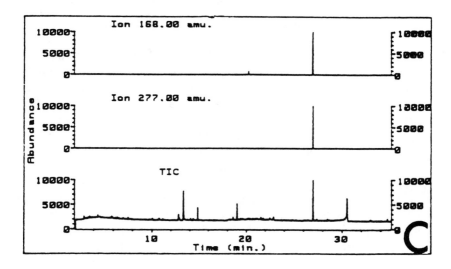

FIGURE 2C.

and scanning the second mass analyzer to obtain a daughter mass spectrum. Analogous to a normal mass spectrum, the daughter mass spectrum can be used for identification of an analyte by standard mass spectral interpretation or by matching the spectrum to an authentic sample.[2] In this way, although only one ion (parent ion) is selected, the spectral information obtained corresponds, generally, to an EI spectrum, which makes this technique particularly powerful for environmental analysis and provides much more information than the conventional GC/MS-EI with SIM.

Since the conversion of the parent ion into specific daughter ions is not complete, and the transmission efficiency of the second mass analyzer is much less than unity, the sensitivity of GC/MS/MS instruments is usually one order of magnitude worse than GC/MS. However, this reduction in sensitivity is compensated by the reduction of the chemical noise in the second analyzer, which often compensates for the reduced sensitivity.[3]

The application of this technique in environmental pesticide analysis has been very scarce and few examples are given elsewhere. In most of the cases, MS/MS has been used without chromatographic separation,[3] e.g., for the analysis of carbamate insecticides.[25] There is not a systematic work in the literature reporting the use of capillary GC/MS/MS for the analysis of pesticides and their corresponding metabolites. GC/MS/MS in the EI mode has been used for the analysis of organophosphorus pesticides[26] and in the PCI mode was employed for the determination of triazine and alachlor herbicides in water.[21]

IV. MISCELLANEOUS GC METHODS

Other hyphenated techniques involving generally GC and MS that are less common are cited below. GC/FT-IR and the more recent GC/FTIR/MS hyphenated technique provide useful information for screening complex environmental matrices; so far they have been applied to the analysis of 2,4-D salts in soil matrices.[27,28]

Organometallic compounds used as pesticides have been assayed by graphite furnace atomic absorption (GFAA) coupled to instruments of GC and LC.[27,29,30] Determination of organotin compounds in environmental samples was accomplished by using GC-GFAA with or without hydride generation.[30] The lowered detection limits thus achieved usually range between 5 and 50 ng and can be lowered to a few picograms by using hydride generation. Finally, the use of diffuse reflectance FTIR (DRIFT-IR) and confirmation by GC/MS has been applied to the determination of the phenoxy acid herbicide 2,4,5-T manufacturing waste.[6]

When techniques of GC/MS are unavailable, other identification methods must be employed. One inexpensive alternative in this respect is the use of linked retention data, parallel flame-photometric and electron-capture detection, and linear temperature retention indices. This approach has been used for the identification of a wide range of organophosphorus, sulfur-containing, and organochlorine pesticides.[31]

V. ON-LINE COUPLING LC/GC

This coupling has gained popularity over the last few years and combines the advantages of LC, that is, the possibility of using LC columns and pumps for preconcentration of pesticides, and the major assets of selective GC detectors. The selectivity of both the mobile and the stationary phase can be varied in LC, which can be exploited for preliminary cleanup of real samples. The LC technique also allows mixtures of components to be split into groups on the basis of chemical classes (group-type separations), and enrichment of very dilute samples. Coupled techniques of LC/GC generally use uncoated and deactivated capillary precolumns (also known as retention gaps), which accommodate the LC effluent while it vaporizes, thereby providing solute preconcentration. In addition, the LC system used can be either conventional or miniaturized. The advantages of microcolumn LC are rather evident. First, the volume of eluent used is reduced, indicating that the solute of interest are diluted in much less eluent. This reduction in peak volume is very important in the case of using reversed-phase eluents in LC, due to the problems caused by polar solvents. In addition, the use of microcolumn LC allows a much larger section of the LC chromatogram to be transferred into the GC system, thus allowing a better quantification of the compounds of interest. For example, the use of microcolumn LC with size exclusion for the elimination of lipid matrices combined on-line with GC was recenty published.[7]

The host of pesticides analyses in waters reported over the last few years
tifies to the growing interest in this approach. Triazines at low parts per
lion levels have been concentrated from drinking water and surface waters
using two precolumns, whether isolated or serially connected and packed
th C-18 and PRP-1, respectively. This is followed by LC analysis using
ferent detection systems (such as UV, fluorescence, and electrochemical[38,40]
a membrane disk containing 500 mg of C-18 material coupled on-line to
/UV.[41]

PHENYLUREA HERBICIDES

Phenylurea herbicides and their corresponding anilines have been pre-
ncentrated from river water by using a platinum phase packed in a short
ncentrator column acting as an aniline filter. Coupling of this precolumn
a C-18 concentrator column permitted subsequent preconcentration of phen-
ureas at the low parts per billion level for UV detection.[37] Detection of
enylureas in river water at 10 ppt level with a 500 ml sample concentration
lume was achieved with UV detection.[39] A similar approach using elec-
ochemical detection was applied to the same herbicides. As expected for an
ectrochemical detector compared with a UV detector, the sensitivities achieved
ere in the sub-parts per billion region and allowed detection of 30 ppt in
rface waters.[42] Phenylurea herbicides have been preconcentrated by using
C-18 column prior to analysis by LC/MS. The system used afforded an
richment factor of 100 to 1000 and characterization by LC/MS in the positive
n mode.[43]

A commercial preconcentration apparatus is shown schematically in Fig-
e 3.[37] This LC system is from Kontron (Zurich, Switzerland) and consists
two model 410 pumps, a model 200 programmer, an MCS 670 column-
vitching apparatus, and a Uvikon 720 variable wavelength UV detector set
241 nm. Separation was achieved by using reversed-phase C-18 columns;
d precolumns are packed with RP-8, RP-18, and 2-amino-1-cyclopentene-
dithiocarboxylic acid (ACDA) supports. Such a preconcentration system
as applied to the determination of phenylurea herbicides and their corre-
onding anilines in water samples. In the determination of phenylureas, there
a problem of discrimination between the analytes and their main biode-
adation products. This discrimination problem can be solved by LC with
e precolumn technology. A special platinum phase packed in a short pre-
olumn acted as an aniline filter for a group separation of the herbicides from
e anilines. The coupling of such a Pt precolumn with a C-18 precolumn
ermits subsequent preconcentration of the herbicides, on-line transfer, and
eparation on a reversed-phase analytical column.

The feasibility of the method was demonstrated for analyzing river water
amples spiked with seven herbicides and seven anilines, as shown in Figure
.[37] The seven anilines (see numbers 1-7 in the chromatogram of Figure 4a)
ere completely removed from the solution by employing the ACDA-Pt filter

The technique offers the possibility of transferring fractic
GC, with injection up to 1 ml, and possibilities for automated
yses. By using LC/GC, atrazine was analyzed in water at the p
level with the aid of a reversed-phase LC column and a so
procedure.[32] This approach has been applied to the analysis
rinated pesticides and herbicides at the parts per trillion le'
sediment,[34] and tobacco[35] matrices followed by electron-cap
The heart-cutting of the LC chromatograms followed by on-lin
analysis was described[35] as a useful multidimensional approacl
contaminated sediment samples. The determinations of atrazii
chlorpyrifos in rat feed, and folpet in hops were accompli
normal-phase LC eluents, such as mixtures of *n*-hexane containi
or ethanol or *n*-heptane with methyl *tert*-butyl ether.[36] For the
of atrazine, LC was connected to a GC autosampler, and cuts c
analyzed using a thermionic detector. In the case of folpet ana
of column-switching valves with two LC columns, one for cl
other as a separation column, were employed in combination
ECD detection. Reversed-phase eluents containing methanol-v
column LC combined with GC were used for the determination c
propachlor in soil.[36] The methods of LC/GC resulted in go
lower detection limits than conventional methods, and partial
manual sample preparation work.

VI. ON-LINE PRECOLUMN LC

Solid-phase extraction procedures applied prior to the s
detection of pesticides in water by on-line liquid chromatograp
are eliciting growing interest. Adsorbents (such as silica, alkylsi
silica, alumina, porous polymers with or without ion exchan
carbon modified materials) are usually packed into short sta
glass columns or cartridges (referred to as concentrator colum
procedures involving a chromatographic column, and can be us
pressures. Solid-phase extraction columns are normally empl
enrichment, cleanup, sample storage, protection of the main
when needed, for derivatization. Trace components in wate
trapped by an appropriate sorbent packed in a concentrator cc
to an analytical column via switching valves, and the pesticide:
directly eluted from the concentrator column to the analytical
suitable mobile phase. Selectivity toward specific compounds c
by coupling different precolumns containing different selective :
as C-18, ion exchangers, or metal-loaded phases) serially. Detail
of the use of on-line concentrator columns in liquid chrom
analysis of pesticides is provided elsewhere.[9,10,37-39]

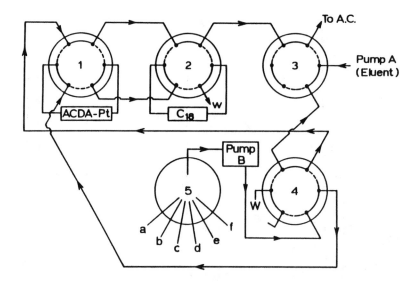

FIGURE 3. Switching-valve configuration for the on-line removal of anilines and preconcentration of phenylurea herbicides prior to LC separation. 1, 2, 3, 4 = High-pressure switching valves; 5 = low-pressure six-port selector valve (a = water, b = sample 1, c = acetonitrile, d = water, e = sample 2, f = acetonitrile); ACDA-Pt and C18 precolumns, 11 × 2 mm I.D., packed with the indicated material; A.C. = analytical column; A and B = LC pumps; W = waste. (Reprinted from Goewie, C. E., Kawkman, P., Frei, R. W., Brinkman, U. A. Th., Maasfeld, W., Seshadri, T., and Kettrup, A. J., *J. Chromatogr.*, 284, 73, 1984. With permission.)

(compare Figure 4a and b), even though this column obviously serves as an efficient filter for the removal of many other (polar) matrix constituents (compare Figure 4b and c). After such a run, the ACDA-Pt precolumn should be washed with 1 ml of acetonitrile to remove all of the sorbed compounds. The applicability of ACDA-Pt allows the removal of at least 0.1 ppm of anilines from aqueous solutions containing 5 to 10 ppb of phenylurea herbicides. In this way, automated sample cleanup and preconcentration of river water samples are achieved by the use of reversed-phase and UV detection.

B. CHLORINATED PHENOXY ACIDS

These acids comprise another group of pesticides of interest for analyses by on-line preconcentration in LC. These compounds, which require derivatization for analysis by GC, have been enriched by using various stationary phases (such as octadecyl silica, nitrile silica, and macroporous polystyrene-divinylbenzene copolymers) from sub-parts per billion level in drinking and surface water with further analysis by LC-UV.[44]

C. CARBAMATE PESTICIDES

These pesticides (such as carbaryl, chlorpropham, propoxur, and

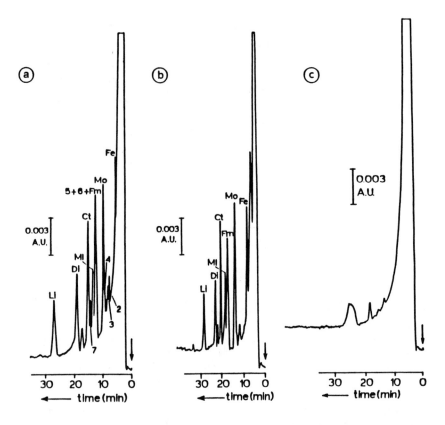

FIGURE 4. Preconcentration of 10 ml of a river water sample spiked with seven herbicides and seven anilines. (a) Preconcentration on a C18 precolumn; (b) as (a), but with insertion of an ACDA-Pt precolumn upstream from the C18 precolumn; (c) river water blank preconcentrated on a C18 precolumn, UV detection at 245 nm. Fe, fenuron; Mo, monuron; Fm, fluometuron; Ct, chlortoluron; Ml, monolinuron; Di, diuron; Li, linuron; peaks 2-7, various anilines. Analytical column, CP-sphere C18 (8 μm), 250 × 4.6 mm I.D. Mobile phase, methanol-water (6:4) at 0.85 ml/min. (Reprinted from Goewie, C. E., Kawkman, P., Frei, R. W., Brinkman, U. A. Th., Maasfeld, W., Seshadri, T., and Kettrup, A., *J. Chromatogr.*, 284, 73, 1984. With permission.)

carbofuran) were preconcentrated from different types of water with C18 sorbent and determined by LC-UV with detection limits in the sub-parts per billion range.[45,46]

VII. POSTCOLUMN SYSTEMS IN LC

The weakest part of LC system is still the UV detector, which is relatively insensitive and requires molecules to possess a chromophore. This restriction has fostered development of chemical reaction methods to enhance detection in LC. Two excellent, recently published reviews showed the great potential of postcolumn LC systems.[47,48] On the other hand, such approaches call for reagent solutions to be added and, hence, for additional pumps to be used. In addition, extra-column peak broadening may result from the reactor. Post-column systems are used in two different ways. One method involves employing ordinary phase separators to extract analytes into an immiscible organic solvent. In this way, aqueous and organic plugs are created that are passed through a phase separator to isolate the organic phase. The pesticides in the organic phase are driven to the detector while the aqueous phase is sent to waste. The other alternative method, more common to the use of postcolumn systems than the previous one, involves postcolumn reactors based either on changes in physicochemical (electrochemical, redox, hydrolytic) properties or photochemical and derivatization reactions (e.g., ion-pair formation and ligand-exchange).

A. POSTCOLUMN SYSTEMS

A typical application of solid-phase reactors is the analysis of *N*-methyl carbamates with on-line hydrolysis and addition of *o*-phthalaldehyde (OPA). The carbamates yield methylamine on hydrolysis, which is converted into a fluorescent compound by addition of OPA in a second reactor. This approach was recently applied to the analysis of carbamate residues at the picogram level in environmental matrices.[49] The same group of pesticides was also assayed by postcolumn photolysis reaction with OPA and fluorescence detection.[50] Other representative examples are the use of oxidation and derivatization reactions with OPA for the determination of glyphosate, one of the most polar herbicides (which can only be extracted in water) with detection by fluorogenic labeling,[51] and the analysis for diquat and paraquat using a postcolumn reaction with sodium hydrosulfide and diode array detection.[52]

B. EXTRACTION SYSTEMS

Postcolumn extraction systems have rarely been applied to pesticide analysis. Such systems have been used for polar pesticides or metabolites, which are difficult to assay under typical LC conditions. Hydroxyatrazine, a polar metabolite of atrazine, was analyzed by ion-pair postcolumn extraction by using 9,10-dimethoxyanthracene-2-sulfonate as a counter ion and fluorescence detection.[53] Fluorescence probes, such as a quaternary ammonium fluorophore, were employed as counter ions in a reaction detector for on-line ion-pair extraction of phenoxy acid herbicides.[54] In this experiment, the herbicides were separated in aqueous methanol at pH 2.5, then deprotonated by the postcolumn addition of a phosphate buffer, and subsequently the ion-pairs were extracted on-line with chloroform-*n*-butanol (80:20) and were monitored by fluorescence detection. Low nanogram levels of detection in drinking water samples were reached.[54] Buffers and ion-pairing agents present in the LC mobile phases are liable to interfere with detection and can be removed by using a postcolumn extraction system to transfer the organic phase to the detector while the inorganic ions remain in the aqueous layer. This procedure

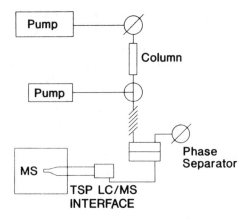

FIGURE 5. Schematic diagram of the postcolumn extraction setup. (Reprinted with permission from Barceló, D., Durand, G., Vreeken, R., de Jong, G. J., and Brinkman, U. A. Th., *Anal. Chem.*, 62, 1696, 1990. Copyright 1990 American Chemical Society.)

has been used for the ion-suppressed extraction of chlorinated phenoxy acid pesticides with on-line MS detection.[55]

A schematic diagram of the experimental setup used for this purpose is given in Figure 5 and consisted of a sandwich-phase separator equipped with two stainless steel blocks with and without a groove and a PTFE disk with a groove. The extraction solvent is delivered by a second high pressure pump. With the phase separator, a purely organic phase can be obtained and directly introduced in the mass spectrometer through a thermospray (TSP) LC/MS interface. This system was used for the analysis of chlorinated phenoxy acids. A LC eluent of acetonitrile-aqueous 0.1 *M* phosphate (pH = 2.5) buffer (50:50) at 1 ml/min was used for the ion-suppressed reversed-phase separation of the chlorinated phenoxy acids. After a postcolumn extraction mixture of dichloromethane-cyclohexane-*n*-butanol (45:45:10) was used as the mobile phase at 1 ml/min, thus extracting the chlorinated phenoxy acids which were subsequently introduced into the system of TSP LC/MS. The extraction efficiencies for 2,4-D, 2,4,5-T, and Silvex were 100, 70, and 60%, respectively.

Figure 6 shows the chromatogram in the full-scan mode obtained for a water sample from the Barcelona harbor spiked at 0.1 ppm with 2,4-D, 2,4,5-T, and Silvex, which has been analyzed with reversed-phase LC in combination with the postcolumn extraction system and with TSP detection in the NI mode. These results clearly demonstrate the usefulness of this system for LC analyses where nonvolatile buffers are needed.[55]

VIII. ON-LINE SFC

Supercritical fluid chromatography (SFC) has been used to analyze various pesticides in the last few years.[56-63] While supercritical fluids are gas-

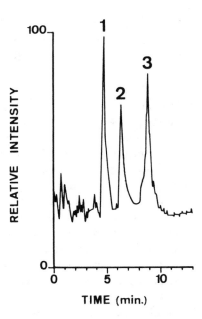

FIGURE 6. Reconstructed ion chromatogram obtained with NI TSP LC/MS of a water sample from Barcelona harbor spiked with 0.1 ppm of (1) 2,4-D and (2) 2,4,5-T. (3) Silvex: stationary phase, 5 μm Lichorsorb RP-18; mobile phase, acetonitrile-water (50:50) + 0.1 M phosphate buffer (pH = 2.5); flow rate, 1 ml/min; extraction solvent, cyclohexane-dichloromethane-*n*-butanol (45:45:10) at 1 ml/min; flow rate into the mass spectrometer, 0.8 ml/min. (Reprinted with permission from Barceló, D., Durand, G., Vreeken, R., de Jong, G. J., and Brinkman, U. A. Th., *Anal. Chem.*, 62, 1696, 1990. Copyright 1990 American Chemical Society.)

like in some aspects and liquid-like in others (they are typically 10 to 100 times less viscous than liquids), they can be used as mobile phases as complementary aids to LC and GC. The main advantages of this technique are the shorter retention times involved in the analysis of moderately polar and thermally labile pesticides of large molecular weights and its compatibility with most detectors of LC and GC (UV, FID, NPD, FTIR, MS). Reported applications of SFC include the analysis of carbamates,[56-60] organophosphorus,[61] and urea[58,62] pesticides.

Since the 1980s, attention has been paid to the development of supercritical fluid extraction (SFE) hyphenated systems.[8] Main advantages of using the on-line combination of SFE with other chromatographic systems are

1. An improved efficiency, as compared with conventional Soxhlet extraction
2. The potential of extracting thermally labile pesticides
3. The ease of separating the analytes from a supercritical fluid
4. The possibility of direct analysis of complex matrices, thus reducing sample manipulation

5.   The compatibility of the method with other on-line chromatographic methods (e.g., GC, TLC, and SFC) in combination with different GC and/or liquid chromatographic detectors (e.g., FID, NPD, MS, and UV/VIS)

Thus, SFE coupled to SFC was successfully used to isolate and analyze carbofuran and its 3-keto and 3-hydroxy metabolites.[63] Other applications include the extraction of pesticides (such as 2,4-D) from soil samples, which are determined with no prior cleanup by SFC with ion mobility detection.[64] Extraction of triazines at a parts per billion level from river sediments was achieved by SFE; then the pesticides were analyzed by LC with UV detection at 225 nm, with recoveries in excess of 90%.[8] The coupling of SFE and SFC was employed for the analysis of sulfonylurea herbicides and metabolites in soil, plant material, and cell culture.[65] In this case, methanol was needed, in combination with carbon dioxide, for the extraction and separation of the analytes studied. The relatively low pressures prevented the decomposition of these thermally labile pesticides, which is a significant advantage over methods of GC that require derivatization prior to analysis.[10] The on-line combination of SFC with MS detection will be discussed in the section on LC/MS, since both combinations have many similarities.

IX. LIQUID CHROMATOGRAPHY/MASS SPECTROMETRY (LC/MS)

The on-line combination liquid chromatography with mass spectrometry (LC/MS) occupies a prominent place in environmental organic analysis and surpasses GC/MS in the analysis of polar pesticides and herbicides in some aspects. Two books[4,5] and two review articles[3,10] that recently have been published provide excellent overviews of LC/MS applications to environmental pesticide analysis. An outstanding review of the different types of interfaces used in this context (including transport systems, direct liquid introduction, thermospray, atmospheric pressure ionization, particle beam, open tubular LC, and continuous flow-fast atom bombardment) was also recently published.[66] Of the different methodologies of LC/MS the thermospray (TSP) interfacing system is probably the most widely used and typically involves reversed-phase columns and volatile buffers, with or without a filament or discharge.

A. THERMOSPRAY LC/MS

This has been applied to the analysis of a variety of pesticides including carbamates,[67-70] organophosphorus compounds,[70-75] pyrethroids,[74] ureas,[73,76,77] chlorinated phenoxy acids,[78] and triazines.[13,77,78] Filament-off and filament-on with positive and negative ion modes are common choices in this context. The filament-off mode is associated with thermospray ionization, while the

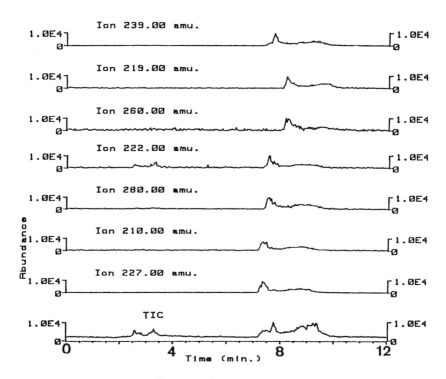

FIGURE 7. Total ion current and selected-ion chromatograms in TSP LC/MS with PI mode of a soil sample spiked with carbamate insecticides at 1 μg/g. Diagnostic ions and compounds identified were at *m/z* 239 (pirimicarb), 219, and 260 (carbaryl); *m/z* 222 and 280 (carbofuran); and *m/z* 210 and 227 (propoxur). Mobile phase, gradient of acetonitrile-water from (30:70) + 0.1 *M* ammonium acetate to (65:35) in 7 min; flow rate, 1 ml/min; column packing, 5 μm LiChrosphere 100 RP-18. (Reprinted from Durand, G., de Bertrand, N., and Barceló, D., *J. Chromatogr.*, 562, 507, 1991. With permission.)

process involved in filament-on is closer to chemical ionization and discharge ionization, which are applied to LC eluents with high water contents and provide further structural information with additional fragments. These two procedures are applied routinely; however, the filament-on alternative is more commonly used on account of its usually higher sensitivity. The choice between the positive or negative ion mode depends on the compound concerned. In any case, the positive ion mode is much more frequently used. It normally yields [M + H]⁺ and/or [M + NH₄]⁺ base peaks. The negative ion mode has been shown to be much more sensitive to rather electronegative compounds (such as chlorinated phenoxy acids) than the positive ion mode; the former yields [M + acetate]⁻ or [M + formate]⁻ base peaks if ammonium acetate or ammonium formate is used as ionizing additive.[76]

Thermospray LC/MS is the most widely applied technique for environmental pesticide analysis. In Figure 7, TSP LC/MS in the PI mode was used for the determination of a mixture of the carbamates propoxur, carbaryl, carbofuran, and pirimicarb at the 1 μg/g level in a soil sample.[68] The ions monitored corresponded to m/z 239 $[M + H]^+$ (pirimicarb); m/z 219 $[M + NH_4]^+$ and m/z 260 $[M + CH_3NH_2CO]^+$ (carbaryl); m/z 222 $[M + NH_4]^+$ and m/z 280 $[M + CH_3NH_2CO]^+$ (carbofuran); and m/z 210 $[M + H]^+$ and m/z 227 $[M + NH_4]^+$ (propoxur). More sophisticated equipment such as on-line LC/MS/MS or continuous fast atom bombardment instruments also permit pesticide characterization. Thus, LC/MS/MS was applied to the analysis of organophosphorus,[75] carbamate,[67,69] and triazine[4] pesticides, and was found to provide more fragmentation than typical thermospray LC/MS.

B. SFC/MS, CZE/MS

SFC/MS[58,60] permits the characterization of different pesticides with electron impact or chemical ionization, which makes the technique potentially useful to environmental analysis. In most of these applications carbon dioxide is used as the supercritical mobile phase and 2-propanol as the polarity modifier.[61] Other novel techniques for analyzing the more polar and amphoteric pesticides include the use of fast atom bombardment,[4,79] as, e.g., for the characterization of glyphosate[4] and paraquat.[79] Another new technique is the use of capillary zone electrophoresis (CZE)/MS,[80] which permits the separation of ionic compounds in an open tube through the application of a voltage gradient. Compounds differentially migrate through the tube as a result of their electrophoresis mobility. An example of commercialy available combinations of CZE-MS is the Sciex IonSpray Atmospheric Pressure Ionization (API) system. Advantages claimed are (1) no pumping of solvent or heat required; (2) mild ambient temperature ionization, thus making the process suitable for labile analytes; (3) instrumental simplicity; and (4) use of MS/MS, thus providing additional structural information. In Figure 8 the determination of phenoxyacetic acid pesticides at the picomole level by using the negative ion mode in SIM CZE/MS is shown.[80]

X. CONCLUSIONS

Several hyphenated methods for pesticide analysis have been reported in this chapter. Determining pesticides by GC/MS is a relatively simple matter, and this will be the methodology of choice for most pesticide residue laboratories. Although three modes of GC/MS operation are generally available (EI, PCI, and NCI), most of the published literature employs EI techniques. While PCI and NCI are good alternative ionization methods, they offer (depending on the compound) better selectivity and/or sensitivity than EI. Selectivity is more important in the case of analyzing environmental samples. When cleanup is poor, e.g., in the determination of pesticides in relatively dirty extracts, GC/MS with PCI and NCI for chlorotriazines and organophosphorus pesticides, respectively, are strongly recommended since diagnostic ions are at higher m/z values and consequently less interference from the matrix will appear in the traces of GC/MS.

NEGATIVE IONS

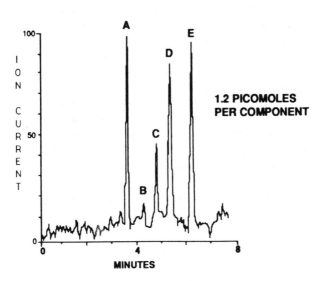

FIGURE 8. Total ion electropherogram in SIM CZE/MS of: (A) 2,4-dichlorophenoxybutyric acid; (B) gibberellic acid; (C) 2,3,6-T; (D) 2,4-D; and (E) *p*-chloromandelic acid. Eluent composition acetonitrile-ammonium acetate 20 m*M* (50:50); hydrostatic injection, 5 cm for 10 s, 10 nl injected of 1.2 picomol per component; 100 μm × 100 cm fused silica column at 33 kV with IonSpray detection 3 kV and liquid junction interface. (Reprinted from *The API Book,* SCIEX, Central Reproduction, Mississauga, Ontario, Canada, 1990, p. 65, Part 6. With permission.)

The use of LC/GC has increased over the last years, and it serves as a hyphenated technique that enhances the sensitivity for pesticides that can be analyzed by GC. Two drawbacks can be mentioned: (1) the need for derivatization, as in GC, when preconcentrating polar herbicides and metabolites that cannot be analyzed directly by GC; and (2) the high cost, which is also common to most of hyphenated techniques. On the other hand, the use of LC/GC/MS offers many possibilities in on-line analysis and confirmation of pesticides in environmental samples. Unfortunately, at present little equipment is commercially available. Methods of LC allow polar and thermally labile pesticides to be assayed without derivatization and direct injection of dirty extracts into water-miscible solvents. The inability of LC to analyze pesticides containing no chromophores or fluorophores is being partly overcome by the increasingly common applications of LC/MS, as most pesticides are responsive to MS and can thus be readily identified. Complementary use of GC/MS and LC/MS should obviously allow monitoring of most pesticide residues.

In general, it should be mentioned that the choice of analytical techniques has broadened enormously over the last few years. Thus, SFC and on-line LC/GC equipment incorporating various detectors (FID, UV, NPD, ECD, MS) or the multielement atomic emission detector for GC and LC are already commercially available and have a promising future in this field.

Future applications of hyphenated systems should concentrate on such relatively unexplored aspects as: (1) the use of alternative MS techniques (e.g., high resolution MS and MS/MS, which although well established and available in many laboratories, are used infrequently), with special emphasis on novel approaches (e.g., micro-LC coupled to continuous-FAB MS or CZE/MS that could be applied to the analysis of glyphosate and/or the "quats"); (2) the coupling of postcolumn extraction systems with TSP LC/MS can be a solution for the analysis of polar and amphoteric pesticides (e.g., for the ion-pair extraction of the "quats" with sulfonate-counter ions); and (3) the potential of coupling SFE with other separation techniques. So far, SFE/GC is the most applied technique, but an interesting hyphenated method that certainly will grow in the near future will be SFE/SFC/MS. The addition of modifiers (e.g., ammonia or nitrous oxide) and modifications of the matrix (e.g., pH change) will certainly help in developing the SFE combination coupled on-line with different chromatographic systems.

REFERENCES

1. **Hirschfeld, T.**, *Anal. Chem.*, 52, 297A, 1980.
2. **Johnson, J. V. and Yost, R. A.**, *Anal. Chem.*, 57, 758A, 1985.
3. **Levsen, K.**, *Org. Mass Spectrom.*, 23, 406, 1988.
4. **Rosen, J. D.**, *Application of New Mass Spectrometric Techniques in Pesticide Chemistry*, John Wiley & Sons, New York, 1987, 1.
5. **Brown, M. A.**, *Liquid Chromatography/Mass Spectrometry Applications in Agricultural, Pharmaceutical and Environmental Chemistry*, American Chemical Society, Washington, D.C., 1990, ACS 420, 1.
6. **Gurka, D. F., Betowski, L. D., Hinners, Th. A., Heithmar, E. M., Titus, R., and Henshaw, J. M.**, *Anal. Chem.*, 60, 454A, 1988.
7. **Sandra, P., David, F., and Redant, G.**, *Organic Micropollutants in the Aquatic Environment*, Angeletti, G. and Bjorseth, A., Eds., Kluwer, Dordrecht, Netherlands, 1991, 102.
8. **Vanndoort, R. W., Chervet, J. P., Lingeman, H., de Jong, G. J., and Brinkman, U. A. Th.**, *J. Chromatogr.*, 505, 45, 1990.
9. **Nielen, M. W. F., Frei, R. W., and Brinkman, U. A. Th.**, in *Selective Sample Handling and Detection in High-Performance Liquid Chromatography*, Frei, R. W. and Zech, K., Eds., Elsevier, Amsterdam, Netherlands, 1988, 5.
10. **Barceló, D.**, *Chromatographia*, 25, 928, 1988.
11. **Stan, H.-J.**, *J. Chromatogr.*, 467, 85, 1989.
12. **Barceló, D., Solé, M., Durand, G., and Albaigés, J.**, *Fresenius J. Anal. Chem.*, 339, 676, 1991.
13. **Durand, G., Forteza, R., and Barceló, D.**, *Chromatographia*, 28, 597, 1989.
14. **Durand, G. and Barceló, D.**, *Anal. Chim. Acta*, 243, 259, 1991.
15. **Lidgard, R. O., Duffield, A. M., and Wells, R. J.**, *Biomed. Environ. Mass Spectrom.*, 13, 677, 1986.
16. **Hajšlová, J., Cuhra, P., Davidek, T., and Davidek, J.**, *J. Chromatogr.*, 479, 243, 1989.

17. **Stan, H.-J. and Kellner, G.,** *Biomed. Mass Spectrom.,* 9, 483, 1982.
18. **Stan, H.-J. and Kellner, G.,** *Biomed. Environ. Mass Spectrom.,* 18, 645, 1989.
19. **Barceló, D., Porte, C., Cid, J., and Albaigés, J.,** *Int. J. Environ. Anal. Chem.,* 38, 199, 1990.
20. **Huang, L. Q. and Mattina, M. J. I.,** *Biomed. Environ. Mass Spectrom.,* 18, 828, 1989.
21. **Rostad, C. E., Pereira, W. E., and Leiker, T. J.,** *Biomed. Environ. Mass Spectrom.,* 18, 820, 1989.
22. **Harrison, A. G.,** *Chemical Ionization Mass Spectrometry,* CRC, Boca Raton, FL, 1983, 1.
23. **Pereira, W. R., Rostad, C. E., and Leiker, T. J.,** *Anal. Chim. Acta,* 228, 69, 1990.
24. **Dougherty, R. C., Roberts, J. D., and Biros, F. J.,** *Anal. Chem.,* 47, 54, 1975.
25. **Hummel, S. Y. and Yost, R. A.,** *Org. Mass Spectrom.,* 21, 785, 1986.
26. **Roach, J. A. G. and Carson, L. J.,** *J. Assoc. Off. Anal. Chem.,* 70, 439, 1987.
27. **Nubbe, M. E., Dean Adams, V., Watts, R. J., and Robinet-Clark, Y. S.,** *J. Water Pollut. Control Fed.,* 60, 773, 1988.
28. **Gurka, D. F., Betowski, L. D., Jones, T. L., Pyle, S. M., Titus, R., Ballard, J. M., Tondeur, Y., and Niederhut, W.,** *J. Chromatogr. Sci.,* 26, 301, 1988.
29. **Lawrence, J. F.,** *Chromatographia,* 24, 45, 1987.
30. **Han, J. S. and Weber, J. H.,** *Anal. Chem.,* 60, 316, 1988.
31. **Stan, H.-J. and Mrowetz, D.,** *J. Chromatogr.,* 279, 173, 1983.
32. **Davies, I. L., Markides, K. E., Lee, M. L., Raynor, M. W., and Bartle, K. D.,** *J. High Resolut. Chromatogr. Chromatogr. Commun.,* 12, 193, 1989.
33. **Noroozian, E., Maris, F. A., Nielen, M. W. F., Frei, R. W., de Jong, G. J., and Brinkman, U. A. Th.,** *J. High Resolut. Chromatogr. Chromatogr. Commun.,* 10, 17, 1987.
34. **Maris, F. A., Noroozian, E., Otten, R. R., van Dijck, R. C. J. M., de Jong, G. J., and Brinkman, U. A. Th.,** *J. High Resolut. Chromatogr. Chromatogr. Commun.,* 11, 197, 1988.
35. **Munari, F. and Grob, K.,** *J. Chromatogr. Sci.,* 28, 61, 1990.
36. **Cortes, H. J.,** in *Multidimensional Chromatography,* Cortes, H. J., Ed., Marcel Dekker, New York, 1990, 251.
37. **Goewie, C. E., Kawkman, P., Frei, R. W., Brinkman, U. A. Th., Maasfeld, W., Seshadri, T., and Kettrup, A.,** *J. Chromatogr.,* 284, 73, 1984.
38. **Subra, P., Hennion, M. C., Rosset, R., and Frei, R. W.,** *Int. J. Environ. Anal. Chem.,* 37, 45, 1989.
39. **Hennion, M. C., Subra, P., Rosset, R., Lamcq, J., Scribe, P., and Saliot, A.,** *Int. J., Environ. Anal. Chem.,* 42, 15, 1990.
40. **Subra, P., Hennion, M. C., Rosset, R., and Frei, R. W.,** *J. Chromatogr.,* 456, 121, 1988.
41. **Brouwer, E. R., Lingeman, H., and Brinkman, U. A. Th.,** *Chromatographia,* 29, 415, 1990.
42. **Nielen, M. W. F., Koomen, G., Frei, R. W., and Brinkman, U. A. Th.,** *J. Liq. Chromatogr.,* 8, 315, 1985.
43. **Maris, F. A., Geerdink, R. B., Frei, R. W., and Brinkman, U. A. Th.,** *J. Chromatogr.,* 323, 113, 1985.
44. **Geerdink, R., Van Balkom, A., and Brouwer, H.-J.,** *J. Chromatogr.,* 481, 275, 1989.
45. **Marvin, C. H., Brindle, I. D., Hall, C. D., and Chiba, M.,** *J. Chromatogr.,* 503, 167, 1990.
46. **Marvin, C. H., Brindle, I. D., Hall, C. D., and Chiba, M.,** *Anal. Chem.,* 62, 1495, 1990.
47. **Brinkman, U. A. Th.,** *Chromatographia,* 24, 190, 1987.

48. Brinkman, U. A. Th., Frei, R. W., and Lingeman, H., *J. Chromatogr.*, 492, 251, 1989.
49. De Kok, A., Hiemstra, M., and Vreeker, C. P., *Chromatographia*, 24, 469, 1987.
50. Miles, C. J. and Anson Moye, H., *Chromatographia*, 24, 628, 1987.
51. Tuinstra, L. G. M. Th. and Kienhuis, P. G. M., *Chromatographia*, 24, 696, 1987.
52. Simon, V. A. and Taylor, A., *J. Chromatogr.*, 479, 153, 1989.
53. Lawrence, J. F., Brinkman, U. A. Th., and Frei, R. W., *J. Chromatogr.*, 185, 473, 1979.
54. De Ruiter, C., Minnaard, W. A., Lingeman, H., Kirk, E. M., Brinkman, U. A. Th., and Otten, R., *Int. J. Environ. Anal. Chem.*, 43, 79, 1991.
55. Barceló, D., Durand, G., Vreeken, R., de Jong, G. J., and Brinkman, U. A. Th., *Anal. Chem.*, 62, 1696, 1990.
56. Kalinoski, H. T., Wright, B. W., and Smith, R. D., *Biomed. Environ. Mass Spectrom.*, 13, 33, 1986.
57. Wright, B. W. and Smith, R. D., *J. High Resolut. Chromatogr. Chromatogr. Commun.*, 8, 8, 1985.
58. Kalinoski, H. T., Udseth, H. R., Wright, B. W., and Smith, R. D., *J. Chromatogr.*, 400, 307, 1987.
59. Berry, A. J., Games, D. E., Mylchreest, I. C., Perkins, J. A., and Pleasance, S., *Biomed. Environ. Mass Spectrom.*, 15, 105, 1988.
60. France, J. E. and Voorhees, K. J., *J. High Resolut. Chromatogr. Chromatogr. Commun.*, 11, 692, 1988.
61. Kalinoski, H. T. and Smith, R. D., *Anal. Chem.*, 60, 529, 1988.
62. Shah, S. and Taylor, L. T., *J. High Resolut. Chromatogr. Chromatogr. Commun.*, 12, 599, 1989.
63. Davies, I. L., Xu, B., Markides, K. E., Bartle, K. D., and Lee, M. L., *J. Microcolumn Sep.*, 1, 71, 1989.
64. Morrissey, M. A. and Hill, H. H., Jr., *J. Chromatogr. Sci.*, 27, 529, 1988.
65. McNally, M. E. P. and Wheeler, J. R., *J. Chromatogr.*, 447, 53, 1988.
66. Tomer, K. B. and Parker, C. E., *J. Chromatogr.*, 492, 189, 1989.
67. Chiu, K. S., Van Langenhove, A. V., and Tanaka, C., *Biomed. Environ. Mass Spectrom.*, 18, 200, 1989.
68. Durand, G., de Bertrand, N., and Barceló, D., *J. Chromatogr.*, 562, 507, 1991.
69. Rudewicz, P. J., *Finnigan Mat Application*, report number 211, 1988, 1.
70. Bellar, T. A. and Budde, W. L., *Anal. Chem.*, 60, 2076, 1988.
71. Durand, G., Sanchez-Baeza, F., Messeguer, A., and Barceló, D., *Biol. Mass Spectrom.*, 20, 3, 1991.
72. Barceló, D., *Biomed. Environ. Mass Spectrom.*, 17, 363, 1988.
73. Barceló, D. and Albaigés, J., *J. Chromatogr.*, 474, 163, 1989.
74. Barceló, D., *LC-GC Mag.*, 6, 324, 1988.
75. Betowski, L. D. and Jones, T. L., *Environ. Sci. Technol.*, 22, 1430, 1988.
76. Barceló, D., *Org. Mass Spectrom.*, 24, 219, 1989.
77. Hammond, I., Moore, K., James, H., and Watts, C., *J. Chromatogr.*, 474, 175, 1989.
78. Barceló, D., *Org. Mass Spectrom.*, 24, 898, 1989.
79. Tondeur, Y., Sovocool, G. W., Mitchum, R. K., Niederhut, W. J., and Donnelly, J. R., *Org. Mass Spectrom.*, 14, 733, 1987.
80. *The API Book*, SCIEX, Central Reproduction, Mississauga, Ontario, Canada, 1990, Part 6.

Chapter 14

LIQUID CHROMATOGRAPHY/MASS SPECTROMETRY MULTIRESIDUE METHODS FOR THERMALLY LABILE PESTICIDES

Lamaat M. Shalaby and Stephen W. George

TABLE OF CONTENTS

I. INTRODUCTION

The demands for analytical methods to detect residues of crop protection chemicals in the environment have increased the need for multiresidue methods. Thermospray liquid chromatography/mass spectrometry (LC/MS) offers the liquid chromatographic separation needed for thermally labile pesticides, high sensitivity, and mass selectivity. This unique combination of attributes makes LC/MS an excellent choice for trace-level environmental multiresidue methods. In recent applications, the technique has been used as a stand-alone, primary method for multiresidue analysis of thermally labile sulfonylurea herbicides and metabolites in crops, soil, and water.[1,2]

Conventional residue methods for pesticides use nonselective detectors, e.g., the ultraviolet (UV) absorbance detector. The only way to eliminate the matrix interferences that are problematic for nonselective detectors is by sample cleanup prior to the determination step by using liquid extraction, solid-phase extraction, or chromatographic separation. These procedures are tedious and time consuming. This problem can be overcome by using the mass spectrometer in the selected ion mode. This provides universally selective detection that minimizes the need for sample cleanup and simplifies method development. Simplification of the sample preparation procedure will increase sample throughput, saving time and reducing solvent usage. In addition, it will improve recovery by reducing analyte losses that normally occur during the cleanup steps. Use of the mass spectrometer for detection offers inherent structure confirmation of the target compounds, which can eliminate the chance for a false positive. False positive results are often a possibility with nonselective detectors when complex matrices, like those found in crop and soil extracts, are encountered. The sensitivity of the mass spectrometer in the selective ion mode allows smaller samples to be extracted. This makes it possible to miniaturize the sample preparation process to increase sample throughput and to reduce solvent consumption.

This chapter describes the application of LC/MS for routine residue analysis of three sulfonylurea herbicides: sulfometuron methyl, [methyl-2-[[[[(4,6-dimethyl-2-pyrimidinyl)-amino]carbonyl]-amino]sulfonyl]benzoate], the active ingredient of the noncrop land herbicide OUST; chlorsulfuron, 2-chloro-N-[(4-methoxy-6-methyl-1 ,3,5-triazin-2-yl)aminocarbonyl]benzenesulfonamide, the active ingredient of the cereal herbicide GLEAN; and bensulfuron methyl, methyl-2-[[[[[(4,6-dimethoxypyrimidin-2-yl)amino]carbonyl]-amino]sulfonyl]methyl]benzoate, the active ingredient of the rice herbicide LONDAX in soil. The chapter also includes an LC/MS method for sulfometuron methyl and a metabolite [methyl-2-(aminosulfonyl)benzoate] in water at the 0.4 ppb level.

The solvents were high performance liquid chromatography (HPLC)-grade acetonitrile, EM OMNISOLV solvent (EM Science, Gibbstown, NJ), and distilled, deionized water obtained from a MILLI-Q water purification system (Millipore Corp., Milford, MA). Ammonium acetate used to prepare the 0.5 *M* solution added postcolumn for thermospray ionization was "Baker Analyzed" Reagent (J. T. Baker, Phillipsburg, NJ). Acetic acid, glacial used to prepare the 0.1 *M* acid mobile-phase was ULTREX "Baker Analyzed" Ultrapure Reagent (J. T. Baker, Phillipsburg, NJ).

B. EQUIPMENT

The HPLC system consisted of a VARIAN model 5560 liquid chromatograph (VARIAN, Instrument Group, Walnut Creek, CA) equipped with a constant-flow pump, a variable wavelength detector, a Rheodyne injector valve (Rheodyne, Inc., Cotati, CA), and an Altech Spherisorb ODS column, 4.6 mm I.D. × 25 cm (Altech/Applied Science, Deerfield, IL).

The mass spectrometer was a Finnigan model TSQ70 (Finnigan MAT, San Jose, CA) triple-stage quadrupole instrument with the ICIS data system. The LC/MS interface was a Finnigan TSP2 thermospray interface with discharge electrode and filament ionization.

A Kratos model Spectroflow 400 (ABI Analytical Kratos Division, Ramsey, NJ) dual piston pulseless HPLC pump was used for postcolumn addition of the 0.5 *M* ammonium acetate solution. A pulseless HPLC pump is necessary with thermospray LC/MS to maintain a stable ion signal.

The LC/MS system was equipped with 2-μm on-line Kel-F A-101X ring filters (Thomson Instrument Co., Newark, DE) located before the LC column and on the Finnigan TSP2 LC/MS interface line prior to the mass spectrometer to prevent clogging of the capillary LC/MS interface line.

Samples were extracted using a THERMOLYNE Maxi-mix Model M16715 (THERMOLYNE Corp., Dubuque, IA) vortex mixer and a Branson Model B-22-4 (Branson Cleaning Equipment Co., Shelton, CT) Ultrasonic Cleaner. The soil extracts were centrifuged on an International Clinical Centrifuge Model CL (International Equipment Co., Needham Heights, MA) centrifuge. Extracts were filtered with GELMAN 0.45-μm Acrodisc-CR filters (GELMAN Sciences, Ann Arbor, MI). Soil extract aliquots were evaporated on an N-EVAP Model 111 Analytical Evaporator (Organomation Assoc., South Berlin, MA) in FALCON 2087 15-mL polypropylene centrifuge tubes (Becton Dickinson Labware, Lincoln Park, NJ). Extract concentrates were filtered with 0.45-μm ACRO LC13 filters (GELMAN Sciences, Ann Arbor, MI).

C. SAMPLE PREPARATION AND EXTRACTION PROCEDURE

Sassafrass soil, which is a sandy loam type of soil, was used to develop this method. Samples were prepared by weighing 10.0 g of soil into 50-mL graduated centrifuge tubes. These samples were fortified with the three compounds, sulfometuron methyl, chlorsulfuron, and bensulfuron methyl, at levels of 0.050, 0.100, and 0.200 ppm. The solvent was evaporated from fortified samples under a stream of nitrogen. An untreated control sample was prepared to check the soil profile by LC/MS and to optimize the LC conditions to separate any matrix interferences from the target compounds.

The samples were extracted with 20 mL of extraction solvent (80% HPLC-grade acetonitrile per 20% MILLI-Q water) and were vortex-mixed for 1 min and then allowed to stand for 10 min. They were then ultrasonicated for 15 min, vortex-mixed for 2 min, centrifuged for 15 min at approximately 1000 rpm, and decanted into separate 50-mL graduated cylinders. The extractions were repeated using 10 mL of fresh extraction solvent. The extracts were collected and filtered. A 5-mL aliquot of each filtered extract was transferred to a 15-mL centrifuge tube for later concentration to less than 1 mL under a nitrogen stream at ambient temperature. Water was added to each concentrate to adjust the final volume to 1 mL prior to LC/MS analysis. No sample cleanup steps were employed prior to the chromatographic analysis.

III. EXPERIMENTAL PARAMETERS FOR LC/MS

Column:	4.6 × 250 mm Altech Spherisorb ODS
LC flow rate:	1.0 mL/min On-column
Mobile phase:	

Time (min)	Acetonitrile (%)	0.1 M Formic acid (%)
0	30	70
15	30	70
18	60	40

Postcolumn addition:	0.5 M Ammonium acetate at 0.3 mL/min
Injection volume:	200-μL Loop
UV detector:	254 nm
Retention times:	Sulfometuron methyl = 19.5 min
	Chlorsulfuron = 21 min
	Bensulfuron methyl = 23.5 min
Selected ions monitored:	Sulfometuron methyl : m/z (124, 233)
	Chlorsulfuron : m/z (141, 358 [MH])
	Bensulfuron methyl : m/z (156, 247 and 411 [MH])
Thermospray probe tip temperature:	120°C
Thermospray MS source temperature:	240°C
Ionization mode:	Thermospray positive ion
Electron multiplier voltage:	1800 V

IV. RESULTS AND DISCUSSION

Figure 1 presents the thermospray positive ion mass spectra for sulfometuron methyl, chlorsulfuron, bensulfuron methyl, and methyl-2-(aminosulfonyl)benzoate generated by full scan LC/MS analyses of the individual test compounds. Sulfonylurea thermospray mass spectra generally contain the protonated molecular ions and protonated fragment ions characteristic of the

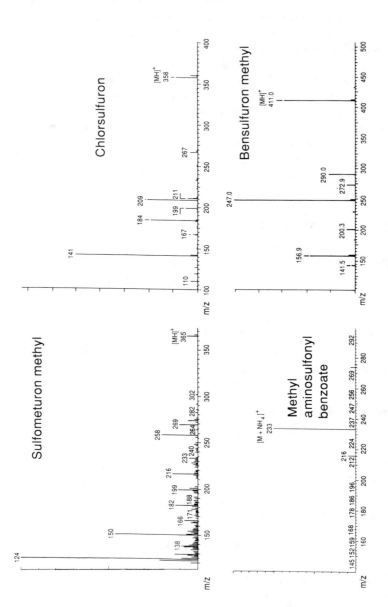

FIGURE 1. The thermospray positive ion mass spectra of sulfometuron methyl, chlorsulfuron, bensulfuron methyl, and methyl-2-(aminosulfonyl) benzoate.

heterocyclic amine, heterocyclic amine urea, and the sulfonamide moieties. The sulfonamide moieties also give ammonium adducts.[3] For trace-level analysis, the most intense ions were selected to quantitate each compound. The ions were m/z 124, 233 for sulfometuron methyl; m/z 141, 358 [MH] for chlorsulfuron; m/z 156, 247, and 411 [MH] for bensulfuron; and m/z 233 [M + NH$_4$] for methyl-2-(aminosulfonyl)benzoate.

The Finnigan LC/MS vaporizer is a 1/32-in O.D. × 0.006-in. I.D. stainless steel tube with a laser-drilled sapphire orifice at its exit end.[4] The solvent and sample are heated within the vaporizer, such that when the liquid exits the vaporizer into the evacuated ion source, it forms a vapor spray in the ion source. The vaporizer acts as a heating element and is heated by the electrical current flowing through it. The temperature of the vaporizer is feedback-controlled, and its operating range is 80 to 130°C. Vaporizer temperature optimization studies were done for sulfonylureas to maximize the ion signal using the same mobile phase used for the LC separation. The optimum temperature was determined to be 120°C, and the ion signal was stable within the 110 to 130°C range.

Our experience with the laser-drilled orifice is that it is clog-resistant. The system has been used continuously for 18 months and we have not experienced any major clogging problem that required replacement of the vaporizer. The thermospray LC/MS interface is maintained by avoiding the use of nonvolatile buffer and rinsing the interface at the end of the day with water and then methanol. During the LC/MS analysis, the back pressure generated from the thermospray vaporization process in the narrow stainless steel line is monitored as the pressure read out on the postcolumn pump. A gain of 20 to 35 bar is generated depending on the solvent flow rate, mobile-phase composition, and the vaporizer temperature. A higher flow rate and amount of water in the mobile phase leads to higher back pressure. Increasing the vaporizer temperature also increases the back pressure. During operation, the back pressure should be constant (+/− 0.5 bar) as long as the three parameters mentioned above are unchanged. An increase in the back pressure may be indicative of a partial blockage of the on-line filter and can generally be corrected by replacing the filter or back flushing the vaporizer tube.

Addition of ammonium acetate solution (0.1 *M*) to the mass spectrometer is required for thermospray ionization.[5] Addition of this buffer on-column could affect the LC retention time, especially for sulfonylurea compounds, where an acidic mobile phase is needed for retention on the LC column.[6] Addition of buffer solution postcolumn at 0.3 mL/min minimized the effect of ammonium acetate on retention times and maintained the performance of the reversed-phase LC column.

The recovery of the three sulfonylureas in soil was determined by comparing the amount detected by LC/MS in the final extract of the fortified samples to the original fortification level. The recoveries were between 88 and 115% for the three compounds, sulfometuron methyl, chlorsulfuron, and

TABLE 1
**Recovery Results of the Multiresidue LC/MS Method for
Sulfometuron Methyl, Chlorsulfuron, and Bensulfuron
Methyl**

Fortification level (ppm)	Average recovery (%)		
	Sulfometuron methyl	Chlorsulfuron	Bensulfuron methyl
0.050	88	101	105
0.100	90	84	115
0.200	92	108	115

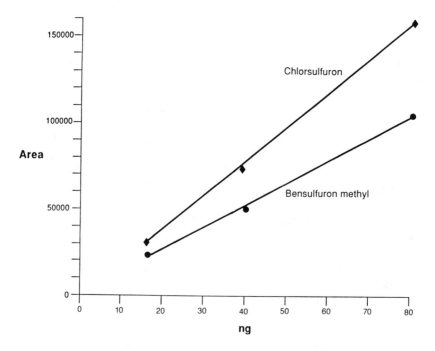

FIGURE 2. Representative calibration curves for chlorsulfuron and bensulfuron methyl.

bensulfuron methyl, at levels of 0.050 to 0.200 ppm in soil. A summary of the recovery results for the three compounds in soil at 0.050, 0.100, and 0.200 ppm is shown in Table 1. The average recovery for the three compounds over the three concentration levels was 99.8%. The detection limit of this LC/MS method was calculated to be 0.010 ppm for sulfometuron methyl and chlorsulfuron and 0.005 ppm for bensulfuron methyl.

Calibration curves were constructed for each test compound based on LC/MS peak areas for a set of standards solution at different levels. Figure 2

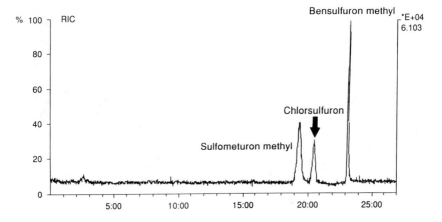

FIGURE 3. The LC/MS total ion chromatogram of 0.1 μg/mL standard solution of sulfometuron methyl, chlorsulfuron, and bensulfuron methyl.

shows representative calibration curves for chlorsulfuron and bensulfuron methyl where the total amount injected into the mass spectrometer was 15 to 80 ng. The calibration curves demonstrated the sensitivity and linear response of the method within the range used for the analysis.

The LC/MS reconstructed ion chromatograms for sulfometuron methyl, chlorsulfuron, and bensulfuron methyl standards (0.1 μg/mL) are presented in Figure 3. The chromatograms indicated a stable baseline, which demonstrated the stability of the ion signal and good performance of the LC/MS vaporizer. Using gradient LC conditions, good separation of the three compounds was obtained. At the same time, the target compounds were separated from the polar components extracted from the soil matrices.

Figure 4 presents the LC/MS total ion chromatogram of an untreated soil sample that was used as a control. The chromatogram provides a typical profile of this soil and shows no interferences at the retention times for the three compounds. The polar components in the soil extract eluted between 2 to 7 min. Gradient LC conditions are selected to separate interfering components from the compounds of interest and allow for diverting the column effluent away from the mass spectrometer for the first 10 min. This procedure is used to keep the mass spectrometer source clean and to prevent the formation of deposits on the interface probe tip. This can be accomplished using a switching valve to divert the column effluent to waste. This step was not applied in this analysis as the amount of polar material extracted was low and did not affect the performance of the interface or the mass spectrometer. With other matrices like crop extracts, which contain sugars and starchy materials, we normally divert those compounds away from the mass spectrometer interface line. Otherwise, deposits will form at the hot vaporizer tip. Such nonvolatile deposits will cause the vaporizer orifice to become irregular

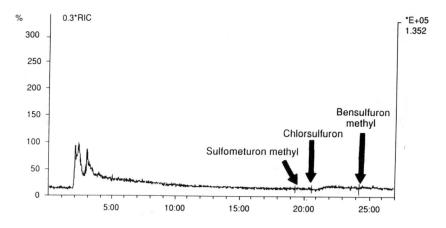

FIGURE 4. The LC/MS total ion chromatogram of an untreated soil extract.

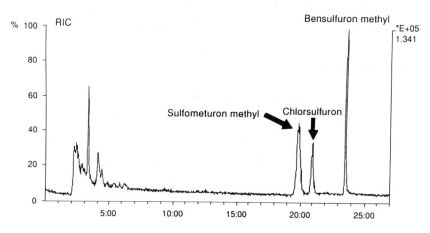

FIGURE 5. The LC/MS total ion chromatogram of 0.5 ppm fortified soil with sulfometuron methyl, chlorsulfuron, and bensulfuron methyl.

and will affect the quality of the thermospray jet in terms of droplet size and spray direction into the source of the mass spectrometer.

Generally, if the LC effluent is diverted away from the mass spectrometer to vent these polar components, the effluent has to be diverted back to the mass spectrometer at least 5 min before the elution of the targeted peaks. This step is required to allow for equilibration of the thermospray reagent plasma in the source. A minimum of 5 min is needed for the temperature and pressure in the ion source to stabilize and to produce a stable ion signal.

Figure 5 presents the LC/MS total ion chromatogram of a 0.05 ppm fortified soil, which is the lowest fortification level. The chromatogram shows good signal-to-noise ratio for the three test compounds and demonstrates that the quantification limits are less than 0.01 ppm for the three compounds.

The most intense ions were selected for quantification to provide good sensitivity. Structure confirmation is based on the detection of compound-specific ions and the relative abundance of these ions compared to the reference standard. Interferences at any of the selected ions can be detected also by comparing the relative ion abundance to the reference standard.

One of the major pathways of sulfonylurea degradation in soil is due to chemical hydrolysis and microbial breakdown.[7] The major hydrolysis pathway in soil is the cleavage of the sulfonylurea bridge to give the corresponding sulfonamide and heterocyclic amine. The thermospray mass spectra for these metabolites contain common ions, as in the mass spectra of the parent compounds. These ions correspond to the common structure moieties that both the parent sulfonylurea and metabolite have. The ions at m/z 124, 141, and 156 correspond to the heterocyclic amine for sulfometuron methyl, chlorsulfuron, and bensulfuron methyl, respectively. On the other hand, the ions at m/z 233, 209, and 247 correspond to the ammonium adducts of the sulfonamide of sulfometuron methyl, chlorsulfuron, and bensulfuron methyl, respectively. Monitoring these common ions with LC/MS will detect both the parent compound and the two major degradates in soil.

To demonstrate this technique, an LC/MS method was developed to detect sulfometuron methyl and its sulfonamide metabolite in water using the selected ions m/z 124 and 233. Both ions would be detected for the parent and only the m/z 233 would be detected for the sulfonamide metabolite. Figure 6 shows the LC/MS total ion chromatogram for a 0.2 ppm standard solution of sulfometuron methyl and metabolite using the same column used in the soil analysis, but the mobile phase was isocratic and consisted of 40% acetonitrile in 0.05 M formic acid.

To validate the method for water analysis, water samples were fortified with sulfometuron methyl and the sulfonamide at the 0.4-ppb level. A 250-mL water sample was extracted three times with 80 mL of methylene chloride after acidification with 1 mL of acetic acid.[8] The methylene chloride extract was then evaporated to dryness in a stream of nitrogen and the dry extract was dissolved in 0.5 mL of 30% acetonitrile in water. Figure 7 shows the selected ion LC/MS chromatograms of m/z 124 and 233 and the total ion chromatogram (bottom trace) for the water extract fortified at 0.4 ppb with the parent, sulfometuron methyl, and its metabolite. The chromatogram indicated good sensitivity. The results of the extraction procedure indicated quantitative recovery.

V. CONCLUSION

The application of thermospray LC/MS with selected ion monitoring has been demonstrated for sulfonylurea herbicides and degradates. LC/MS provides a sensitive and selective method to determine sulfometuron methyl, chlorsulfuron, and bensulfuron methyl simultaneously in soil at levels down to 0.05 ppm. In addition, it is applicable to multiresidue analysis and provides

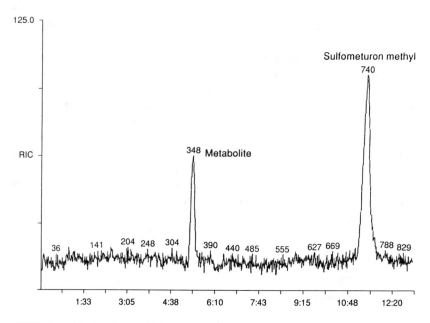

FIGURE 6. The LC/MS total ion chromatogram of 0.2 μg/mL standard solution of sulfometuron methyl and metabolite.

a simultaneous structure confirmation. The LC/MS selectivity allows for simple sample preparation technique and eliminates cleanup steps, which increases sample throughput to 8 to 12 samples per day. Method development of residue LC/MS analysis is much faster than conventional nonselective techniques. The one-step extraction procedure and the elimination of cleanup steps minimize analyte losses, which enhances recovery. A typical residue method for one analyte in soil using LC/UV takes a few months to develop and validate. On the other hand, an alternative LC/MS method will take only a few weeks and can include more than one herbicide and metabolite. These unique advantages make the technique cost-effective and a good investment. The continuing improvement and proliferation of commercial LC/MS instruments and the increase in the number of LC/MS methods that are validated for trace-level analysis should lead to more general acceptance of the technique. LC/MS will be a valuable tool in the efforts to meet the growing need for multiresidue environmental analyses.

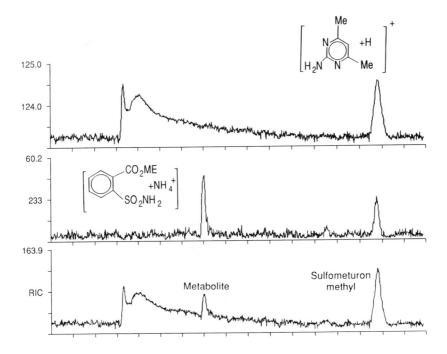

FIGURE 7. The LC/MS selected ion chromatograms of an extract of fortified water at 0.4 ppb with sulfometuron methyl and metabolite.

REFERENCES

1. **Shalaby, L. M. and George, S. W.,** in *Liquid Chromatography/Mass Spectrometry: Applications in Agricultural, Pharmaceutical, and Environmental Chemistry,* Brown, M. A., Ed., (ACS Symposium Series), No. 420, American Chemical Society, Washington, D.C., 1990, 75.
2. **Shalaby, L. M.,** Proc. 7th IUPAC Meeting, IUPAC, Hamburg, 1990, 276.
3. **Shalaby, L. M. and Reiser, R. W.,** in *Mass Spectrometry of Biological Materials,* McEwen, C. N. and Larsen, B. S., Eds., Marcel Dekker, New York, 1990, 379.
4. Finnigan TSQ70 Operator's Manual, TSP2 option, Finnigan MAT, San Jose, CA, 1989.
5. **Garteiz, D. A. and Vestal, M. L.,** *LC Magazine,* 3(4), 334, 1985.
6. **Shalaby, L. M.,** in *Pesticide Chemistry: Chemical Analysis,* Rosen, J. B., Ed., Wiley-Interscience, New York, 1987, 161.
7. **Beyer, E. M., Duffy, M. J., Hay, J. V., and Schlueter, D. D.,** in *Herbicide Chemistry, Degradation, and Mode of Action,* Vol. 3, Marcel Dekker, New York, 1987, chap. 3.
8. **Wheeler, J. R.,** Private communication, Du Pont Agricultural Products, Wilmington, DE, 1989.

Chapter 15

IMMUNOCHEMICAL METHODS FOR PESTICIDE ANALYSIS

J. Paul Aston, David W. Britton, Michael J. Wraith, and Alan S. Wright

TABLE OF CONTENTS

I. INTRODUCTION

The last decade has witnessed a phenomenal expansion in the application of immunoassay technology in environmental monitoring.This expansion has been prompted mainly by a need for simple, rapid procedures to cope with escalating demands for pesticide analysis. The principle aim was to exploit the specific molecular recognition and binding properties of antibodies for their analytes to reduce and, hopefully, eliminate the preparative work which is so often the most costly and exacting step of conventional analytical methods such as gas chromatography/mass spectrometry. Moreover, the overall simplicity and versatility of immunoassay technology held the promise of field applications which would be of particular value in environmental monitoring studies.

Initial applications of immunoassay to measure pesticides met with only limited success. For example, radioimmunoassays (RIA) developed for malathion, DDT, and parathion[1-3] failed to provide reproducibly quantitative results. However, technological advances, particularly the introduction of the enzyme-linked immunosorbent assay (ELISA), have led to the development of successful immunoassays for a broad range of pesticides including organochlorines,[4-6] organophosphates,[7-9] pyrethroids,[10,11] benzoyl phenylurea insect growth regulators,[12] triazine herbicides,[13-17] and sulfonylurea herbicides.[18] Some of these methods are commercially available and are used routinely in many laboratories.[19,20]

II. IMMUNOANALYTIC PRINCIPLES

Immunoanalysis depends on the reversible, noncovalent binding of analyte to the binding sites of specially tailored antibodies in a reaction which obeys the law of mass action. The selective binding affinity of the antibody for its analyte may be exploited to develop highly sensitive and specific immunoanalytical systems which offer the prospect of monitoring analytes even where present in complex mixtures, for example, environmental samples.

A. ANTIBODY PRODUCTION

Molecules capable of inducing an immune response in experimental animals have molecular weights greater than 1000 Da.Small molecules such as pesticides do not usually elicit antibody production, but may be rendered immunogenic by covalent attachment to high molecular weight protein carriers, e.g., bovine serum albumin, keyhole limpet hemocyanin, or chicken gamma globulin (hapten-protein conjugates). If the pesticide molecule does not contain a functional group suitable for conjugation to proteins it may be necessary to synthesize a derivative possessing an appropriate functionality, e.g., an amino or carboxyl group, in order to prepare the immunogen. This topic has been reviewed recently.[21]

Extensive use has been made of carbodiimide reactions to couple haptens containing carboxyl groups to proteins.[22] Heterobifunctional cross-linking agents such as maleimide succinyl ester and, more recently, N-(aminobenzoyloxy)succinimide[23] have been employed successfully to link amino groups of haptens to phenolic groups of tyrosine residues or imidazole groups of histidine residues of bovine serum albumin (Figure 1).

In certain instances it may be advantageous to include a spacer arm between the hapten and protein in order to prevent the masking of important antigenic determinants (epitopes) by the bulky protein moiety.[24,25] However, spacer attachment close to key structural features should be avoided as this may limit their recognition by antibodies. Ideally a spacer arm should be located distally to potentially important or unique antigenic determinants, thereby serving to maximize exposure of that determinant, thus optimizing the specificity of the immune response.

Injection of a laboratory animal with a hapten-protein conjugate (immunogen) induces an immune response. This consists of a proliferation of antibody-secreting cells (B lymphocytes), each of which produces a unique antibody (immunoglobulin) directed to a specific epitope or group of epitopes on the immunogen. If successful, a proportion of these antibodies will specifically recognize and combine with epitopes on the hapten. Thus, a mixture of antibodies accumulates in the bloodstream of the animal and many pesticide immunoassays utilize polyclonal antisera derived in this way. Although each constituent antibody has a unique specificity and binding affinity for any given epitope, the characteristics of the antiserum reflect the composite properties of the mixture.

More recently the value of monoclonal antibodies in pesticide monitoring has been recognized.[9] Monoclonal antibodies are the molecules formed by a single clone of B lymphocyte cells and are uniform in terms of structure and function. These antibody-secreting clones cannot normally be cultured indefinitely *in vitro*. However, in 1975 Kohler and Milstein[26] overcame this problem by fusing antibody-secreting cells from an immunized mouse with malignant mouse myeloma cells which had been adapted to permanent and vigorous growth in culture. The fused product, or hybrid, retained the phenotypic characteristics of the lymphocyte for specific antibody secretion while the parental myeloma cell conferred on the hybrid cell the ability to grow indefinitely in culture or as a tumor *in vivo*. Hybridoma technology is now firmly established and numerous strategies are in use.

The use of carefully designed screening procedures enables the selection of homogeneous antibody of predetermined specificity. Evaluation of specificity is crucially important in assessing the value of a particular antibody for the development of immunoanalytical methods. This is especially true, for example, where there is a need to monitor pesticides in heterogenous samples containing potentially cross-reacting metabolites or other structurally related chemicals. For example, many novel pesticides operate by inhibiting

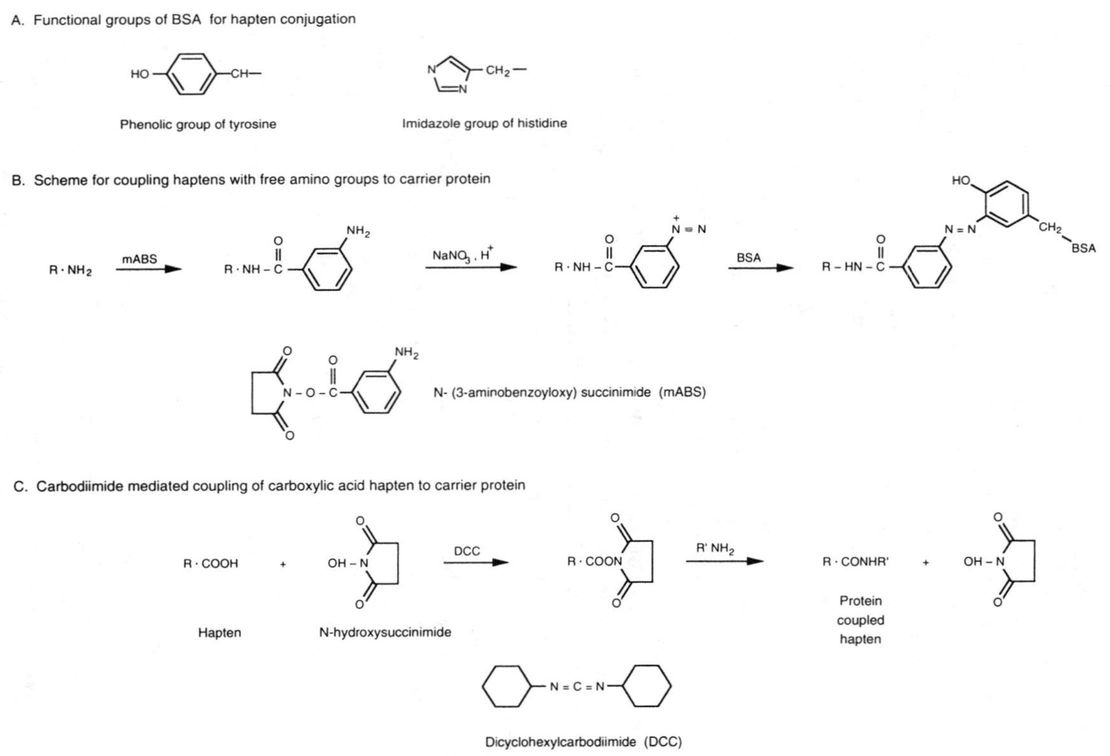

FIGURE 1. N'-(Aminobenzoyloxy)succinimide links amino groups of haptens to phenolic groups of tyrosine residues or imidazole groups of histidine residues of bovine serum albumin.

specific enzymes in the target species and often are structural analogues of naturally occurring substrates. Thus, trace levels of pesticide may have to be detected in the presence of a large molar excess of structurally related substances.[27] Judicious screening may permit the selection of unique monoclonal antibodies possessing the high specificity required to meet such challenging objectives.

Invariably, the binding affinities of monoclonal antibodies are reported to be lower than their polyclonal counterparts, particularly for large multi-epitope antigens.[28] There is evidence that the overall effectiveness of a polyclonal antiserum may reflect synergism between the constituent antibodies.[29] Cooperativity between antibodies for distinct epitopes on an antigen may be associated with antibody bivalency and the formation of multimeric structures and may considerably increase the avidity of the mixture.[30] However, such synergism is less likely to occur in the case of small antigens and we have produced several monoclonal antibodies that display greater affinity than corresponding polyclonal antisera for some haptens with molecular weights in the range of 300 to 500 Da.

The availability, and, therefore, the long-term use of a polyclonal antiserum, is limited by the amount of serum which may be obtained from an animal bleed, the extent and consistency of the subsequent responses to repeated immunizations, and the life-span of the animal. In contrast, a hybridoma cell line provides a potentially unlimited supply of homogeneous antibody reagent and this may be an important consideration, particularly where large quantities of antibody are required for a protracted period, for example, in immunoaffinity chromatography. For pesticide monitoring it seems likely that monoclonal antibodies offer the greater potential of satisfying the requirements of immunoanalytical performance, particularly the key aspect of specificity, and meeting the demands of regulatory authorities for standardized reagents and methods.

B. IMMUNOASSAY DESIGN
It is convenient to classify immunoassays into two categories.[31]

1. Competitive Assays
These are based on the use of a limiting concentration of antibody reagent in which, analyte and derivatized, i.e., labeled or bound, analyte molecules compete with each other for a limited number of antibody binding sites. The relative proportion of free analyte combining with antibody is a function of its concentration (Figure 2).

2. Noncompetitive (Immunometric) Assays
This assay format utilizes an excess of labeled antibody reagent which serves to convert analyte to a labeled derivative emitting a signal which is directly proportional to the concentration of analyte present[32] (Figure 2).

Competitive immunoassay

$$Ag^* + Ab + Ag \;\; \rightleftharpoons \;\; Ag^*Ab + Ag\,Ab$$

Where Ag^* represents a fixed concentration of derivatised
(bound or labelled) analyte and [Ab] is limiting

The binding of Ab to Ag^* is a function of the concentration of Ag

Immunometric assay

$$Ag + Ab^* \;\; \rightleftharpoons \;\; Ag\,Ab^*$$

Where Ab^* = Labelled antibody reagent and $[Ab^*]$ is in excess

The binding of Ag to Ab^* is a function of the concentration of Ag

FIGURE 2.   Examples of competitive and noncompetitive (immunometric) assays.

The latter format offers the theoretical prospect of greater assay sensitivity, enhanced specificity, a linear response for a considerable dose range, and shorter reaction times compared with the competitive immunoassay. However, the immunometric assay configuration depends on the simultaneous recognition of at least two distinct binding sites on the analyte and requires large amounts of purified antibody reagent. Furthermore, pesticide residues and their metabolic products are usually small molecules (molecular weight 100 to 1000 Da) and therefore simultaneous binding to more than one antibody is precluded by steric hindrance. This factor has restricted the application of the immunometric (antibody-reagent excess) configuration to the immunoassay of pesticides. As a consequence, pesticide immunoassay has relied almost entirely on the competitive assay format. The ELISA, in which free analyte competes with a bound form of the analyte in the presence of a limiting concentration of antibody, has largely superseded RIA and is now by far the more widely used approach.[33] In contrast to the immunometric assay, this system does not require the use of highly purified antibody. The reaction is monitored by measuring the amount of antibody which combines with the bound form of the analyte. This is achieved by means of an enzyme-labeled second antibody directed against the antianalyte antibody. Enzymes used for this purpose include alkaline phosphatase, horseradish peroxidase (HRP), and beta galactosidase. Quantitation is based on the measurement of the rate of formation of the enzyme-catalyzed reaction product, which is inversely proportional to the concentration of analyte.

    Enzyme-multiplied immunoassay technology (EMIT) has also been used in pesticide immunoassay.[34] EMIT is a form of competitive immunoassay which requires no separation step. In this system, analyte and an analyte-enzyme conjugate compete for binding to a limiting amount of antianalyte antibody. Antibody binding to the enzyme-conjugate fraction results in the modulation, usually inactivation, of its enzymic activity to an extent which is related to the concentration of analyte.

An alternative competitive immunoassay format based on a labeled antihapten monoclonal antibody has been used successfully to measure haptens of clinical interest at subnanomolar concentrations[35] and may have a role in the immunoassay of pesticides. In this approach, an analyte (hapten)-protein conjugate competes with analyte for binding to labeled antibody (Figure 3). Immune complexes formed between the conjugate and the labeled antibody are monitored by binding to an excess of immobilized second antibody directed to the protein portion of the conjugate.

A single-antibody immunometric assay which detects a small hapten (digoxin) with good precision at the microgram per liter level has been reported.[36] In this format, excess labeled antibody is incubated with analyte and unreacted labeled antibody is then removed by using antigen immobilized on a solid support (Figure 4). Quantitation of the fraction not bound to the solid support provides a direct measure of analyte concentration. This procedure has been automated in the form of an affinity column-mediated immunometric assay. However, possible drawbacks to this approach could be poor assay sensitivity and reproducibility, especially if the reaction kinetics result in rapid dissociation of the labeled antibody-hapten complex.

III. TRENDS IN PESTICIDE IMMUNOASSAY

There have been several important advances in pesticide immunoassays since their introduction over 15 years ago[37] (Table 1). Initial applications centered on RIA and proved capable of measuring pesticides in complex matrices, for example, ground water, serum, and a range of crop extracts.[5,7,38,44] Further benefits included significant reductions in sample volumes and increased sample throughput compared with conventional analytical methods. However, practical limitations of RIA, for example, assay speed, the potential hazards associated with the use of radioisotopes, and the requirement of complex instrumentation, militate against its application in field-test screening. The application of nonisotopic immunoassay has gradually superseded the use of radioisotopes. Probably the most significant advance in immunoassay, in terms of speed and simplicity, for pesticide field assays came with the introduction of enzyme-linked immunoassays.[33] Enzyme immunoassays are now becoming established routinely and with increased usage, improvements in assay performance are becoming increasingly evident, largely due to a focus on immunoanalytic principles and hapten-protein conjugate design leading to the production of very effective antibodies. These advances have permitted the reproducible monitoring of pesticides at levels which satisfy

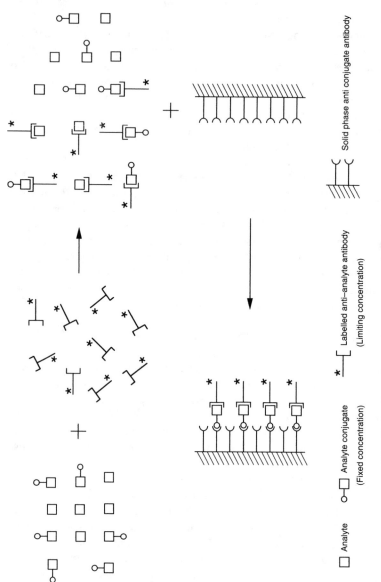

FIGURE 3. Competitive immunoassay using labeled antianalyte antibody.

the requirements of regulatory authorities and which are, invariably, below the detection limits reported using physicochemical approaches.

Not surprisingly, higher sensitivity immunoassays have been achieved using high affinity polyclonal antisera which, however, are of finite supply. A good example of the effective use of a polyclonal antiserum has been reported by Jung et al.[43] These authors describe a variant form of a competitive ELISA for the fungicide fenpropimorph and the herbicide molinate. This assay is based on competition between hapten and hapten-enzyme conjugate for binding to a specific antibody; the resulting immune complex is then "captured" using immobilized antibody reagent which is directed to the specific analyte antibody. The authors report advantages both in antibody reagent conservation and assay sensitivity.

Recently, immunoassays based on monoclonal antibodies have been developed for a range of pesticides including the herbicides paraquat[45] and atrazine[16] and the insecticides paraoxon,[9] permethrin,[11] and heptachlor[6] (Table 1). The herbicide monoclonal antibody immunoassays are approximately an order of magnitude less sensitive than their polyclonal based counterparts. However, similar limits of detection have been reported using monoclonal and polyclonal antibodies for the insecticide permethrin[11] and the structurally related cypermethrin plant metabolite, 3-phenoxybenzoic acid,[10] respectively. Moreover, in a comparison of polyclonal and monoclonal immunoassays for the detection of the herbicide picloram in river water, urine, soil, and plant extracts, the authors concluded that a monoclonal antibody-based ELISA was more sensitive, accurate, and precise than assay formats which used a polyclonal antiserum.[50]

The performance of an immunoassay is a function of the affinity and specificity of the antibody reagent. It is envisaged that careful design, not only of haptens and hapten-conjugates, but also of the chemicals employed to screen or select the appropriate monoclonal antibodies, will provide the precision necessary to tailor antibodies to meet most, if not all, requirements in pesticide monitoring.

IV. EMERGING CONCEPTS

A. IMMUNOAFFINITY CHROMATOGRAPHY

One of the more difficult problems encountered in pesticide monitoring is the initial sample preparation step where pesticide residues must be isolated from complex environmental samples. The unique specificity of an antibody-antigen binding reaction may prove to be extremely useful in addressing this problem. Affinity chromatography is a separation procedure which selectively isolates specific molecules from complex mixtures by adsorption using an immobilized secondary molecule (ligand).[54] Immunoaffinity chromatography, as the term suggests, exploits the specific interaction between an antibody and its antigen to purify and concentrate the antigen from a dilute solution.[55]

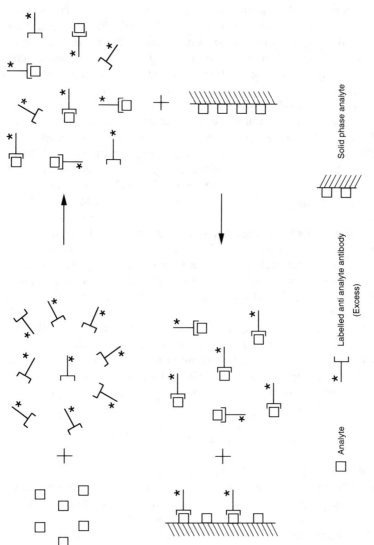

FIGURE 4. Single antibody immunometric assay.

TABLE 1
Pesticide Immunoassays

Pesticide	Chemical structure	Detection limit (ng ml^{-1})	Assay format	Antibody reagent[a]	Ref.
Benomyl		0.1	FIA[b]	P	37
		1.25	RIA	P	38
		0.12	ELISA	P	39
Metalaxyl		0.063	ELISA	P	40
Fungicides					
Iprodione		0.2	ELISA	P	41

TABLE 1 (continued)
Pesticide Immunoassays

Pesticide	Chemical structure	Detection limit (ng ml^{-1})	Assay format	Antibody reagent[a]	Ref.
Triadimefon		1	ELISA	P	42
Fenpropimorph		0.013	ELISA	P	43
Paraquat		0.1 0.8 0.1	RIA ELISA ELISA	P M P	44 45 46
2,4-Dichlorophenoxyacetic acid		0.10	RIA	P	5

Herbicides

Compound	Structure	Concentration	Method	P/M	Ref.
Diclofop-methyl		23 45	EIA FIA	P P	47 47
Chlorosulfuron		0.1	ELISA	P	18
Terbutryn		25	ELISA	P	13
Cyanazine		0.5	ELISA	P	17
Atrazine		.011 1×10^{-4} 0.5 0.1	ELISA EIA EIA ELISA	P P P M	14 14 15 16

TABLE 1 (continued)
Pesticide Immunoassays

Pesticide	Chemical structure	Detection limit (ng ml^{-1})	Assay format	Antibody reagent[a]	Ref.
Molinate		3	ELISA	P	48
Alachlor		0.1	ELISA	P	49
Picloram		5 1	ELISA ELISA	P M	50 50
Clomazone		10	ELISA	P	51

Insecticides

Name	Structure		Method	P/M	Ref.
Parathion	O_2N—C₆H₄—O—$P(OCH_2CH_3)_2$ (S=)	4	RIA	P	7
Paraoxon	O_2N—C₆H₄—O—$P(OCH_2CH_3)_2$ (O=)	.028	ELISA	P	8
		1×10^3	ELISA	M	9
Diflubenzuron	F₂-C₆H₃—$CO.NHCO.NH$—C₆H₄—Cl	2	ELISA	P	12
Permethrin	$Cl_2C=CH$—CH< (CH₃)(CH₃) cyclopropane —$CO.COCH_2$—O—C₆H₄—O—C₆H₅	50	ELISA	M	11
Endosulfan	Cl-bicyclic chlorinated structure with —O—SO—O— (SO_2)	3	ELISA	P	52

TABLE 1 (continued)
Pesticide Immunoassays

Pesticide	Chemical structure	Detection limit (ng ml^{-1})	Assay format	Antibody reagent[a]	Ref.
Avermectins	c	0.5	ELISA	M	53
Heptachlor		10	ELISA	M	6

[a] P, polyclonal antiserum; M, monoclonal antibody.
[b] Competitive binding fluorescence polarization immunoassay.
[c] 16-Membered lactones containing an α-L-oleandrosyl-α-L-oleandrosyl disaccharide attached to the lactone ring through the C13-hydroxy group.

Typically, antibody (ligand) is covalently coupled to an insoluble support matrix, for example, agarose or sepharose, and packed into a column through which the antigen-containing sample is passed. Antigen is specifically retained on the column whereas other soluble components of the mobile phase pass through without binding.

Immunopurification systems are usually designed to isolate one specific analyte. However, an antibody reagent selected to exhibit a broad, or class, specificity, that is, a capacity to recognize a structural feature common to a class of chemicals, may permit the isolation of a class of pesticides or metabolic products which, in some instances, may be even more useful.

Results of preliminary studies in our laboratory have demonstrated the principle of class-specific immunopurification. Mice were immunized with the hapten 6-(2,4-dinitroanilino)hexanoic acid and monoclonal antibodies were selected by screening for class specificity toward the 2,4-dinitrophenyl moiety. One of these antibodies, covalently coupled to a solid support (cyanogen bromide-activated sepharose) proved to be very effective in selectively adsorbing chemicals possessing a 2,4-dinitrophenyl group from an aqueous cocktail containing low concentrations of related chemicals variously substituted in the aromatic ring with one or two nitro groups. These findings underline the importance of carefully planned screening strategies in selecting antibodies possessing the necessary (desired) specificity.

B. ENZYME IMMUNOCHROMATOGRAPHY

An immunoassay procedure that obviates a requirement for analytical instruments, separation steps, and wash cycles would be of particular benefit for field-test monitoring. Enzyme immunochromatography is a quantitative immunoassay technique comprising a reactangular strip of filter paper coated with analyte-specific antibody which is inserted into a solution of analyte, analyte-HRP conjugate, and the enzyme glucose oxidase.[56] The latter enzyme becomes distributed uniformly throughout the length of the test strip by capillary migration, whereas analyte and analyte-HRP conjugate are immunospecifically bound to the surface of the strip and migrate according to their relative concentrations. The test strip is then immersed in a developer solution of glucose and a chromogenic substrate for HRP. In the presence of peroxide, formed by the reaction of glucose and glucose oxidase, HRP catalyzes the oxidation of its substrate to an insoluble colored product which adheres to the surface of the test strip (Figure 5). The procedure is rapid and the position of the colored migration front is related to the concentration of analyte.

C. IMMUNOPROBES AND IMMUNOSENSORS

The requirement for rapid, simple, and continuous monitoring procedures in clinical medicine has provided the incentive for the development of biosensors.[57] Similar devices could prove highly effective for field testing and batch screening of pesticides. For example, immunoprobes comprising an

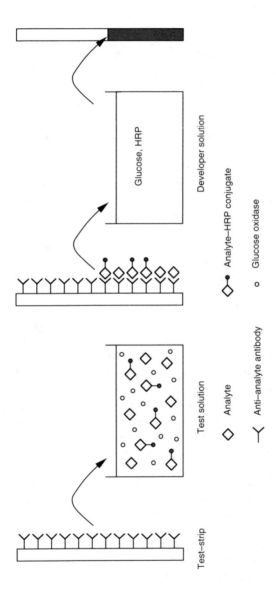

FIGURE 5. Principle of enzyme immunochromatography.

antibody-coated solid support could be developed and used to monitor pesticide levels by direct immersion in an aqueous sample. The extent of binding by analyte provides a measure of the concentration of the analyte which is determined by introducing the probe into the appropriate developer solution. Alternatively, advances in biosensor technology raise the possibility of developing immunosensors in which quantitation relies on a physicochemical change triggered by the specific binding of analyte to the antibody. In this case, the immunological event is detected by an optical or electrical transducer.[58] The exciting prospect of multianalyte (compound- or class-specific) immunosensors capable of monitoring trace levels of several analytes rapidly could, during the next decade, revolutionize pesticide analysis and environmental monitoring.

REFERENCES

1. **Haas, G. J. and Guardia, S. J.**, *Proc. Soc. Exp. Biol. Med.*, 129, 546, 1968.
2. **Centeno, E. R. and Johnson, W. J.**, *Fed. Proc.*, 26, 704, 1967.
3. **Hammock, B. D. and Mumma, R. O.**, in *Pesticide Analytical Methodology*, Harver, J., Jr. and Zweig, G., Eds, American Chemical Society, Washington, D.C., 1980, 321.
4. **Langone, J. J. and Van Vunakis, H.**, *Res. Commun. Chem. Pathol. Pharmacol.*, 10, 163, 1975.
5. **Rinder, D. F. and Fleeker, J. R.**, *Bull. Environ. Contam. Toxicol.*, 26, 375, 1981.
6. **Stanker, L. H., Watkins, B., Vanderlaan, M., Ellis, R., and Rajan, J.**, in *Immunoassays for Trace Chemical Analysis*, ACS Symposium Series No. 451, Vanderlaan, M., Stanker, L. H., Watkins, B. E., and Roberts, D. W., Eds., American Chemical Society, Washington, D.C., 1990, 108.
7. **Ercegovich, C. D., Vallejo, R. P., Gettig, R. R., Woods, L., Bogus, E. R., and Mumma, R. O.**, *J. Agric. Food Chem.*, 29, 559, 1981.
8. **Hunter, K. W. and Lenz, D. E.**, *Life Sci.*, 30, 355, 1982.
9. **Brimfield, A. A., Lenz, D. E., Graham, C., and Hunter, K. W.**, *J. Agric. Food Chem.*, 33, 1237, 1985.
10. **Wraith, M. J., Hitchings, E. J., Woodbridge, A. P., Cole, E. R., and Roberts, T. R.**, Proc. 6th Int. Congr. Pesticide Chemistry, Poster Session, Ottawa, Canada, 1986.
11. **Stanker, L. H., Bigbee, C., Van Emon, J., Watkins, B., Jensen, R. H., Morris, C., and Vanderlaan, M.**, *J. Agric. Food Chem.*, 37, 834, 1989.
12. **Wie, S. I. and Hammock, B. D.**, *J. Agric. Food Chem.*, 32, 1294, 1984.
13. **Huber, S. J. and Hock, B.**, *Z. Pflanzenkr. Pflanzenschutz*, 92, 147, 1985.
14. **Huber, S. J.**, *Chemosphere*, 14, 1795, 1985.
15. **Bushway, R. J., Perkins, B., Savage, S. A., Lekousi, S. J., and Ferguson, B. S.**, *Bull. Environ. Contam. Toxicol.*, 40, 647, 1988.
16. **Karu, A. E., Harrison, R. D., Schmidt, D. J., Clarkson, C. E., Grassman, J., Goodrow, M. H., Lucas, A., Hammock, B. D., Van Emon, J. M., and White, R. J.**, in *Immunoassays for Trace Chemical Analysis*, ACS Symposium Series No. 451, Vanderlaan, M., Stanker, L. H., Watkins, B. E., and Roberts, D. W., Eds., American Chemical Society, Washington, D.C., 1990, 59.
17. **Robotti, K. M., Sharp, J. K., Ehrmann, P. R., Brown, L. J., and Hermann, B. W.**, 192nd American Chemical Society National Meeting, Anaheim, California, 1986.

18. **Kelley, M. M., Zahnow, E. W., Petersen, W. C., and Toy, S. T.,** *J. Agric. Food Chem.,* 33, 962, 1985.

19. **Vanderlaan, M., Watkins, B. E., and Stanker, L.,** *Environ. Sci. Technol.,* 22, 247, 1988.

20. **Jung, F., Gee, S. J., Harrison, R. O., Goodrow, M. H., Karu, A. E., Braun, A. L., Li, Q. X., and Hammock, B. D.,** *Pestic. Sci.,* 26, 303, 1989.

21. **Harrison, R. O., Goodrow, M. H., Gee, S. J., and Hammock, B. D.,** in *Immunoassays for Trace Chemical Analysis,* ACS Symposium Series No. 451, Vanderlaan, M., Stanker, L. H., Watkins, B. E., and Roberts, D. W., Eds., American Chemical Society, Washington, D.C., 1990, chap. 2, 14.

22. **Erlanger, B. F.,** in *Methods in Enzymology,* Vol. 70, Van Vunakis, H. and Lagone, J. L., Eds., Academic Press, London, 1980, chap. 4, 85.

23. **Fujiwara, K., Matsumoto, N., Kitagawa, T., and Inouye, K.,** *J. Immunol. Methods,* 134, 227, 1990.

24. **Vallejo, R. P., Bogus, E. R., and Mumma, R. O.,** *J. Agric. Food Chem.,* 30, 572, 1982.

25. **Goodrow, M. H., Harrison, R. O., and Hammock, B. D.,** *J. Agric. Food Chem.,* 38, 990, 1990.

26. **Kohler, G. and Milstein, C.,** *Nature,* 256, 495, 1975.

27. **Vanderlaan, M., Stanker, L., and Watkins, B.,** in *Immunoassays for Trace Chemical Analysis,* ACS Symposium Series No. 451, Vanderlaan, M., Stanker, L. H., Watkins, B. E., and Roberts, D. W., Eds., American Chemical Society, Washington, D.C., 1990, chap. 1, 2.

28. **Siddle, K.,** in *Alternative Immunoassays,* Collins, W. P., Ed., John Wiley & Sons, New York, 1985, chap. 3, 13.

29. **Ehrlich, P. H., Moyle, W. R., and Moustafa, Z. A.,** *J. Immunol.,* 131, 1906, 1983.

30. **Jackson, A. P., Siddle, K., and Thompson, R. J.,** *Biochem. J.,* 215, 505, 1983.

31. **Ekins, R. P.,** in *Alternative Immunoassays,* Collins, W. P., Ed., John Wiley & Sons, New York, 1985, chap. 13, 219.

32. **Miles, L. E. M. and Hales, C. N.,** *Nature,* 219, 186, 1968.

33. **Engvall, E. and Perlmann, P.,** *Immunochemistry,* 8, 871, 1971.

34. **Ferguson, B. S.,** in Proc. 6th Int. Congr. Pesticide Chemistry, Poster Session, Ottawa, Canada, 1986.

35. **Sturgess, M. L., Weeks, I., Mpoko, C. N., Laing, I., and Woodhead, J. S.,** *Clin. Chem.,* 32, 532, 1986.

36. **Freytag, J. W., Lau, H. P., and Wadsley, J. J.,** *Clin. Chem.,* 30, 1494, 1984.

37. **Lukens, H. R., Williams, C. B., Levison, S. B., Dandliker, W. B., and Murayama, D.,** *Environ. Sci. Technol.,* 11, 292, 1977.

38. **Newsome, W. H. and Shields, J. B.,** *J. Agric. Food Chem.,* 29, 220, 1981.

39. **Newsome, W. H. and Collins, P. G.,** *J. Assoc. Off. Anal. Chem.,* 70, 1025, 1987.

40. **Newsome, W. H.,** *J. Agric. Food Chem.,* 33, 528, 1985.

41. **Newsome, W. H.,** in *Pesticide Science and Biotechnology,* Greenhalgh, R. and Roberts, T. R., Eds., Blackwell Scientific, Oxford, 1986, 349.

42. **Newsome, W. H.,** *Bull. Environ. Contam. Toxicol.,* 36, 9, 1986.

43. **Jung, F., Meyer, H. H. D., and Hamm, R. T.,** *J. Agric. Food Chem.,* 37, 1183, 1989.

44. **Fatori, D. and Hunter, W. M.,** *Clin. Chim. Acta,* 100, 81, 1980.

45. **Niewola, Z., Hayward, C., Symington, B. A., and Robson, R. T.,** *Clin. Chim. Acta,* 148, 149, 1985.

46. **Van Emon, J. M., Hammock, B. D., and Seiber, J. N.,** *Anal. Chem.,* 58, 1866, 1986.

47. **Schwalbe, M., Dorn, E., and Beyermann, K.,** *J. Agric. Food Chem.,* 32, 734, 1984.

48. **Gee, S. J., Miyamoto, T., Goodrow, M. H., Buster, D., and Hammock, B. D.,** *J. Agric. Food Chem.,* 36, 863, 1988.

49. **Rittenburg, J. H., Grothaus, G. D., Fitzpatrick, D. A., and Lankow, R. K.,** in *Immunoassays for Trace Chemical Analysis,* ACS Symposium Series No. 451, Vanderlaan, M., Stanker, L. H., Watkins, B. E., and Roberts, D. W., Eds., American Chemical Society, Washington, D.C., 1990, chap. 3, 28.

50. **Deschamps, R. J. A. and Hall, C. J.,** in *Immunochemical Methods for Environmental Analysis,* ACS Symposium Series No. 442, Van Emon, J. M. and Mumma, R. O., Eds., American Chemical Society, Washington, D.C., 1990, chap. 8, 66.

51. **Dargar, R. V., Tymonko, J. M., and Van Der Werf, P.,** in *Immunochemical Methods for Environmental Analysis,* ACS Symposium Series No. 442, Van Emon, J. M. and Mumma, R. O., Eds., American Chemical Society, Washington, D.C., 1990, chap. 14, 170.

52. **Dreher, R. M. and Podratzki, B.,** *J. Agric. Food Chem.,* 36, 1072, 1988.

53. **Schmidt, D. J., Clarkson, C. E., Swanson, T. E., Egger, M. L., Carbon, R. E., Van Emon, J. M., and Karu, A. E.,** *J. Agric. Food Chem.,* 38, 1763, 1990.

54. **Lerman, L. S.,** *Proc. Natl. Acad. Sci. U.S.A.,* 39, 232, 1953.

55. **Goding, J. W.,** in *Monoclonal Antibodies: Principles and Practice,* 2nd ed., Academic Press, London, 1986.

56. **Zuk, R. F., Ginsberg, V. K., Houts, T., Rabble, J., Merrick, H., Ullman, E. F., Fischer, M. M., Sizto, C. C., Stiso, S. N., and Litman, D. J.,** *Clin. Chem.,* 31, 1144, 1985.

57. **Ekins, R. P.,** in *Clinical Applications of Monoclonal Antibodies,* Hubbard, R. and Marks, V., Eds., Plenum Press, New York, 1988, 41.

58. **North, R. J.,** *Trends in Biotechnol.,* 3, 180, 1985.

Chapter 16

FUTURE TRENDS IN PESTICIDE RESIDUE METHODOLOGY

Thomas Cairns

TABLE OF CONTENTS

I. INTRODUCTION

There can be no doubt that chemical technology has greatly advanced the global standard of living through increased food production but with concurrent potential threats to the health of society. Health-oriented government agencies responsible for the protection of the public from possible adverse effects such as chemical residues face a formidable task.

The driving force — Intensive legislative activity throughout the world over the last two decades has culminated in a dramatic increase in the role of analytical chemistry in the protection of the public health. Trace analysis has quickly become the buzz word of the discipline. Clearly, this emphasis placed on trace analysis mandates technological advancements at a faster pace to overcome difficult analytical problems such as PCBs, dioxins, etc. More importantly, however, is the key role analytical chemistry should play in providing reliable data via surveillance of air, water, food, etc. to ensure the full vigor of consumer protection. Perhaps the most disturbing pressures are those inadvertently applied by toxicological studies and risk assessment. Both these disciplines frequently use extrapolation methodologies and the relevant "safe" levels are often below the current level of detection.

The analytical response — In the 1991 biannual review on pesticides by Sherma in *Analytical Chemistry,* over 400 key references to pesticide analysis were reported. This overwhelming evidence provided by these reports has clearly indicated the rapidly developing response of analytical chemists to residue problems. While the main thrust of an analytical scheme, multi-residue or target compounds, is still gas chromatography, various new technologies have entered the arena for serious consideration as routine methods. Each of these new approaches is geared to one ultimate goal — more efficient sample analysis. After all, the pressure has now been applied to collect more and more residue information with less resources.

II. THE STATE OF CURRENT PESTICIDE ANALYSIS

This volume has attempted to bring together a global consensus of our present knowledge and evaluation of current methodologies. Each author has presented an objective and concise report of selected topics that reflect those areas where sufficient progress has been made during the last decade. All too often there is a mild form of myopia that does not allow the scientist to view the panorama of possibilities and encourage expansion beyond their narrow range of research. The rapidly expanding research on pesticide analysis represented in this book stands as testimony to the multifaceted approaches being attempted to resolve difficult analytical problems.

A. EXTRACTION AND CLEANUP

In Part I, the first three chapters demonstrated three major innovative developments in response to the need for improvements in the extraction and cleanup portion of a complete protocol. The development of microextraction methods by Steinwandter has addressed the need for reducing the cost of analysis as well as the reduction in the generation of hazardous waste. While both objectives are noble goals, there is still the long road to interlaboratory validation before worldwide acceptance. But the movement has started! In the case of solid-phase partition column chromatography using bonded-phase and solid-phase partition columns by Hopper, a real attempt has been made to replace conventional liquid-liquid partition to streamline existing procedures to save both time and labor. Richter in Chapter 3 has provided convincing documentation that supercritical fluid extraction is indeed a viable alternative to the more conventional sample extraction techniques. While not presented as the ultimate solution to faster, cheaper, and more efficient extractions, the technique is ideally suited to multiresidue methods.

B. MULTIRESIDUE APPROACHES

Like the unending search for the Holy Grail, the main analytical thrust in pesticide analysis continues to be the development of a more efficient multiresidue procedure. In Part II, six chapters have outlined the major advances in this field of interest. While there is still considerable interest in tailoring conventional methods to selected target groups such as fruits and vegetables (Holland and Malcolm), carbamates (Krause), organophosphorus (Barceló and Lawrence), and organonitrogen (Luchtefeld), the most exciting areas are the application of two-dimensional capillary gas chromatography (Stan) and headspace methods for dithiocarbamates (Hill).

C. EMERGING TECHNOLOGIES

Perhaps the most exciting developments in pesticide analysis are the pioneering applications of new analytical techniques to hopefully improve existing methods. In Part III there are six chapters that address the hope for the future — fiber optic spectroscopy (Grey), enzyme-linked competitive immunoassay (Nugent), ion trap mass spectrometry (Mattern and Rosen), hyphenated methods (Barceló), liquid chromatography/mass spectrometry (Shalaby and George), and immunochemical methods (Aston, Britton, Wraith, and Wright). Each author has described an alternative approach to conventional detection methods in a serious attempt to stress the advantages and disadvantages of adopting such routines.

III. THE FUTURE FOR PESTICIDE ANALYSIS

Clearly the problem of pesticide analysis can be dissected into three major portions for research development — extraction and cleanup, separation, and detection. Each area has been receiving intense examination as evidenced by the extensive literature published each year. Before acceptance as routine methods, the long extensive process of validation and interlaboratory studies may have to be performed. Therefore, which techniques show the greatest promise for future routine application? Attempting to forecast technological acceptance may seem a risky business in an analytical field where residue levels are constantly being lowered as chemicals are reevaluated or introduced. With the complexity of the overall three component concept to pesticide analysis, certain predictions can be made with some certainty of conviction and scientific belief.

A. EXTRACTION

While a great deal of attention has been paid to organic partitioning for multiresidue procedures, there are a large number of pesticides and their metabolites that remain in the aqueous layer. Therefore, it would seem reasonable that a shift in emphasis might occur in the very near future to extend the range of compounds extracted, i.e., the aqueous layer, causing liquid chromatography to play a progressively more important role in pesticide analysis. Certainly the role of supercritical fluid extraction and various miniaturization processes will increase dramatically to help resolve both extraction efficiency and cost and avoidance of hazardous waste generation. These advancements, however, must be supported by recovery studies before widespread acceptance. Both these technologies, however, have the highest probability of becoming routine procedures before the end of the decade.

B. SEPARATION TECHNOLOGIES

For over two decades gas chromatography has been the main support in the separation of various pesticides. More recently, the use of capillary columns and megabore capillary columns have replaced the traditional packed columns, but have brought with that change the need for cleanup before analysis. Many multiresidue procedures have ignored the need for sample cleanup because packed columns afforded the luxury of removing the first inch or so of the packing material after repeated use and replacement and conditioning overnight. Besides, any cleanup procedures involved extra time and cost through clear demonstration of recovery data through the selected matrix. Therefore, it is understandable why most multiresidue procedures have been devised to make them independent of extensive sample cleanup. Sadly, there are no direct substitutes to gas chromatography and liquid chromatography being developed, only extensions and extrapolations of these two basic chromatographic techniques.

C. DETECTION TECHNOLOGIES

There can be no doubt that mass spectrometry, in conjunction with both gas chromatography and liquid chromatography, will continue to simultaneously provide both detection and confirmation of presence. By far the most exciting development in analytical chemistry is the commercial appearance of the ion trap mass spectrometer. Combined with a temperature-programmed injector and megabore column, the basic elements are present to support a multiresidue detection system in one injection compared with the multi-injections required by element sensitive detectors, e.g., N, P, Cl, S, etc. With a capital investment far lower than a conventional quadrupole mass spectrometer, the ion trap technology will quickly gain acceptance.

Second in importance is the development of immunoassay procedures for individual pesticides and metabolites. The strength of this technology is in the area of surveillance monitoring prior to a multiresidue procedure for quantification. Such assays are expected to gain in popularity as soon as sufficient assays are available to monitor a wide range of compound types. Immunoassay procedures are not expected to replace the more traditional approaches to pesticides, but rather to provide an alternative route to scrutiny of certain pesticides or groups of compounds.

Yet another evolving technique that is gaining respect from analytical chemists involved in pesticide analysis is the atomic emission detector. Since the performance of this detector is not degraded by crude extracts without sample cleanup, it offers two distinct advantages: (1) additional specific elements in the compound can be detected beyond the existing element-sensitive detectors, e.g., differentiation of halogens; and (2) the elimination of the need for sample cleanup simplifies the multiresidue approach. The first papers to address the application of atomic emission detectors to pesticide residue analysis are expected to appear in the literature and the conclusions are promising.

IV. THE LAST WORD

The concepts of multiresidue pesticide analysis and now its rapid growth have coincided with the intense controversy over risk assessment. Admittedly, pesticide analysis is still a young and developing science. However, the fresh ideas being brought to bear on difficult analytical problems are strong evidence that giant strides are about to be made in improving the process. No one single approach or chemist will give birth to the universal solution. Primo Levi, the great Italian writer and chemist, commented, *the history of technology demonstrates that when it is faced by new problems, scientific education and precision are necessary but insufficient. Two other virtues are needed, and these are experience and inventive imagination.*

Index

INDEX

C